Vaulting Ambition

Vaulting Ambition

Sociobiology and the Quest for Human Nature

Philip Kitcher

The MIT Press
Cambridge, Massachusetts
London, England

This book was set in Palatino by Achorn Graphic Services, Inc., and printed and bound by The Murray Printing Co. in the United States of America.

Library of Congress Cataloging in Publication Data

Kitcher, Philip, 1947–
Vaulting ambition.

Bibliography: p.
Includes index.
1. Sociobiology. I. Title.
GN365.9.K58 1985 304.5 85-7229
ISBN 0-262-11109-8

To the memory of my mother
in thanks for all that she was

Contents

Preface

In my optimistic moments I hope that the chapters that follow will put an end to the controversies that have beset sociobiology in the past decade. My principal goal has been to explain as clearly as possible what sociobiology is, how it relates to evolutionary theory, and how the ambitious claims that have attracted so much public attention rest on shoddy analysis and flimsy argument. This aim is not entirely negative. If readers are provided a clear view of the nature of sociobiology and of the methodological standards that the discipline ought to meet, premature speculations about human nature may be jettisoned but the exciting research into the behavior of nonhuman animals may be appreciated for what it has achieved. Perhaps the picture I offer may be useful in the vigorous pursuit of that research. Perhaps it may even help us to envisage the *future* development of an approach to human behavior that makes genuine use of biological insights.

This book is longer than I expected it to be, and longer than I wanted it to be. Its length results from my efforts to be complete. I have tried to do justice to the main approaches to human sociobiology in a way that will avoid the charge that my criticisms have been selective, and this has forced me to consider a large number of examples of sociobiological analysis. I have also attempted to make the book self-contained. Interested readers without any prior background in biology or in the philosophy of science will find, I hope, that the technical ideas are explained and that they do not have to look elsewhere to understand the main issues and arguments.

A brief plan of the chapters may serve as a guide for identifying points of principal interest. The introduction is intended to raise the questions that stem from the political implications of some sociobiological claims and to give those questions their due. The next four chapters set the stage for later critical analysis. Chapter 1 addresses the main proposals and arguments made in the most popular version of sociobiology—that introduced by E. O. Wilson in his early sociobiological writings. Chapter 2 offers an overview of the main ideas of contemporary evolutionary theory and of the philosophical issues that arise in connection with that theory. Chapter 3 surveys the new developments in the theory that have sparked the study of animal behavior. In chapter 4, I draw on the material of the previous chapters to explain what sociobiology is, how it relates to evolutionary theory, and how the controversial proposals about human beings relate to the sociobiology of nonhuman animals.

Chapters 5 through 8 offer a systematic criticism of the most popular kinds of sociobiological claims about humans. Chapter 5 compares the analyses of aspects of human behavior with some of the best work on the behavior of nonhuman animals. Chapter 6 tries to expose some common methodological errors that beset sociobiological arguments about human nature. Chapter 7 is slightly more technical, focusing on the legitimacy of the adaptationist program, a program that underlies much work in sociobiology and in evolutionary theory generally. In chapter 8, I attempt to focus the argument by considering how the objections of the prior chapters apply to four main examples.

Chapter 9 is devoted to an analysis of the different approach to human sociobiology pursued by Richard Alexander and his coworkers. In chapter 10, I consider Wilson's recent work with Charles Lumsden, which is intended to provide a theory of gene-culture coevolution. Because of the technicalities introduced in the Lumsden-Wilson approach, this chapter is, of necessity, more demanding on the reader than most of the earlier discussions. However, I hope to have made the aims and claims of the theory of gene-culture coevolution more accessible than they have previously been. Chapter 11 addresses the sociobiological arguments about human altruism, human freedom, and ethics.

Some chapters contain arguments whose rigorous formulation requires a certain amount of mathematical symbolism. Although the algebra is not usually very difficult, I have tried to avoid losing the reader in a thicket of symbols by setting the mathematical discussion off from the main body of the text. Those who would prefer to skip the mathematics can find a qualitative version of the argument in the text. Those who want the details can consult technical discussions A through O.

In writing this book, I have incurred many debts. A large number of people have given me valuable comments, criticisms, and suggestions about parts of the manuscript. I am particularly grateful to John Beatty, William Charlesworth, Eric Charnov, Jim Curtsinger, Norman Dahl, John Dupré, Douglas Futuyma, Frank McKinney, Craig Packer,

Anne Pusey, and Rolf Sartorius. Needless to say, it should not be assumed that these people agree with what I have written.

Five people deserve special mention because of the extensive help and advice they have given me. From Peter Abrams and Joan Herbers I have received valuable suggestions about many points of detail and about several issues of general importance. Richard Lewontin gave me numerous comments on an earlier draft, and his remarks have been extremely valuable in improving the final version. Elliott Sober's careful reading and his line-by-line comments have enhanced my discussion of almost every issue. Patricia Kitcher's unerring eye for cloudy argument and obscure formulations has helped me to make this book far clearer and more readable than it would otherwise have been. I am deeply appreciative of the time and thought that all these people have invested. I hope they will think it was worthwhile.

My research on evolutionary biology and on sociobiology was begun while I held a fellowship from the American Council of Learned Societies. I am grateful to the Council for its support, and to the Museum of Comparative Zoology at Harvard University for its hospitality. The book could not have been written without the benefit of a year off, provided by a Fellowship for College Teachers from the National Endowment for the Humanities. I would like to thank the Endowment for awarding me the fellowship and the University of Minnesota for support in the summer of 1983.

Finally, my thanks to Sally Lieberman and Betsy Anderson for their help in preparing the manuscript.

Vaulting Ambition

A Bicycle Is Not Enough

When I was growing up on the South Coast of England in the 1950s, I was haunted by a vision of judgment. The image was not of the final division, familiar from Michelangelo, with stern judge at top and equal numbers of happy sheep and groveling goats below. My worries, like those of most of my contemporaries, were thoroughly secular and mundane. Those of us whose families were not rich enough to sidestep the state educational system knew that judgment awaited us at age eleven. An examination would separate the academic sheep from the academic goats. We did not want to find ourselves among the goats.

For those who failed the famous British eleven-plus—about fifty percent—judgment was virtually final. Institutions suited to their perceived abilities awaited them. These establishments tried, usually unsuccessfully, to combine sound discipline with the inculcation of mechanical skills. Once committed to them, few of my contemporaries would return to the company of the educational elect. Perhaps one or two pupils per school per year succeeded in showing that they had been wrongly categorized. Occasionally, newspaper reports celebrated the achievement of a child who had initially failed the eleven-plus but had managed to obtain the credentials required for entering university. The exceptions contrasted with the overwhelming mass of cases, the thousands of students who, at age eleven, were stamped as failures and debarred from the routes that led to the more rewarding positions in British society.

Although the educational system continued to devise tests for comparing the merits of those who were initially successful, the original division made the greatest impact on the lives of British children. Many working-class and lower-middle-class parents placed enormous emphasis on preparing for the examination, hoping that their children would be able to enjoy positions of higher status and greater earning power than they themselves enjoyed. The pressure from parents was often unavailing. Not only was a child's performance on a given day to be taken as the index of aptitude, but one third of the examination was devoted to a test that was carefully divorced from prior study and preparation. My friends and classmates knew that part of the exam was to evaluate our "general intelligence." Since few of us knew what "general intelligence" was—or whether we had any—our apprehension about the eleven-plus was increased by the air of mystery surrounding it, by the idea that here was an arcane enigma, only to be revealed to us on the day when our futures would be decided.

Luckily, the system of final educational division at age eleven is a thing of the past—at least in most areas of Britain. In its twenty years of hegemony that system probably stifled the aspirations of thousands of lower-class children. Yet it was designed with the best intentions. Its architects conceived of it as providing the economically disadvantaged with genuine educational opportunities, as enabling the sons and daughters of miners and milkmen, laborers and lorry drivers, to compete on an equal basis for access to Britain's most elite academic institutions. Only gradually was it recognized that children from poorer homes were disproportionately unsuccessful both in passing the eleven-plus and in surmounting later hurdles in the educational system. That growing recognition was a major factor in the long campaign to dislodge the idea of assessing the intellectual abilities of children, once and for all, around the age of eleven.

Where had the idea come from? What motivated a system whose potential harm now seems so obvious? Those who designed and administered the arrangement were enlightened people, concerned to do their best for British children, and their concern drove them to adopt and apply the established psychological wisdom of the day. In a series of publications, Sir Cyril Burt, educational psychologist and adviser to the London County Council, had reported his results on the measurement of general intelligence, showing to everyone's satisfaction that there is such a thing as general intelligence, that it can be assessed by a particular type of test, and that beyond the age of eleven a child's measured general intelligence does not vary significantly. Quite evidently, then, a good policy would be to devise the appropriate tests, require children to take them around the age of eleven, and match the children with educational programs suited to their abilities. So the decree went forth that all the children should be measured—and measured we were.

Burt's claims were dubious. His lifelong acceptance of his psychological and educational views stemmed from faith in the existence of "general intelligence"—and little more. Ultimately, that faith would lead him to announce phony findings. The extensive studies showing correlations in general intelligence between identical (more exactly, monozygotic) twins reared apart, studies that amassed ever more data for precisely the same conclusions, were fabrications, apparently designed to protect Burt's pet idea. In the 1970s Leon Kamin, a Princeton psychologist, spotted a remarkable property of Burt's results: some of the correlations persisted unaltered, to the third decimal place, even when the size of the sample changed quite dramatically (Kamin 1976; Lewontin, Rose, and Kamin 1984, chapter 5). Soon the entire fabric of "supporting evidence" began to unravel. Not only was it discovered that Burt's assessments of IQ, in both parents and children, often depended on his "personal interviews" of those concerned, but some of the "informants" who had supplied him with data were unmasked as fictions.

My aim is not to vex Sir Cyril's ghost but to draw a moral. Those who frame social policies often look to the findings of the sciences for guidance. If the studies they consult are incorrect, then the mistakes may reverberate through the lives of millions. Dogmatic faith, wedded to deliberate fraud, is only the most glaring (and surely the least common) source of socially costly scientific error. From the point of view of those who ultimately suffer, however, the source does not greatly matter. For those of my friends whose talents were left to moulder because of a benighted educational policy, it would hardly be a consolation to know the error arose from carelessness rather than deliberate deception. The moral is transparent. When scientific claims bear on matters of social policy, the standards of evidence and of self-criticism must be extremely high.

Like previous investigations of the heritability of intelligence (Jensen 1969) and of the alleged residues of human evolutionary history (Ardrey 1966; Morris 1967), contemporary work in the biology of behavior runs the risk of being used to support social injustice. Human sociobiology has been portrayed by its vocal critics as a doctrine that perpetuates inequitable divisions on the basis of sex, race, and class. Their reaction can readily be traced to the content of E. O. Wilson's monumental book *Sociobiology: The New Synthesis* (1975a) and to the acclaim that this volume quickly received. Wilson begins his final chapter, entitled "Man: From Sociobiology to Sociology," with a sentence calculated to pique the interest of anyone looking for a guide to scientific social policy: "Let us now consider man in the free spirit of natural history, as though we were zoologists from another planet completing a catalog of social species on Earth." Wilson goes on to draw some sobering conclusions. Humans emerge as xenophobic, deceitful, aggressive, and "absurdly easy to indoctrinate." These are alleged to be intrinsic features of our nature, and Wilson suggests that certain social institutions also have deep biological roots. After appealing to anthropological findings to support the idea that women are frequently used in barter in preliterate societies, controlled and exchanged by the competitive, dominant males, Wilson continues by comparing industrial societies to our hunting-and-gathering cousins.

> The building block of nearly all human societies is the nuclear family. . . . The populace of an American industrial city, no less than a band of hunter-gatherers in the Australian desert, is organized around this unit. In both cases the family moves between regional communities, maintaining complex ties with primary kin by means of visits (or telephone calls and letters) and the exchange of gifts. During the day the women and children remain in the residential area while the men forage for game or its symbolic equivalent in the form of money. The males cooperate in bands to hunt or deal with neighboring groups. (1975a, 553)

Here, a highly distinguished scientist suggests that there is a biological basis for political institutions that many people would like to alter. Small wonder that those concerned with eradicating inequities of sex, race, and class are enraged.

"Rage" is exactly the term for the reaction of some of Wilson's critics. For some of his politically sensitive contemporaries, the suggestions of a new biological approach to human social behavior appeared—in one, but only one, sense—as a red flag, provoking them to respond with a violence that undercut their own arguments. Replying to a laudatory review of *Sociobiology* that had appeared in the *New York Review of Books*, a group later known as the "Sociobiology Study Group of Science for the People" discussed Wilson's proposal in the context of the social Darwinism used by John D. Rockefeller, Sr., to justify the rapacious excesses (or vigorous freedom) of U.S. capitalism and of the theories of genes and behavior underlying Hitler's eugenic policies. Wilson's opponents did not mince their words:

> [Wilson] purports to take a more solidly scientific approach using a wealth of new information. We think that this information has little relevance to human behavior, and the supposedly objective, scientific approach in reality conceals political assumptions. Thus we are presented with yet another defense of the status quo as an inevitable consequence of "human nature." (Allen et al. 1975, 261)

Harvard professors rarely relish being treated as kissing cousins to industrial robber barons and fascist dictators. Wilson's subsequent discussions of his views about human nature and human social relations adopt the tone of quiet but firm regard for uncomfortable truths.

The strident opposition receives a polite response, but Wilson and his recent co-worker Charles Lumsden make clear their belief that the critics have deserted the standards of science in deference to political ideology.

> The flaw in their argument, pointed out by many subsequent writers during the debate, is the assumption that scientific discovery should be judged on its possible political consequences rather than on whether it is true or false. That mode of reasoning led earlier to pseudo-genetics in Nazi Germany and Lysenkoism in the Soviet Union. (Lumsden and Wilson 1983a, 40)

The theme of devoted pursuit of scientific truth is reinforced by the tone of Wilson's Pulitzer Prize winning book, *On Human Nature* (1978). Here we find no brusque announcements to the effect that certain social arrangements are biologically inevitable. Instead, Wilson proposes that our nature sets constraints on our social systems, so that "there are costs, which no one yet can measure," in adopting schemes for social reform that would honor the demands of those who clamor for justice. The message has not changed, but the medium has been exquisitely refined.

Some of Wilson's fellow sociobiologists are less diplomatic. In a book that bears testimonials from both Ardrey and Wilson, David Barash throws down the gauntlet.

> We will explore the biological underpinning of human nepotism and altruism, the basis for families as well as a surprising and sure-fire prescription for conflict. We will analyze parental behaviors, the underlying selfishness of our behavior toward others, even our own children, and take a hard-headed (and undoubtedly unpopular) look at the evolutionary biology of differences between men and women, including a theory for the biological basis of the double standard. (Barash 1979, 2–3)

Home truths about *Homo* may infuriate some readers, but Barash writes with the air of one who dares us to look the facts about ourselves in the face, adjusting our political views to what we see, not to what we want to see. His colleague, Pierre van den Berghe, also warms to the idea of confronting social reformers with uncomfortable results. He will have no part of the "currently fashionable brand of professorial liberalism" or the "kow-towing to feminism" (van den Berghe 1979, 2). For when we take a good look at the data, uncontaminated by "ideological passions," we find support for the "conventional wisdom." It turns out that the family is here to stay, that women are better with the kiddies, that men are turned off by aggressive females, and that politics is a man's game (1979, 195–197). In what might have been a valentine for Phyllis Schlafly, he declares that "neither the National Organization for Women nor the Equal Rights Amendment will change the biological bedrock of asymmetrical parental investment" (196).

So the initial political critique of human sociobiology is met with the straightforward theme of service to truth. Sometimes the expression is tinged with regretful sympathy for ideals of social justice (Wilson), at other times with a zeal to *épater les féministes* (van den Berghe). Yet there are two reasons for thinking that the pledge of devotion to the truth comes a little too easily. In the first place, it is far from clear that sociobiologists appreciate the political implications of the views they promulgate. Those implications become clear when a *New York Times* series on equal rights for women concludes with a serious discussion of the limits that biology might set to women's aspirations (see Gould 1980a, 263) and when the new right in Britain and France announces its enthusiasm for the project of human sociobiology. Second, and more important, those who are angered by the claims of recent human sociobiology do not think of themselves as engaged in a desperate fight to hide uncomfortable truths. The critics believe that, in deference to the authority of eminent scientists, politically harmful falsehoods are likely to be accepted on the basis of inadequate and misleading evidence.

The political impact of sociobiology cannot be measured simply by isolating the claims that offend utopian sensibilities, the assertions of sexual asymmetry and the emphasis on a competitive scramble for positions of dominance. Much of the sociobiological literature is marked by a general approach to human behavior that can easily be used to support harmful views. Because they stress the genetic basis of behavior, many sociobiologists *seem* to be endorsing a strategy of linking behavioral differences to genetic differences, and this strategy encourages the denigration of particular racial and social groups. When we discover differences in test performance between blacks and whites, between urban poor and wealthy suburbanites, a general presumption in favor of a genetic basis for behavioral differences can be used to buttress the suggestion that there is a strong genetic component to the difference in performance. This drastic step is easily taken in advance of any specific studies of the relevant behavior. Recent proclamations to the effect that emphasis on a genetic *basis* for behavior does not entail commitment to the idea that genes *determine* behavior (see van den Berghe 1979, 29; Barash 1977, 39ff.) do not forestall the use of sociobiology in the cause of racism or other forms of social injustice. So long as there is no clear distinction of those cases in which the sociobiologists take the existence of genetic constraints on human social behavior to be well established, so long as ambitious speculations about the involvement of the genes in all aspects of human affairs are mixed with the examples for which the sociobiologists believe that they have firm evidence, the vagueness of the term "genetic basis" and the vast scope that is announced for human sociobiology will combine to provide a weapon for those who wish to perceive differences in behavior as impervious to social change.

One of my first tasks will be to introduce some clarity into the murk that surrounds talk of "genetic determinism." For the moment I shall simply conclude that the early critics of sociobiology were correct in regarding human sociobiology as potentially damaging to the cause of social justice. Their vehemence in making the point, however, fostered the impression that sociobiology was being rejected for that reason alone. Defenders of human sociobiology were handed the reply along with the criticisms: politics cannot be allowed to dictate to science. Wilson develops this obvious reply:

> Human sociobiology should be pursued and its findings weighed as the best means we have of tracing the evolutionary history of the mind. In the difficult journey ahead, during which our ultimate guide must be our deepest and, at present, least understood feelings, surely we cannot afford an ignorance of history. (1975b, 50)

> Knowledge humanely acquired and widely shared, related to human needs but kept free of political censorship, is the real science for the people. (1976, 302; also Lumsden and Wilson 1983a, 41; Alexander 1979, 136–137)

So, because free inquiry is intrinsically good and because it generates the knowledge we need to help with social problems, research in human sociobiology should continue.

Who could disagree? Certainly, if the issue is whether we should censor efforts to inquire into the biological basis of human behavior, all reasonable people should join in a shout of denial. Any social policy that would proscribe areas of scientific research threatening to yield unpalatable conclusions is plainly misguided. One does not help the oppressed by indulging in fantasies about their capacities and designing programs for them in accordance with utopian hopes. But the critics of sociobiology are not committed to a frantic struggle to silence the messenger. One of the most prominent members of the Sociobiology Study Group presents the point clearly:

> Scientific truth, as we understand it, must be our primary criterion. We live with several unpleasant biological truths, death being the most undeniable and ineluctable. If genetic determinism is true, we will learn to live with it as well. But I reiterate my statement that no evidence exists to support it, that the crude versions of past centuries have been conclusively disproved, and that its continued popularity is a function of social prejudice among those who benefit most from the status quo. (Gould 1977, 258)

Several of the points Gould makes are hotly debated by advocates of human sociobiology: they will resist the imputation of genetic determinism and the suggestion that political motives play a role in the popularity of scientific claims. Postponing discussion of these points, we can use Gould's remarks to show that the sociobiologists have failed to defend where their critics attack.

Gould's reference to evidence, and to "truth as we understand it," makes it clear that the issue is miscast by talking about censorship and the proscription of research. Opponents of sociobiology are concerned primarily with the acceptance of hypotheses about the springs of human behavior, not with the pursuit of research. (In a case where the hasty acceptance of hypotheses spawns a particular research program that threatens to yield conclusions both socially harmful and ill supported, they may also argue against the implementation of such a program. The argument derives from a more fundamental opposition to the hasty acceptance of hypotheses.) Gould and his colleagues maintain that the central claims of human sociobiology have been adopted on inadequate grounds, so that what Wilson perceives as the "true science for the people" is really a collection of harmful mistakes, sloppily acquired but unfortunately widely disseminated.

We come to the central theme of this book: *The dispute about human sociobiology is a dispute about evidence.* Friends of sociobiology see the "new synthesis" as an exciting piece of science, resting soundly on evidence and promising a wealth of new insights, including some that are relevant to human needs. To critical eyes, however, the same body of doctrine seems a mass of unfounded speculation, mischievous in covering socially harmful suggestions with the trappings and authority of science. From this perspective it might appear that the political considerations could be left behind. After all, if all reasonable people agree that we must accept hypotheses as the evidence dictates, then the fact that the hypotheses under study can readily be connected with political controversies can be disregarded. The issue reduces to a question about truth, pure and simple.

Lady Bracknell's reminder is apposite—the truth is rarely pure and never simple. Everybody ought to agree that, *given sufficient evidence* for some hypothesis about humans, we should accept that hypothesis whatever its political implications. But the question of what counts as sufficient evidence is not independent of the political consequences. If the costs of being wrong are sufficiently high, then it is reasonable and responsible to ask for more evidence than is demanded in situations where mistakes are relatively innocuous. In the free-for-all of scientific research, ideas are often tossed out, tentatively accepted, and only subsequently subjected to genuinely rigorous tests. Arguably, the practice of bold overgeneralization contributes to the efficient working of science as a community enterprise: hypotheses for which there is "some evidence" or, perhaps, "reasonably good evidence" become part of the public fund of ideas, are integrated with other hypotheses, criticized, refined, and sometimes discarded. Yet when the hypotheses in question bear on human concerns, the exchange cannot be quite so cavalier. If a single scientist, or even the whole community of scientists, comes to adopt an incorrect view of the origins of a distant galaxy, an inadequate model of foraging behavior in ants, or a crazy explanation of the extinction of the dinosaurs, then the mistake will not prove tragic. By contrast, if we are wrong about the bases of human social behavior, if we abandon the goal of a fair distribution of the benefits and burdens of society because we accept faulty hypotheses about ourselves and our evolutionary history, then the consequences of a scientific mistake may be grave indeed.

These conclusions do not rest on misty sentimentality or unrealistic standards of evidence but on fundamental ideas about rational decision. A familiar principle of decision making is that agents should act so as to maximize expected utility. The rationality of adopting, using, and recommending a scientific hypothesis thus depends not merely on the probability that the hypothesis is true, given the available evidence, but on the costs and benefits of adopting it (or failing to adopt it) if it is true and on the costs and benefits of adopting it (or failing to adopt it) if it is false. The abstract principle is familiar to us from many concrete cases. Drug manufacturers rightly insist on higher standards of evidence when there are potentially dangerous consequences from marketing a new product.

The genuine worry behind the political criticism of sociobiology is that, while claims about nonhuman social behavior may be carefully and rigorously defended, the sociobiologists appear to descend to wild speculation precisely where they should be most cautious. (Gould expresses admiration for Wilson's nonhuman sociobiology

[1977, 252]; however, even nonhuman sociobiology has its critics.) Appreciating the worry, we can formulate the central questions about the program of human sociobiology, the questions that I shall consider in this book: How good is the evidence for sociobiology? How good is the evidence in the field of human sociobiology? How good is the evidence for certain headline-grabbing claims about human nature made by various practitioners of sociobiology?

I shall begin by trying to present as clearly as possible the line of reasoning that has led Wilson and others to their conclusions about human nature. This is no trivial task. Wilson has been forced to complain again and again that his position has been misunderstood; the trouble lies in the elusive character of his argument. To give a serious evaluation of the discipline of sociobiology and of the prominent subdiscipline that deals with humans, we shall have to achieve a clear view of contemporary evolutionary theory and of some methodological questions that arise with respect to the theory of evolution.

Violent rhetoric must give way to painstaking assessment, and there is no substitute for beginning at the beginning. We shall need to understand the biological ideas that are central to contemporary thinking about evolution and the recent theoretical contributions that have inspired talk of a "new synthesis." Armed with this understanding, we can proceed to questions that ultimately belong to my own specialty, the philosophy of science: What is sociobiology? What is its relation to modern evolutionary theory? By providing clear answers to these questions, I hope to show that we should resist the idea of sociobiology as a unitary theory, something that must be accepted or rejected as a whole and that can only be called into question by abandoning large parts of contemporary biology. The challenge issued by the ambitious advocates of human sociobiology turns on a false dilemma; we can honor our scientific conscience without hurling brickbats at those currently struggling to obtain their political rights.

Although the focus of my investigation will be the methodological status of sociobiology, it would be wrong to lose sight of the political implications of the inquiry. In considering the merits of sociobiological claims, we must always be aware that the evidence must not be simply suggestive but must justify us in adopting beliefs that will affect our perception and treatment of other people. Close methodological scrutiny is appropriate precisely because of the consequences for social policies and individual lives if we turn out to be mistaken. As Scott Fitzgerald's narrator reminds us at the end of *The Great Gatsby*, carelessness that results in the destruction or diminution of human life is unforgivable. There is no guarantee that our beliefs about ourselves will be correct, no matter how carefully we weigh the evidence. But the more extensive our inquiry, the more secure we are against error. That, at least, is the hope of human rationality.

In the early 1970s, on a visit to England, I went to see a distant cousin. One of her children had just failed the eleven-plus—the old system of final judgment lingered on in the bastion of Conservatism in which I spent much of my youth and in which my cousin lives. Like many children before her, the girl had been promised a new bicycle if she passed the eleven-plus. Like many parents before them, her mother and father had given her the bicycle anyway. The daughter was visibly depressed. She felt that she had failed her parents, and she was not looking forward to the beginning of the school year when she, together with the other "failures," would transfer to a new school. Still, the bicycle was there, a small consolation to her and a token of her parents' continued support. As she wobbled down the sidewalk (the bicycle was somewhat too big for her), pride in her new possession temporarily overcame her sense of inadequacy. As I watched her, I remembered many of the children I had known, and the ways in which the educational system had narrowed their horizons at an early age. Those whose aspirations have been mangled and whose lives have been reduced through the application of misguided science direct us to look closely at any theorizing that might lead us to further mistakes. Their descendants deserve better. A bicycle is not enough.

Chapter 1
From Nature Up

Homage to Aesop

Admirers of particular animals sometimes see in their favorites the human condition writ small. Eighteenth-century naturalists studied the animal creation for testimony to the wisdom and beneficence of its author. Their twentieth-century successors are more likely to draw conclusions about a less exalted subject. For contemporary students of animal behavior, the ladder of evidence does not reach all the way from nature up to nature's god. Yet the observations of animal activity are seen as offering general morals: the Uganda kob and the well-trained pigeon reveal human aspirations in different dress. Cautionary fables are there for those who have the wit to read them.

Sociobiology is popularly regarded as a program launched in 1975 with the publication of Wilson's book. So conceived, it stands in a long tradition of attempts to discern the elements of human nature in the behavior of nonhuman animals and thus to justify the ways of man to man. Advocates of the program pride themselves on having transcended the naive attempts to derive grand conclusions about human nature from detailed observation of a single species or from scattered investigations into selected types of behavior in a motley of species. Not for them the casual assumption that explanations that work today for pigeons will work tomorrow for people. Not for them the collection of scraps of natural history to provide suggestive analogy. Sociobiology has put system and science in the place of parochial vision and animal anecdote. Contemporary evolutionary theory has supplied a new philosophers' stone, one that enables Wilson and his followers to turn ethological dross into sociobiological gold.

Wilson explicitly links his "new synthesis" to the faltering but meritorious efforts of his predecessors. After commending popular books by Konrad Lorenz, Robert Ardrey, Desmond Morris, Lionel Tiger, and Robin Fox for their "great style and vigor," he continues,

> Their efforts were salutary in calling attention to man's status as a biological species adapted to particular environments. . . . But

> their particular handling of the problem tended to be inefficient and misleading. They selected one plausible hypothesis or another based on a review of a small sample of animal species, then advocated the explanation to the limit. (Wilson 1975a, 551; see also 28–29, 287)

Those who are fortunate enough to draw on refinements of evolutionary theory that have made possible a science of human social behavior can afford to sympathize with their pioneering predecessors who were not so lucky.

The central message of one program in human sociobiology, a program that derives from some of Wilson's writings (notably 1975a, 1978), is that the integration of evolutionary insights with careful observations of animal behavior yields a particular theory of human nature. Much of the controversy about sociobiology has concerned the credentials of this program. The harsh response of the Sociobiology Study Group of Science for the People was intended to devastate the program. Yet the message rings on in the pages of such committed early Wilsonians as David Barash (1977, 1979) and Pierre van den Berghe (1979). Whoever is not for the program is against Darwin.

Despite its appropriation of the term "sociobiology," Wilson's new synthesis does not appeal to all those who want to call themselves "sociobiologists." There are sociobiologists and sociobiologists and sociobiologists. Some are card-carrying followers of the program begun in *Sociobiology* and *On Human Nature*. Others, most prominently Wilson himself, have gone on to new heights. Still others, writers like Richard Alexander and Napoleon Chagnon, maintain that there are important implications of evolutionary theory for the study of human nature and human society, while diverging in important ways from the analysis favored by Wilson and his followers. In addition to these, there remains a large group of scientists, probably a majority of those interested in the behavior of nonhuman animals, for whom the controversies that swirl about sociobiology provoke profound discomfort. For people in this group, the evolution of animal behavior is interesting in its own right. They have no wish to play Aesop, and they distrust the idea that the theory of evolution offers any direct insight into human nature. They worry that lurid advertisements for programs like Wilson's will cause sound biology to be viewed as politically dubious. Some would even prefer to use another label to describe their work (Hinde 1982, 151–153).

It will be useful to have a term for the enterprises that promise important insights into human nature. *Pop sociobiology*, as I shall call it, consists in appealing to recent ideas about the evolution of animal behavior in order to advance grand claims about human nature and human social institutions. I use the term as an abbreviation for "popular sociobiology"; the name seems appropriate because the work that falls under this rubric not only is what is commonly thought of as sociobiology but is deliberately designed to command popular attention. Pop sociobiology is practiced by such people as Wilson, Alexander, Robert Trivers, Richard Dawkins, Barash, van den Berghe, and Chagnon. Some of these people (the first four, for example) also engage in biological investigations into the evolution of nonhuman behavior. One of the tasks of the chapters that follow will be to differentiate their additions to pop sociobiology from their contributions to the biology of nonhuman behavior.

Pop sociobiology should be distinguished from both the subdiscipline of evolutionary theory that studies the behavior of nonhuman animals and a possible future discipline that might employ ideas from evolutionary theory in investigating human social behavior. Pop sociobiology is a particular historical movement—more exactly, a cluster of related historical movements—a collection of ideas, arguments, and conclusions that have emerged in recent years. I shall postpone to the very end of this book the question whether it is possible to develop a genuine science of human behavior that draws on the insights of evolutionary biology. For the moment I am only concerned to note that the pop sociobiologists should not be granted a monopoly. There is no a priori reason to believe that any serious biological study of human behavior must go the way of pop sociobiology. For the present, however, pop sociobiology dominates this area of inquiry.

There are three major rival programs within pop sociobiology. Although there are important affinities among them, each deserves separate treatment. First, and most widely known among general readers, is the *early Wilson* program, whose central texts are *Sociobiology* and *On Human Nature*. Second is the program inaugurated by Wilson and Lumsden in *Genes, Mind, and Culture*. Finally, perhaps the most influential sociobiological program among practicing social scientists is the enterprise recommended by Richard Alexander. By considering the proposals of each of these three approaches, I hope to offer a clear diagnosis of the state of pop sociobiology.

Like any group of scientists, sociobiologists have their differences. But the variation among students of the social behavior of animals is not simply a routine range of disagreements about specific points. The primary division is between the advocates of pop sociobiology and those who explicitly distrust any grand theorizing about human nature (see Maynard Smith 1982b, 3). The caution of the latter often strikes more ambitious sociobiologists and their supporters as a failure of nerve. Thus Michael Ruse chides Maynard Smith for failing to see that "analogies work two ways"—if Maynard Smith is content to apply methods originally introduced for studying human decision making to the behavior of nonhumans, then, Ruse contends, he ought to be prepared to apply his conclusions about nonhuman behavior to humans as well (Ruse 1979, 147). The anthropologist Chagnon is also puzzled by Maynard Smith's reluctance to "apply" sociobiology to humans (1982, 292). For him, for Ruse, for Wilson, and for a host of other writers, sociobiology is conceived as a general doctrine, which flows ineluctably from evolutionary theory and which yields profound consequences for our understanding of human behavior. There is no separating any such thing as "pop sociobiology." There is only sociobiology, practised with more or less consistency and intellectual fortitude.

I believe that this conception is radically incorrect and that it has proved seriously misleading. Insofar as there is a subject, "sociobiology," which flows from evolutionary theory, it is not a general doctrine. Insofar as there is a general doctrine that challenges us with important claims about people and their institutions, it does not flow from evolutionary theory. If we are to avoid losing blooming babies with unwholesome bath water, it will be necessary to make some important distinctions and to offer a better picture of the varieties of sociobiology. Because of its importance in the political controversies and because it offers a way into many of the central issues, I shall start by trying to characterize the early Wilson program.

Wilson's Ladder

No one who picks up a copy of *On Human Nature* can avoid the advertisement. The cover of the hardbound edition announces that the book "begins a new phase in the most important intellectual controversy of this generation: Is human behavior controlled by the species' biological heritage? Does this heritage limit human destiny?" Once inside, the vision blurs. It seems that conclusions about human limitations are supposed to follow from premises about the evolution of behavior, but the structure of the argument is elusive. Sometimes a brief sentence makes the intent plain: "Polygyny and sexual differences in temperament can be predicted by a straightforward deduction from the general theory of evolution" (1978, 138). Yet how does the "straightforward deduction" go?

Unsympathetic critics can easily devise versions of their own, demolish the arguments they have constructed, and move on to wage new intellectual battles elsewhere. Wilson evidently envisages a ladder that will enable him to ascend from studies of nature up to controversial claims about human nature. Let us briefly consider how he justifies his remarks about "polygyny and sexual differences in temperament." Wilson devotes several pages to discussing the differences between male and female behavior that we might expect to find, given that human behavior is a product of evolution. He suggests that evolution under natural selection would favor the differences that are found in most human societies. Putting evolutionary expectations together with claims about the widespread occurrence of the differences, Wilson arrives at the conclusion that there are genetic constraints on gender roles.

When we try to reconstruct the argument, we have to fill in some lacunae. There is a version of Wilson's ladder that will apparently accommodate our example, and it has figured prominently in discussions of Wilson's ideas. It runs as follows.

Wilson's Ladder (Naive Version)

1. Evolutionary theory yields results to the effect that certain forms of behavior maximize fitness.
2. Because these forms of behavior are found in many groups of animals, we are entitled to conclude that they have been fashioned under natural selection.
3. Because natural selection acts on genes, we may conclude that there are genes for the forms of behavior in question.
4. Because there are genes for these forms of behavior, the forms of behavior cannot be altered by manipulating the environment.

When the argument is stated so baldly, almost anybody will disavow it. There are passages in Wilson's writings—even in his most recent books—in which he seems to come close to embracing the ideas of the naive version. In many of his remarks, however, especially on occasions on which he is taking care to guard himself against criticism, Wilson will have no truck with anything so crude. His articles are full of weary attempts to dissociate himself from some of the doctrines that the naive version ascribes to him. I shall take these remarks seriously, in the hope of seeing if a more refined version is available.

To discover how Wilson's ladder is really constructed, I suggest that we start with the naive version and consider how to modify it in the light of Wilson's disclaimers. This strategy is forced on us because of Wilson's preference for "pungency and simplicity of style" (1978 jacket blurb) over logical explicitness. Protestations about the ways in which he has been misunderstood provide the best clues we have to the character of the intended argument.

What is wrong with the naive version of the ladder? Plenty. Its conception of the deliverances of evolutionary theory is suspect, its assumption that optimal behavior signals a history of selection deserves scrutiny, and its talk of genes "for" forms of behavior will make any practicing geneticist wince. But the most glaring error comes at the end. Enshrined among the commonplaces of contemporary biology is a principle that every beginning student learns—the characteristics of an organism are the result of the interplay between the genes of the organism and the environment in which it develops. We do not live by our genes alone. Wilson's critics have forced him again and again to announce his devotion to this commonplace. His detractors read the announcements as a smokescreen. There must be a secret denial behind the public acceptance, they allege, because if Wilson really meant what he says he means, then his version of pop sociobiology would become trivial. Without a commitment to "genetic determinism" pop sociobiology may be full of sound and fury, but it signifies nothing.

Our attempt to find a better version of Wilson's ladder can profitably begin with this vexed issue. Unless we can dissolve some of the myths that surround the idea of "genetic determinism," then it is likely that we shall bequeath "the most important intellectual controversy of our generation" to generations without end.

The Iron Hand Meets the Empty Mind

The organization of organisms is deceptively simple. Animal bodies are composed of cells. The nuclei of the cells provide a home for the chromosomes. Genes are segments of chromosomes. If we select a particular animal we can, in principle, identify the collection of genes that distinguishes it from any other animal (with the possible exception of siblings formed from the same fertilized egg). This complement of genes, the animal's *genotype*, will be found in almost all the cells that together make up the animal. The exceptions (barring mutations) will be the sex cells—the sperm or ova—produced by the animal. Typically, these contain only half of the animal's chromosomal material, and thus about half of the animal's genes.

Reproduction transmits the parental genes to a new organism. Animals that reproduce sexually pass on half their genes to their offspring. Asexual organisms do better at making faithful copies of themselves. Barring mutation, their offspring are perfect replicas of the parent. (Whether this should count as success for the single parent is a delicate question in contemporary evolutionary theory. See Williams 1975; Maynard Smith 1978; Stanley 1980.)

Memories of Mendel dominate popular ideas of the process of reproduction and foster the belief in genetic determination of readily observable traits. Mendel believed that there was a one-to-one correlation between genes and observable characteristics. Using notions that were not available to him, we can explicate his ideas as follows. The chromosomes of a sexually reproducing animal typically pair up just before the division in which the gametes are formed. Each chromosome has a mate (the chromosome *homologous* with it) with which it pairs, and, if the process of division goes smoothly, one member from each pair goes to a gamete. (For the moment I shall ignore complications.) Let a *locus* be any region of chromosomal material at which exactly one gene occurs. The different genes that can occur at a particular locus are called *alleles.* If we look at corresponding loci on homologous chromosomes, we find a pair of alleles. One way to formulate Mendel's idea is to say that the combination of alleles present at a pair of corresponding loci determines the form of the characteristic that is governed by that pair of loci. So, to revert to a classic example, Mendel envisaged a locus for seed color in garden peas. Depending on the alleles present at that locus and the corresponding locus, the color of the seeds set by the plant would be yellow or green.

Contemporary geneticists know more than Mendel did, and they recognize that simple kinds of connections between genes and observable traits are extremely rare. The totality of the characteristics of an organism, the organism's *phenotype,* is the product of a complicated interaction between the genotype of the organism and the environment(s) in which it develops. Occasionally we can isolate traits that are dependent on only one locus and for which environmental effects are negligible. (Terminological note: Here and hereafter I use "locus" to refer to a pair of corresponding regions on homologous chromosomes; this agrees with the usual practice of geneticists.) Typically, however, many genes combine to affect the characteristics we observe, and their action can be perturbed by changes in the environment. Eye color in the fruit fly is controlled by an array of genes scattered across all the chromosomes. Raise your fly at an unusual temperature and you will find that the eye color manifested is like that found in flies with unusual genotypes. Grow plants with identical genotypes in different soils, and you will discover a wide range of variation in height, vigor, and quality of flowers and fruits.

Mendel's successors have made clear to us why we should find so complicated a story. Genes are segments of chromosomes, and chromosomes are composed of DNA. *Structural* genes are chunks of DNA that direct the formation of proteins. *Regulatory* genes are stretches of DNA that control the times and rates at which the structural genes operate. As an embryo develops, there is a sequence of cell divisions—itself ultimately directed by the reactions that occur in the individual cells—and a series of chemical combinations in each cell. Depending on the internal state of a cell, certain genes will be "switched on." The products of the genes, the proteins, will react with one another and with other molecules within the cell. New genes may become active, previously productive genes may go into retirement. As the cell's internal state changes, its relations with other cells may alter, through cellular motion, through changes of shape, or through cell division. In the process new contacts may be made with other cells. Molecules may be transported across cell membranes, yielding a novel chemical state within the cell. Further genes may become active. Finally, as the result of a long and complex series of exquisitely timed reactions, we have an organism with certain observable features. If we now focus on one feature, asking ourselves how that feature has been affected by the genes and the environment, it is easy to see that there are many possible ways for the end product to be altered. Typically there will be a host of gene products that have to be available at just the right times. In environments where the developing organism is stimulated in certain ways, it will not be able to obtain the molecules it needs to continue the ordinary sequence of cellular divisions, motions, and interactions. (This is especially obvious in very dramatic cases, as when a developing mammal is deprived of food and water; but such dramatic cases are only the tip of the iceberg.) Given certain sorts of gene mutations, or simply an unusual combination of genes, particular molecules may be unavailable at the stages at which they are needed. So we can point to a host of genes and a host of environmental factors, and claim that, had any one of them been appropriately different, the final result would have been changed. The shoe might have been lost for want of any of a large number of nails.

Developing organisms are buffered against catastrophe, and our appreciation of the fine timing with which reactions occur in the growth of an organism should be tempered by recognition that backup systems are often available if matters go awry. Even if the normal causal sequence breaks down, the organism may still contrive to reach its usual end state by following an alternative route. Nevertheless, the moral of the last paragraph stands. The organism comes to be as it is because of a complex interaction. If some of its properties are stable under relatively large changes in environment or genetic constitution, that is often because, under different circumstances, different complex sequences of reactions would generate the same trait. The availability of alternative routes in no way detracts from the actual causal efficacy of a host of genes and environmental factors. Consider an analogy. Colonel Custard died because Major Mango shot him. Had Mango missed, the colonel would have drunk the poisoned martini on the table before him, prepared by Private Prune. Mango's counsel would be ill advised to plead that the accused's actions were not causally responsible for the death on the grounds that Custard would have been killed anyway. Custard went out with a bang, not a gurgle.

We do not literally pass our phenotypic characteristics on to our progeny. The idea of eyes and noses—or, more pertinently, talents and dispositions—being handed down across the generations is a myth. Out of the recognition that it is a myth comes the "most important intellectual controversy of our generation." We give our children particular protein makers. What exactly does the gift entail?

The ideas that I have been reviewing are commonplace. None of those who participated in the fierce debate about the merits of Wilson's early pop sociobiology questions the picture of development as involving complex interaction between genes and environment. Yet it is a convenient tactic to portray one's opponents as denying the commonplace. Tactical convenience breeds caricatures, and the true debate is never joined.

As we have seen, those inspired by Wilson's early pop sociobiology view themselves as identifying the limits that human genes place on human behavior and on the development of human social systems. (Reminder: Early Wilsonians are not the only pop sociobiologists, and their self-image should not be attributed to those, such as Richard Alexander, who have different ideas.) Their claims frequently give offence, for they appear to foreclose the possibility of the kinds of society hoped for by those who suffer most from present social arrangements. Even in the mollifying terms of *On Human Nature*, we are told that there may be "unmeasurable costs" involved in trying to implement certain ideas of social justice. Critics are quick to react. The conclusions that quicken the conservative pulse, they assert, are obtained at the cost of denying the commonplace. Wilson and his followers have fallen into a well-known trap in theorizing about human nature. Some critics even believe that the lapse is no accident but reflects the way in which dominant ideology shapes scientific research (Lewontin, Rose, and Kamin 1984).

Pop sociobiology, in its early Wilsonian form, revives a familiar flop, the tawdry drama of genetic determinism. So, at least, claim the critics, whose favorite metaphor is the iron hand of the genes. Everybody acknowledges that our genotypes set limits to the ways in which we can behave; all human genotypes are such that, whatever environmental manipulations we make, humans will never be able to fly simply by flapping their arms. Wilson is charged with confusing this innocuous idea with a stronger genetic constraint and thus supposing that there are genes that direct—or determine—specific pieces of behavior, no matter what the environment.

Gould makes the criticism with characteristic lucidity in an influential review of *Sociobiology*:

> Wilson's primary aim, as I read him, is to suggest that Darwinian theory might reformulate the human sciences just as it previously transformed so many other biological disciplines. But Darwinian processes cannot operate without genes to select. Unless the "interesting" properties of human behavior are under specific genetic control, sociology need fear no invasion of its turf. By interesting, I refer to the subjects sociologists and anthropologists fight about most often—aggression, social stratification, and differences in behavior between men and women. If genes only specify that we are large enough to live in a world of gravitational forces, need to rest our bodies by sleeping, and do not photosynthesize, then the realm of genetic determinism will be relatively uninspiring. (1977, 253)

Wilson is aware of the elementary fact that phenotypes are the product of an interaction between genes and environment, and Gould is aware that Wilson is so aware. Gould quotes Wilson's remark that "the genes have given away most of their sovereignty" (1975a, 550). Moreover, Wilson's subsequent writings abound with explicit disavowals of the view that genes determine human behavior and with metaphors intended to convey his ideas about the relation between genes and behavior. Yet critics galore follow Gould in contending that Wilson cannot be serious (see, for example, Lewontin, Rose, and Kamin 1984). Their assessment rests on the kind of argument Gould provides in the passage I have quoted.

I am not concerned with fathoming Wilson's exact intentions. The crucial issue is not whether Wilson believes in the position Gould ascribes to him, whether he really holds the view that he professes when pressed by his critics, or whether he oscillates between the two according to the phases of the moon. I am interested in trying to find the best argument behind the early Wilsonian version of pop sociobiology. Thus we can abandon speculative psychology in favor of attention to matters of logic. Does Gould's reasoning show that, on pain of trivializing his enterprise, Wilson is logically committed to the theses about genetic determination that he disavows?

No. The connections in this area are much more elusive than Gould's remarks suggest. Even if we cannot suppose that biology is the key to all human behavior, the recognition that genes are causally relevant to the development of behavior might make massive changes in some social scientific circles. Scientific revolutions are sometimes born of awareness that certain extra variables need to be considered. More important, there is a non sequitur. True enough, showing that genes control our inability to photosynthesize does not a revolution make. It does not follow that revealing how the genes limit our range of possible forms of behavior in the area of sexual relations would be equally boring. The Wilson program does not depend on genetic determinism for its excitement.

I shall develop the point in detail in the next section. For the moment, let us consider a simple way in which Wilson could offer provocative conclusions without embracing the doctrines Gould attributes to him. It is possible to argue that male propensities for parental care are not genetically determined: there are some environments in which males grow up to be loving and conscientious parents. Similarly, it is possible to deny that females are genetically bound to reject promiscuity: there are environments in which females develop dispositions to great sexual freedom. There is no inconsistency in now claiming that our genes preclude the combination. If the environments that dispose males to parental care do not overlap the environments that prompt females to promiscuity, then, without any crude commitment to genetic determinism (of the kind that Gould envisages), pop sociobiologists can still maintain that they advance revolutionary conclusions. Although the example is an artificial one, it is not entirely divorced from the sociobiological literature. At the end of a scholarly article on monogamy in mammals, Devra Kleiman suggests that the ideals of some feminists—increased male parental care, increased female sexual freedom—may be unattainable. The reason? They are "biologically inconsistent" (Kleiman 1977, 62).

The foes of sociobiology invent a myth, the myth of the iron hand of the genes. The myth cannot be regarded as integral to Wilson's version of pop sociobiology. The champions of pop sociobiology avail themselves of similar tactics, however. The blistering attack of the Sociobiology Study Group provoked Wilson to a quick defense, and in quick defense he concocted his own myth. The critics, he claimed, believe in the "infinite malleability" of human beings. "They postulate that human beings need only decide on the kind of society they wish, and then find ways to bring it into being" (Caplan 1978, 292).

So Wilson and his followers appropriate for themselves the sensible position that phenotypes—including behavioral characteristics—result from an interaction between genes and environment (see, for example, Barash 1977, 39–43). Opponents are assigned the myth of the blank mind, and in recent theoretical developments much attention has been lavished on the problem of showing that blank minds would be eliminated in the course of evolution (Lumsden and Wilson 1981).

What initially appears as a furious debate quickly dissolves into a tempest in a teapot. Pop sociobiologists and their opponents agree that genes and environment together determine phenotype, and that is the end of the matter. This conclusion ought to be disquieting. How have intelligent people managed to convince themselves that they have deep differences? I think that the illusion of a particular type of disagreement can easily be replaced by an illusion of agreement. For nearly a decade the iron hand of the gene has wrestled with the blank mind. Nobody believes in the iron hand of the gene, and nobody believes in the blank mind. Everybody honors the picture of the inheritance of genes and the complicated development of phenotypes that I have outlined in this section. Yet there is still an important divergence of opinion, a debate not about "genetic determinism" or "cultural determinism," a debate not readily captured in a single formula. To understand the early Wilson version of pop sociobiology and the position of its critics, we have to move beyond the public postures.

Fixed Proteins and Protean Organisms

Proteus, the legendary sea god, could assume any form he chose and consequently enjoyed great advantages in achieving his goals. The apparent threat of Wilson's pop sociobiology lies in its denial that we can mimic Proteus. Because of the genes that we have inherited from our ancestors, we are not sufficiently flexible to attain our social ends. However we vary the environment, we cannot create Utopia.

The threat can be made precise by borrowing one of the fundamental ideas of quantitative genetics, the notion of a *norm of reaction.* Suppose that we are interested in some property that admits of degrees—the height of a plant, for example. It would be folly to suppose that plants with a particular genotype always have a particular height. The composition of the soil in which they grow is plainly relevant to the heights they eventually achieve. We know enough about the requirements of plants to provide a convenient representation of the effects of the genotype. We can draw a graph that plots the height of a plant with the given genotype against the critical environmental variables. Our graph displays the norm of reaction of the given genotype. Suppose, for the sake of simplicity, that the only crucial factor is the amount of water the plant receives each day. Then our graph will reveal the height of the plant for different values of the "watering index," the number of liters that the plant receives per day. Obviously it would be more realistic to consider a number of other environmental variables—the acidity of the soil, the nitrogen and phosphorus content, the amount of sunlight, and so forth. Taking such factors into account would deprive us of the possibility of giving a simple two-dimensional representation of the dependence, but it is still possible to imagine a higher-dimensional generalization of the same basic idea.

Let us take a similar approach to the observable characteristics of any organism. Suppose that the genotype of the organism is fixed. The considerations of the last section make it clear that that, by itself, does not determine a unique phenotype that the organism will inevitably manifest. However, we can ask for the way in which the phenotype will vary across all possible environments (or, perhaps, all possible environments in which the organism can survive). By analogy with the notion of a norm of reaction, we can associate with the genotype a function that, for any possible environment, assigns the phenotype the organism will manifest in that environment. We expect that different genotypes will be associated with different functions and that we shall be able to compare the effects of different genotypes by looking at the functions associated with them.

In its simplest form the disagreement that lurks behind the rhetoric about genetic and cultural determinism is a disagreement about the forms of the functions when the genotypes we consider are human genotypes and the phenotypic properties in whose variation we are interested are the kinds of properties that anthropologists squabble about. To a first approximation, Wilson and his followers believe that the values of the functions vary relatively little and that they do so only when the environment is quite drastically altered. The critics maintain that the values of the functions are quite responsive to changes in environmental variables. Each side may justly claim to have absorbed the commonplace story about genes and development. There is still a genuine difference, deriving from alternative articulations of the story.

The norm of reaction of a genotype is a function that assigns a phenotypic value to each appropriate argument. An appropriate argument is some combination of the critical environmental variables—the amount of water added, the acidity of the soil, the amount of sunlight, and so forth. For plant geneticists, norms of reaction reduce (in the literal sense of "reduce") much more complex and unmanageable mappings. In principle, we could consider the function that takes any possible plant environment onto the height of the plant, but nothing would be gained by doing so. We organize the set of possible environments by picking out the critical variables and focusing our attention on the changes that result from varying just these factors. When we compare two plants with different genotypes, we do so by looking at the two functions that assign heights to different combinations of the critical environmental variables. Plant genetics has no need to differentiate environments in which kindly gardeners sing lullabies to their budding shoots. At least, not yet.

Human behavior is another matter. In this domain we may speculate about environmental variables that may be relevant, but it would be rash to assume that we already know how to identify all the critical factors. So the task of investigating human behavioral flexibility must be approached by way of the "unreduced" mapping that takes a possible environment onto some measure of the behavior in which we are interested. The disputes that underlie the sterile exchange about "genetic determinism" often concern the merits of attempts to argue that a particular reduction of the environmental variables effectively represents this complex mapping. The parties on one side claim that we can organize the vast collection of possible environments by concentrating solely on certain environmental variables; any possible variation in phenotype is supposed to be available by modifying some of the selected environmental variables. Their opponents contend that the simplification overlooks possibilities of phenotypic variation that would only be revealed by altering different features of the environment.

To appreciate the character of the disputes, let us begin with a hypothetical example. Suppose that it is alleged that women are by nature more disposed to spend time in child rearing than men. What could this claim mean?

Consider any female genotype. With respect to this genotype, there is a mapping that assigns to each possible environment a measure of the willingness to spend time in child rearing. (For the sake of simplicity let us assume that actual time spent in child rearing is an appropriate measure. This is obviously implausible, but our present concern is to understand what a claim about human nature might *mean*, not to assess its truth.) By averaging out the assigned values for different female genotypes, we can construct a composite mapping that represents the dependence of the disposition to spend time in child rearing on possible environments for some kind of "average woman." The details of the construction are as follows. Fix any environment. Take a particular female genotype and find the value of the measure of the willingness to spend time in child rearing for that genotype in that environment. Repeat the process for all other female genotypes in the same environment. Average the values obtained. This average value is now the value for the composite mapping in that environment. Repeat the procedure for each possible environment.

Now one obvious interpretation of our hypothetical claim is the suggestion that the value of the mapping for the "average female" is always greater than the value for the "average male." On this strong construal the claim would be analyzed as follows:

> (A) For any possible environment, the value of the child-rearing propensity is greater for the "average woman" than it is for the "average man."

There is a simple way to present (A). Let us say that a state is *precluded* for a given genotype (or collection of genotypes) if there is no possible environment in which that genotype (or collection of genotypes) attains that state. Then (A) is simply the claim that, for the composite mappings drawn from male and female human genotypes, the state in which males have a propensity to spend time in child rearing that is greater than or equal to that of females is a precluded state.

(A) is not the only construal of our hypothetical claim, however, even if we interpret it as a proposal about average males and average females. A weaker suggestion is that we can indeed achieve a state in which males have an equal enthusiasm for child rearing, but that this can only be done at considerable cost. (Perhaps it can only be attained in situations in which all parents are extremely reluctant to care for their children. A possible example is the sorry state of the Ik, studied by Colin Turnbull [1972].) The only possible environments in which the state is attained are highly undesirable. Here is an analysis of the weaker proposal:

> (B) There is a collection of desirable properties (desiderata) such that any possible environment in which the value of the child-rearing propensity is equal for the "average male" and the "average female" is an environment in which at least one of the desiderata is absent.

In other words, (B) tells us that the state in which equal propensities to rear children are accompanied by all our cherished human institutions is precluded.

I think that statements like (A) and (B) make explicit what most pop sociobiologists have in mind when they talk about limits set by human nature. It should be obvious that claims like these are neither unexciting nor committed to the simplistic version of "genetic determinism" often ascribed to Wilson and his followers. It should be equally obvious that there are many alternative analyses that might be offered. Our envisaged composite functions could be compared in other ways. We could weight genotypes or weight environments. We could also resist the idea of contrasting "average female" and "average male" values in favor of a direct comparison between individuals. I shall not explore these alternatives, not only because I doubt that they represent the intentions of any flesh-and-blood pop sociobiologist, but also because the considerations relevant to discussing (A) and (B) seem to me to be equally pertinent to other possible construals.

My confidence that (A) and (B) represent what prominent pop sociobiologists have in mind is based on what they say. Describing the aims of his version of sociobiology, Barash writes, "the process of evolution, operating on human beings, has produced a creature for whom certain behaviors just don't go at all, whereas others go very well indeed" (1979, 11). Wilson is more canny, opting for (B) rather than the more provocative (A): "There is a cost, which no one yet can measure, awaiting the society that moves either from juridical equality of opportunity between the sexes to a statistical equality of their performance in the professions, or back [*sic*] toward deliberate sexual discrimination" (1978, 147). Talk of composite mappings, possible environments, and precluded states may seem artificial in contrast with the plain idiom in which Barash and Wilson announce their conclusions. Yet the unnatural idiom serves its purpose in enabling us to see clearly how the pop sociobiological view is compatible with conventional wisdom about gene-environment interaction.

My reformulation also reveals what Herculean labors await those who hope to arrive at conclusions about the limits of human nature. Comparison of actual behavior is not enough. We cannot compare the overall behavior of two functions by looking at their values for a single argument or for arguments within a small interval: if we only consider values between 0 and 1, x is greater than x^2; this does not show us that the value of the former function is always greater than that of the latter. Yet, as Wilson's critics have repeatedly pointed out, there is a long and dismal history of drawing grand conclusions from just such comparisons (see Lewontin 1976; Gould 1981). A quick look at actual behavior and at behavioral differences among groups has all too frequently served to buttress hypotheses about the fixity of human institutions and the impossibility of eradicating inequalities among races or classes. Plant breeders who inferred the qualities of rival strains from consideration of relative vigor in a single environment, or from casual inspection of a collection of environments, would have a pronounced tendency to go rapidly out of business. By contrast, their imitators in the behavioral sciences usually seem to thrive.

Ironically, it is the immensity of our ignorance about the environmental influences on human behavior that enables behavioral scientists to practice methods that would doom their plant-breeding cousins. In the human case we lack the lore that enables a cautious plant breeder to arrive at a justified assessment of the relative merits of particular strains. We have no representation of the collection of possible environments that will reduce it to a space of manageable dimensions. We know, in a dim and unsystematic way, that features of child rearing and of cultural history can make profound differences in the behavior of individual human beings. What eludes us is the detail, the behavioral counterpart of the adjustments of pH or the nitrogen content.

If Wilson's ladder is to enable us to climb from nature to self-knowledge, then it must surmount the old problem of the casual comparison. We shall have to achieve some clear view of the kinds of changes in environment that would be critical for changes in various kinds of human behavior. We shall have to be given evidence that the forms of behavior and social institutions alleged to be stable—or, perhaps, modifiable only at great cost—really do remain constant when the crucial variables vary. Wilson's version of pop sociobiology has so far emerged as an *intelligible* program, and we have been able to understand its conclusions without assuming them to depend on denying the commonplace. However, to understand it is one thing. To see if it is plausible is quite another.

The One and the Many

The time has come to make a distinction that I have so far cheerfully blurred. Wilson and his followers are interested in deriving conclusions about human nature, about limits on the behavior of individuals or, perhaps, limits on their dispositions to behave in various ways. Yet this is not their only concern. Pop sociobiologists are devoted to certain human institutions: home, family, and maternal care, to name but three. They hope to show that such institutions will be permanent features of our social condition, that they are grounded in behavioral characteristics that are, in their turn, extremely stable.

Just as the passage from the genotype to the phenotype is fraught with complications, so too there is no easy bridge between the behavior (or behavioral propensities) of individuals and the character of the society to which they belong. The first point to recognize is that the aspirations and attitudes of individuals are typically shaped by the institutions of prior generations. (This is a leading theme of Bock 1980; Wilson attempts to come to terms with it in his recent work on gene-culture coevolution.) Second, societal institutions and societal attitudes need not mirror the aspirations of individuals. Nations may be aggressive despite the fact that a majority of their citizens are peace loving—witness the Germany of the 1930s. There may be institutions promoting inequalities of race, sex, and class, even though individuals would prefer to treat one another as equals. To emphasize these points is not to invoke some mysterious "Force of Culture" that is responsible both for shaping the ideas of developing individuals and for distorting the societal expression of those ideas. It is simply to recall the obvious facts that human social environments reflect human history and that, when groups of people interact, the arrangements they reach may be wildly at odds with their individual preferences.

Thomas Schelling has provided some beautiful illustrations of the latter point (Schelling 1978). Some of them involve everyday occurrences, small irritants in our social lives. Most people have found themselves on an expressway clogged with traffic because the drivers ahead have slowed down to peek at an accident on the opposite side. We would willingly forgo the chance to take a look if everyone else would do likewise. The other people in the traffic jam share this preference. However, when we arrive at the scene, there is no further cost to ourselves in indulging our curiosity, and so we make our own small contribution to delaying those who are behind us.

There are other examples that concern matters of importance. Consider the assorting of individuals according to race or class. Suppose that members of Group I have a range of tolerance for living in the same area as members of Group II. The most tolerant people in Group I are happy with a situation in which they are outnumbered 2:1. The least tolerant will be unhappy if they have any Group II neighbors at all. The median members of Group I are happy if the ratio is 1:1. Group II people have exactly the same distribution of attitudes toward Group I people. Now suppose that there are many more people in Group I than in Group II. How will people distribute themselves into neighborhoods? (We are suppressing all kinds of complications that arise in realistic situations—ability to afford particular kinds of housing, and so forth.) Much depends on the initial conditions and the details of the dynamics of movement. However, it is relatively easy to reach a situation in which all members of Group I live together and all members of Group II live together. (See Schelling 1978, 157ff., for details.) This result is quite compatible with the assumption that most members of both groups are either very happy or only mildly unhappy with a situation in which the ratio is 1:1.

We can now begin to see that there are two distinct ways in which goals of social justice might be protected against the pop sociobiological charge that they fall afoul of human nature. The more obvious is to suggest that our present social structure and our cultural history are important determinants of the forms of behavior and the attitudes that individual human beings develop. Implementing this strategy would lead to the type of debate envisaged in the last section. Pop sociobiologists would insist that the forms of behavior are relatively resistant to modification through adjustment of our social environments. Their opponents would claim that appropriate social changes can alter the forms of behavior and the underlying propensities.

Less obvious is the strategy of denying that our present social arrangements accurately reflect our individual preferences and propensities. It is possible to concede that those individual attitudes are relatively invulnerable to change through modification of the social environment, but still to deny that our institutions are unalterable. Just as genotypes do not determine phenotypes, so too, individual propensities to behavior do not determine the character of a society.

To revert to our example, the clustering of people by race need not reflect any individual racial prejudice. It is possible that most of the people in a society would prefer to live in racially mixed neighborhoods, but that accidents of initial distribution and initial movements should produce a collection of racially homogeneous neighborhoods. The social institutions that arise from a collection of individual propensities may be crucially shaped by an accident of history. The social system begins to go in a particular direction, and once it does, it has its own momentum. (In appreciating this point, we should not lose sight of the other important fact, that individual propensities are themselves shaped by social arrangements. Thus the dynamics is even more complicated than our simple example indicates.)

As in our study of gene-environment interactions, we can give a more precise analysis of the type of relationship envisaged here. Suppose for the sake of the present discussion that the propensities of individuals are not altered by changes in the social environment. We want to understand the societal implications of the individual propensities of the people who make up the society. We regard these people as interacting with one another to refashion the institutions of the society into which they are born. So we associate with a particular set of people having particular characteristics a mapping that assigns to each possible social context the social situation that would result from the response of those people to that context.

The idea behind the view that our present social arrangements reflect accidents of history can now be stated quite simply. Given certain initial social contexts, a collection of people with common behavioral propensities will develop certain social institutions; once in place, these social institutions will belong to the initial social contexts of subsequent generations and will be stable. There are alternative social contexts, however, in which alternative social institutions develop and are equally stable. Consider a simple example (suggested to me by Elliott Sober): we want to use the same language to communicate with one another; it is an accident of history that we use English and not Chinese.

The accidents of history put us into one social tradition—one sequence of social contexts—rather than another, and we can only remove ourselves from this tradition by conscious social engineering. By engaging in social engineering we reach an alternative stable set of social institutions. Since our common behavioral attitudes are compatible with both social traditions, we are not compelled to regard our present social arrangements as the only ones that are possible for us. This, I suggest, is the major claim of those of Wilson's critics who have emphasized the role of history and culture (Sahlins 1976; Bock 1980). That claim is intelligible, it is by no means obviously false, and it is not committed to the existence of dubious entities (mysterious "Forces of Culture").

Thoughtful people should not wed themselves to either one of the two strategies I have envisaged as responses to Wilson's early pop sociobiology. They should insist both that changes in social environment can effect changes in behavioral propensities and that the social environment itself is the product not only of individual attitudes but of the prior social arrangements that have been developed. The most extreme versions of pop sociobiology contend that human genotypes preclude certain forms of behavior and that our actual behavioral dispositions preclude certain social arrangements. We ought to be suspicious about both types of claims.

In recognizing the need for an extra step in defending some sociobiological conclusions, the need to justify projecting the structure of society from the behavioral inclinations of individuals, I have touched on a problem that has moved Wilson to go beyond his early program. (Wilson notes that this move was inspired by the challenges of Bock [1980], Harris [1979], and Sahlins [1976]. See Lumsden and

Wilson 1983a, 44.) The theory of gene-culture coevolution, developed by Lumsden and Wilson, is an attempt to respond to the kind of difficulty that I have been describing. Lumsden and Wilson hope to show how the institutions of a society are determined by the behavioral propensities of the members of the society. In chapter 10 we shall consider the extent to which their hopes are well-founded.

For the time being, however, I want to set to one side the problem of the one and the many in favor of the more straightforward approach of Wilson's early program. Even before we ask whether the flaws Wilson now perceives in that program are indeed corrected by his most recent efforts (see Lumsden and Wilson 1983a, 47–50), we should consider whether there are other defects that ought to be addressed. Even if our goal is simply to understand the genetic limits on the behavior of individual human beings, can we reasonably hope to attain it in the way that Wilson suggests?

Short Cuts and Blind Alleys

My aim has been to clarify the kinds of conclusions at which pop sociobiology is directed. I have been trying to find the points against which the top end of Wilson's ladder is supposed to rest. However, once we have identified the target area, it is natural to wonder about the route. Why should a search for conclusions about the genetic limits on human behavior begin with evolutionary theory and with the behavior of nonhumans? Why not take a simple short cut and investigate ourselves, as we are, here and now?

There is, of course, a science of the genetics of human behavior, and it has some notable discoveries to its credit. The most convincing results are instances in which pathological conditions have behavioral effects, as in the case of various metabolic disorders and defects in color vision (see Ehrman and Parsons 1981, 281–285, 288–291). These hardly provide a basis for the grand conclusions after which pop sociobiologists hanker. Because the types of behavior most susceptible to rigorous genetic analysis are not those that pop sociobiology finds most interesting, the writings of pop sociobiologists do not brim over with technical reports from human behavioral genetics—although Wilson and his followers are not too proud to advertise any promising suggestions when the moment seems right (see Wilson 1978, 47). As in the early days of classical genetics, when the fruit fly was the geneticist's best friend and when mutants were gifts of the population cage, not artifacts of X-ray bombardment, the kinds of genetic systems that are best understood in human beings are those in which variation produces markedly deleterious effects. Human geneticists can look sadly on as their colleagues employ the impressive arsenal of classical and molecular techniques developed in this century. Their own hands are tied.

Orwellian fantasies aside, it is considered poor form to subject people to the kinds of procedures that are used in rigorous genetic analysis of nonhuman animals. Breeding pure lines, rearing us in controlled environments, irradiating us to induce mutations, inserting genetic markers, and like tactics are clearly out of the question. Moreover, even if we were to be grossly insensitive to the ethical considerations, the long wait from birth to reproduction makes us poor subjects for classical genetic analysis. The hope of human behavioral genetics is that, without interference, it will be possible to trace the features of the genetic components in human behavior from the unsystematic collection of human genotypes and environments that are actually given to us. One prominent and familiar method is to investigate cases in which monozygotic twins (twins who originate from the same fertilized egg and who thus usually have the same genotype) are reared in different environments. Yet, while geneticists may yearn for a world in which monozygotic twins are born in profusion and in which they are reared completely apart in radically different environments, that world is not ours. In consequence, the task of achieving justified views about the genetics of human behavior is difficult and painstaking.

If we are not prepared to wait for the slow and cautious accumulation of conclusions by classical methods, for the patient survey of interactions between genes and environments as they haphazardly occur, for the development of biochemical techniques and of the tools of molecular biology, if we want a grand theory of human nature and we want it now, what can we do? There is no short and direct route to constructing the function that, for a fixed genotype, maps possible environments onto a behavioral phenotype. With respect to interesting human characteristics—such as the notorious example of human intelligence—there are well-known short cuts that end in blind alleys (see Block and Dworkin 1976a). The ambitious student of human behavior needs something new.

Frustrated by the cautious plodding of behavioral genetics, pop sociobiologists, like Ardrey, Lorenz, and Morris before them, turn to our evolutionary history. We have the genes we do because we inherited them from our ancestors. Perhaps we can learn something about them by investigating animal behavior, by understanding how it is adapted, by appreciating the selective forces that have been at work in evolution. But how exactly can we learn? For Ardrey and Morris it is simply a matter of seeing suggestive analogies. Wilson and his followers are more systematic. They try to build a ladder from nature up.

The central issue in the sociobiology controversy is whether there is a firm ladder that will take pop sociobiologists where they want to go. The need for the ladder is clear. We have managed to identify the intended terminus. It is now time to return to the origin.

Chapter 2

The Rules of the Games

Darwin and After

"But, oh Lord!, how little do we know and have known to be so advanced in knowledge by one theory. If we thought ourselves knowing dogs before you revealed Natural Selection, what d--d ignorant ones we must surely be now we do know that law." So wrote Joseph Hooker to his friend Charles Darwin, for whose theory of evolution by natural selection he was one of the earliest and most stalwart campaigners (F. Darwin 1903, I, 135). Hooker's comment captures the double achievement of Darwin's work: on the one hand its tremendous advancement of biological knowledge, on the other its definition of large areas of previously unsuspected ignorance. Pop sociobiologists believe that they have made a new advance in the tradition begun by Darwin. They claim to build on the body of results amassed by Darwin and his successors and to venture into areas where evolutionary biology has previously been forced to confess its ignorance.

To understand and evaluate the contentions of pop sociobiology, it is necessary to know the rules of the game—more exactly, to know the rules of two related games. First, we need an appreciation of the main ideas of evolutionary theory. Second, we must consider the methodological questions that have swirled around the theory from the beginning. Serious philosophical discussion presupposes biological familiarity. So I shall begin with a look at the theory of evolution.

There is no better place to start than with Darwin, with his ideas and with his problems. In the early chapters of the *Origin of Species* we find four principles on which the theory of evolution by natural selection apparently rests.

> *The Principle of Variation*
> Organisms are different. At any stage in the history of a species, there will be variation among the members of the species.

> *The Principle of the Struggle for Existence*
> Organisms compete with one another. At any stage in the history of a species, more organisms are born than can survive to reproduce up to their full potential.
>
> *The Principle of Variation in Fitness*
> Organisms differ in ways that affect their competitive abilities. Some organisms have characteristics that better enable them to survive and reproduce than others.
>
> *The Strong Principle of Inheritance*
> Organisms transmit a large majority of their characteristics to their descendants.

From these four principles Darwin draws what appears to be his major conclusion. If organisms vary, and if they do so in ways that affect their abilities to compete in a struggle for limited resources, then those with superior competitive abilities are likely to survive longer and produce more offspring. If most characteristics are transmitted in reproduction, then the fortunate offspring will inherit the superior abilities of their parents. Hence

> *The Principle of Natural Selection*
> Typically, in the history of a species, those characteristics that endow their bearers with superior competitive abilities will become more prevalent in subsequent generations.

Here we might think we have it. Darwin's theory is encapsulated in the four fundamental principles and the major consequences we have drawn from them. Contemporary neo-Darwinism is a purified version of the same ideas, obtained with the cleansing help of modern genetics and mathematical formulations.

Not so. The reconstruction just given bypasses Darwin's achievement. The principles that I have assembled were not novel. What was revolutionary was the use that Darwin made of them. (For a detailed defense of this interpretation, see Kitcher 1985.)

Darwin proposed that we could explain vast numbers of biological phenomena by viewing the histories of groups of organisms as processes in which, over long stretches of time, large changes in characteristics occur. These processes are governed by the principles of competition, variation in fitness, and inheritance. In consequence, groups of organisms (populations, species, genera, and so forth) are modified by natural selection—though, for Darwin, they are not modified by selection alone. Phenomena that had previously seemed to be beyond the scope of science were brought into the domain of biology, to be explained by tracing histories of descent with modification, throughout which the chief, but not the only agent, has been natural selection.

What were these phenomena? Consider first the topic that seems to have moved Darwin most, biogeographical distribution. As naturalists became aware of the distribution of plants and animals around the globe, numerous puzzles arose. Why does Australia contain so rich a collection of marsupials? Why do we find woodpeckers where no tree grows? Why do the Galapagos Islands contain several distinct species of finch, recognizably similar to one another and to the finches of the South American continent?

Darwin provided a way of approaching such questions. We are to see the present distribution of plants and animals as the outcome of a long history of descent with modification and dispersal. In some particularly perplexing cases he could sketch explanations. The woodpeckers without trees to peck are modified descendants of tree-dwelling ancestors who have dispersed and colonized a new habitat. Australian marsupials are the descendants of primitive marsupials who were once widely spread around the globe. In the later struggle with placental mammals, most marsupials were eliminated. Luckily, the successful placentals were unable to reach Australia, so that some marsupials survived in a refuge.

Darwin was also able to illuminate other issues. No longer was the order of the fossil record to be perceived in terms of a succession of inexplicable waves of creation punctuated by equally inexplicable cataclysms. The evolutionary perspective could also suggest accounts of the structural and physiological similarities among contemporary organisms and the anatomical affinities of those organisms with fossils. Darwin proposed that the details of both the succession of fossils and the similarities among organisms are to be understood by revealing the histories of descent with modification. His revolutionary insight was that the principles assembled above, principles that almost all of his contemporaries accepted, could be applied to explain the details of life.

So far, natural selection has hardly figured in the discussion. Darwin's approach to the phenomena of biogeography, the order of the fossil record, and the affinities among past and present organisms usually requires us only to relate histories of descent with modification. We are to understand the phenomena by recognizing a pattern of relationship among ancestors and descendants. We can avoid speculating about the causes that have given rise to the pattern. True, we must be justified in thinking that the modifications we hypothesize were possible, and this will require us to have some ideas about the ways in which evolutionary changes can be effected. However, for the purposes so far envisaged, we do not have to match our favored agents of change to the modifications actually described in our histories.

Matters are different when we turn to a different type of biological question, one that Darwin took so seriously that he declared his theory would be unsatisfactory unless he could answer it. His undergraduate years at Cambridge reading Paley's *Natural Theology* left their mark on him, and he required that his theory should show "how the innumerable species inhabiting this world have been modified, so as to acquire that perfection of structure and coadaptation which most justly excites our admiration" (Darwin 1859, 3). The key to the problem of adaptation was to be natural selection.

Organisms sometimes "excite our admiration" by striking us as perfectly suited to their environments. Darwin suggests that we understand the presence of admirable characteristics by tracing a particular type of history of descent. We start with a group of remote ancestors who lacked the trait. A few variant organisms arose, possessing the beneficial characteristic (quite possibly in a rudimentary or primitive form). The characteristic gave them an advantage in the struggle for limited resources. As a result they left more descendants, and the trait spread. Ultimately, all members of the group came to have it.

Darwin was quick to appreciate the fact that understanding adaptations as products of selection removes any sense of wonder at those features of the organic world that might offend delicate sensibilities—the behavior of parasitic wasps, for example (Darwin 1859, 472). He also recognized that traits that emerge from evolution under selection might sometimes fail to square with our ideas of good design. Some adaptations are imperfect, clumsy solutions, achieved with whatever materials are available. The orchids provide a beautiful collection of examples. Darwin showed how their baroque internal structures have been shaped by selection against self-pollination, acting on the morphology of ancestral flowers (Darwin 1862; see also Ghiselin 1969, chapter 6, and Gould 1980b).

Despite Darwin's revolutionary insights and his detailed suggestions about numerous plants and animals, Hooker's judgment was apt. There was much for Darwin's successors to do. No single study could do more than touch a minute part of the diversity of nature. Furthermore, large theoretical questions remained to be confronted. What is the basis of the variation in the members of a species? How do new variants arise? To what extent is heritability the rule? What kinds of factors operate in the struggle to survive and reproduce? How exactly do advantageous characteristics spread through a population? Contemporary evolutionary theory extends Darwin's insights by grafting onto his framework answers to these major questions.

The Missing Gene

A major character was absent from Darwin's script, and the absence caused him much trouble. Darwin had no systematic account of heredity, and his more ingenious critics diverted themselves by suggesting inconsistencies among the accounts of variation, selection, and inheritance that evolutionary theory seemed to require (see, for example, Jenkin 1867). We can safely admire their ingenuity, in confidence that we know how to resolve the difficulties. Darwin could not be quite so serene.

Genetics gives us a perspective on some of the large questions about the evolutionary process. Consider first Darwin's emphasis on the variation among organisms. In any population of organisms (that is, any group of organisms whose members have a greater propensity to mate with one another than with organisms outside the group) some organisms will differ with respect to their observable properties but not with respect to underlying genes. Flies carrying the same alleles may look strikingly different if they are exposed to different temperatures at critical points in their development. Variation of this kind is of no account to natural selection. Where there is no difference in genes, there is nothing for a competitively superior organism to pass on to its progeny. Darwin's approach rests on the claim that typical populations are populations on which selection can act—that is, there is typically genetic variation among the individuals in a population.

Immediately, a simple view of the workings of evolution commands our attention. Imagine that two alleles are initially present at a locus in a population. In the environment inhabited by members of the population, individuals who have two copies of one of the alleles (homozygotes for that allele) invariably develop characteristics that lower their relative ability to reproduce. Their more fortunate rivals are the homozygotes for the other allele and those organisms—the heterozygotes—who have one allele of each kind. (If the heterozygote *Aa* has the same phenotype as one of the homozygotes *AA*, then *A* is said to be dominant with respect to *a*. Alternatively, *a* is recessive with respect to *A*. In the present example I am assuming that the favorable allele is dominant.) Successful organisms are more likely to carry, and to transmit, copies of the favored allele. With each generation the frequency of the allele in the population increases. Ultimately, the rival allele is eliminated.

As we shall see shortly, this simple story represents only one among many possible evolutionary scenarios. (The point is worth remembering when we ponder pop sociobiological discussions of the course of evolution.) Nevertheless, the story helps us to see how the concepts of genetics refine Darwin's ideas. We can think of evolutionary changes, by and large, as changes in the frequency of alleles in a population. (There are a few exceptional cases, but for present purposes we can ignore them.) Selection can accomplish evolutionary change: those whose genes promote characteristics that are advantageous in the struggle to survive and reproduce are rewarded through the transmission of their genes to the next generation.

Evolutionary change does not work by selection alone. If we think of a population as a large pool of genes, and if we think of evolutionary changes in terms of the representation of various alleles in the pool, then it is clear that there are many ways for the frequencies to change. New alleles, or new copies of alleles already present, may arise through mutation. Allele frequencies may be altered through the arrival of new organisms from elsewhere or through the departure of some members of the population.

Moreover, the intricacies of the process through which genes are transmitted complicates the action of selection. Recall that the gametes are formed in a cell division in which chromosomes align themselves in pairs. Before the chromosomes are passed on to the gametes, there is an exchange of genetic material. Typically, part of one chromosome changes places with the corresponding part of the homologous chromosome. In this process (recombination) new chromosomes (recombinant chromosomes) with new combinations of genes are formed. Recombination is likely to split alleles that are relatively far apart on the same chromosome, whereas alleles that occupy neighboring loci (alleles that are closely linked) are likely to remain together. Linkage can affect the operation of selection because alleles that keep good company benefit from the selection of their neighbors.

Finally, a host of extrinsic factors may affect evolutionary change. Actual populations are finite, and it is all too easy for alleles to be lost through the accidents of death and mating. Organisms with the best available characteristics can find themselves in the wrong places at the wrong times, and as a result they may forfeit the chance to pass on their genes.

We arrive at a general, qualitative picture of the evolutionary process. New alleles arise by mutation. New configurations of genes arise through recombination and through the process of reproduction. As a result of interaction between genes and environment, organisms come to have different phenotypic properties. Given the environment in which the organisms live, some properties are more advantageous to their bearers in the struggle to survive and reproduce. Where natural selection operates untrammeled, the genes carried by organisms possessing advantageous properties are likely to achieve greater representation in subsequent generations of the population. The process can be critically affected by other factors. Mutation, migration, and linkage can subvert the workings of selection. Accidental contingencies can alter the fate of an evolving population. The river floods too soon, and superplant or superanimal, bearer of wondrous genes, is no more.

The qualitative picture helps us to understand arguments in contemporary evolutionary theory and sociobiology. A more precise account of evolution is even more helpful. Let us take a quick look at the ways in which a little mathematics brings exactness to our qualitative conclusions about the genetics of populations.

A First Course in Auto Mechanics

Richard Lewontin calls theoretical population genetics "the auto mechanics of evolution." The name is apt. Like auto mechanics, population genetics is both essential and unlovely: essential because evolutionary changes are (almost always) changes in gene frequencies, unlovely because the equations governing the changes typically do not admit of elegant solution.

According to legend, the inelegant subject was born in an elegant setting. G. H. Hardy, great mathematician and cricket lover, used to watch the university matches at Cambridge in company with one of the early Mendelians, R. C. Punnett. During one match Punnett posed the following puzzle: what would occur if the individuals in a population, possessing two alleles at assigned frequencies, mated at random? The problem is not difficult. It is the type of thing a mathematician of Hardy's stature could be expected to solve between innings, and, indeed, the question had already been answered independently by a German physician, Wilhelm Weinberg. The result is the Hardy-Weinberg law.

Suppose we have a population so large that it can be considered as infinite and that, at a particular locus, the only alleles present in the population are A and a. Individuals in the population mate at random, in the sense that, for any member of the population, any member of the opposite sex is equally likely to be picked as a mate. The initial relative frequencies of A and a in both sexes are p and q, where $p + q = 1$. The Hardy-Weinberg Law states that under these conditions the population will reach an equilibrium distribution of the allelic combinations *AA*, *Aa*, and *aa* in one generation. The distribution will be $p^2(AA)$, $2pq(Aa)$, and $q^2(aa)$.

Why is this? Assume that the allelic combinations are originally distributed in any way that is compatible with the assigned relative frequencies. When the founding mothers and fathers produce their sex cells (gametes), each gamete contains either the allele *A* or the allele *a*. The relative frequency of *A*-bearing sperm will be p, that of *A*-bearing ova will be p, that of *a*-bearing ova will be q, and that of *a*-bearing sperm will be q. (Consider each event of gamete formation. The chance that the ovum formed contains *A* will be the frequency of *A* alleles among the females, and so forth.) When a sperm and an egg unite to form a zygote, there are four possible combinations:

A-bearing sperm	× *A*-bearing egg	produces *AA* zygote
A-bearing sperm	× *a*-bearing egg	produces *Aa* zygote
a-bearing sperm	× *A*-bearing egg	produces *Aa* zygote
a-bearing sperm	× *a*-bearing egg	produces *aa* zygote

Elementary considerations of probability tell us that the chance of a union between a particular kind of sperm and a particular kind of egg is the product of the relative frequencies of those kinds of sperms and eggs. When we apply this result to the situation under study, we can conclude that for the next generation the probability of being *AA* is p^2, the probability of being *Aa* is $2pq$, and the probability of being *aa* is q^2. This distribution results no matter what the initial distribution of allelic combinations, provided only that they conform to the condition on the relative gene frequencies. So, in particular, it will result when the organisms of the latest generation mate to produce grand-offspring for their parents. Hence, the distribution is an equilibrium.

The Hardy-Weinberg law is analogous to the law of inertia in Newtonian dynamics. It tells us what happens to a population in which no evolutionary change is occurring, in which "no forces intervene." (Elliott Sober [1984] gives a lucid discussion of this analogy.) Obviously, there are numerous ways in which the frequencies of genes and allelic combinations could be affected: we need only amend some of the assumptions that were made in the argument for a constant distribution. The sources of evolutionary change that figure in the qualitative picture of the last section can now be given their exact due. If some of the zygotes (the fertilized eggs) develop, grow up, and wander away, if those who stay at home are joined at mating time by interlopers from elsewhere, or if both things occur, then the rates at which allele frequencies and distributions of allelic combinations change will be determined by the rates of immigration and emigration. Similarly, we can appreciate the effects of mutation and hope to relate them exactly to the rates at which various types of mutation occur.

There are other evolutionary "forces" that we might not have anticipated. If there were a propensity for heterozygotes to transmit one allele rather than another to their gametes, then that would affect the frequencies of alleles and allelic combinations in the next generation. Likewise, retraction of our claims about random mating would subvert the calculations. If like systematically attracts like, or systematically attracts unlike, the resulting distribution of alleles among genotypes will be different in identifiable ways.

Evolutionary theorists who are interested in understanding the genetic basis of directional change focus primarily on natural selection. While the forces mentioned in the last paragraph may overwhelm selection in particular cases, Darwin's preferred candidate continues to figure as a major cause of evolutionary change. (However, there are important, unresolved questions about the importance of random changes in the evolutionary process. See chapter 7 for some discussion of this issue.) Natural selection is relatively easy to integrate with the approach adopted in deriving the Hardy-Weinberg law.

Suppose that the bearers of different allelic combinations do not all have equal chances of surviving and reproducing. For the sake of simplicity assume that the organisms under study that survive to reproductive age all have the same chance of finding mates, and assume that all are equally fertile. Selection acts only by virtue of differences in the probability of survival to reproductive age. Consider an initial distribution of allelic pairs among the zygotes as follows:

AA	Aa	aa
p^2	$2pq$	q^2

The distribution of allelic pairs at the time of reproduction will be

AA	Aa	aa
$w_{AA}p^2$	$w_{Aa}2pq$	$w_{aa}q^2$

where w_{aa} is the probability that an aa zygote will survive to reproductive age (similarly for w_{AA} and w_{Aa}). The organisms now mate to produce a new generation. The frequency of A gametes in the mating process will be

$$p' = \frac{w_{AA}p^2 + \frac{1}{2}w_{Aa}2pq}{w_{AA}p^2 + w_{Aa}2pq + w_{aa}q^2} = \frac{p(pw_{AA} + qw_{Aa})}{\bar{w}}$$

where $\bar{w} = w_{AA}p^2 + w_{Aa}2pq + w_{aa}q^2$. The frequency of A alleles among the zygotes of the next generation will also be p', and, as in our derivation of the distribution of allelic pairs in the Hardy-Weinberg case, the frequency of AA combinations among the zygotes will be p'^2.

Evolutionary change is change in gene frequencies. The change in frequency of the A allele is easily calculated:

$$\begin{aligned}\Delta p &= p' - p \\ &= (1/\bar{w})\,[p(pw_{AA} + qw_{Aa}) - p(w_{AA}p^2 + w_{Aa}2pq + w_{aa}q^2)].\end{aligned}$$

After a little algebraic manipulation this reduces to

$$\Delta p = (pq/\bar{w})[(w_{AA} - w_{Aa})p + (w_{Aa} - w_{aa})q].$$

If the probabilities of survival from zygote to adult remain constant from generation to generation, then the equation just derived can be iterated to compute the long-term trajectory of evolutionary change.

One relatively obvious point that emerges from our equation is that the change in gene frequencies would be no different if all the survival probabilities (w_{AA} and so forth) were multiplied by the same constant. What matters is the ratios of the probabilities to one another. We could have achieved the same result by proceeding—as population geneticists typically do proceed—with the greatest value of the ws normalized to 1 and the other values expressed as proportions of the maximal value.

This is as we might expect. The character of evolutionary change depends on the relative merits of the organisms that are struggling for existence, not on any absolute measure of their success. Suppose that there are two populations, the Cushies and the Grotties, and that, for each genotype, the survival probability for that genotype among the Cushies is a thousand times the survival probability for the genotype among the Grotties. Even though the Grotties may be rapidly dying out, the relative frequency of each allele in the Grotties will be just the same as the relative frequency in the Cushies. If $\Delta p > 0$, then A will be increasing in frequency relative to a in both populations, despite the fact that among the Grotties the total number of organisms, and therefore the total number of alleles, is dwindling from generation to generation.

Other anticipated results also follow. If $w_{AA} > w_{Aa} > w_{aa}$, then Δp will be positive. That is, if AA organisms are on average more viable than Aa organisms, and if Aa organisms are on average more viable than aa organisms, then the A allele will increase in relative frequency. Reverse the inequalities and you get the opposite conclusion. A more interesting consequence is obtained by considering the situation in which w_{Aa} is greater than both w_{aa} and w_{AA}. A little reflection will reveal that there is not only a solution to the equation $\Delta p = 0$, but, if the value of p should rise above the solution value $\hat{p}$, then in these circumstances Δp will be negative. If p falls below $\hat{p}$, then, by contrast, Δp will be positive. This means that there is a stable equilibrium of gene frequencies in cases where the heterozygote is more viable than either homozygote.

A well-known illustration is provided by the occurrence of sickle-cell anemia in some human populations. The AA homozygotes have normal hemoglobin. The aa homozygotes suffer sickle-cell anemia. Heterozygotes have sufficient normal hemoglobin to function normally; in some African environments they obtain a bonus through having a resistance to malaria that the normal homozygotes lack. In areas where malaria-carrying mosquitoes are prevalent, realistic figures for w_{AA}, w_{Aa}, and w_{aa} are 0.9, 1, and 0.04, respectively. (These figures are normalized in the way indicated above.) Substituting in our equation yields the conclusion that the equilibrium value of p, $\hat{p}$, is about 0.91. It is worth noting that the a allele would persist in the population, albeit at a lower frequency, even if it were strictly lethal, that is, even if w_{aa} were 0.

I derived the equations for change in gene frequencies by making some simplifying assumptions. Organisms of the three genotypes were supposed to be, on average, equally likely to mate and equally fertile. An obvious way to obtain a more general theory is to retain the equation and reinterpret the symbols w_{AA}, w_{Aa}, and w_{aa} that occur in it. Darwin was interested in the different abilities of organisms to pass on their characteristics to later generations. Perhaps we can capture his idea by assigning weights to the allelic pairs so as to reflect their capacities for transmitting the constituent genes. Instead of regarding the symbols w_{AA} and so forth as denoting survival probabilities, we can view them as standing for all those factors that affect gene transmission. If the AA organisms have a high probability of survival, then that increases the value of w_{AA}; if they have a high fertility (so that any single act of mating produces many offspring), that too increases the value of w_{AA}; if they have a low chance of mating, then the value of w_{AA} is decreased. We take w_{AA} to be a composite index of the factors that are important in survival and reproduction. It is the *absolute fitness* of the allelic pair AA. Of course, the direction of evolution is determined by the relation among the absolute fitnesses of the available combinations. We can proceed in practice by considering relative fitnesses. (For further discussion of the general notion of fitness, see the next section.)

A major part of the auto mechanics of evolution is the mathematical study of how different constellations of fitnesses affect the transmission of genes. Population geneticists try to fathom the genetic systems that occur in the course of evolution, the conditions to which they are subject, and on this basis they attempt to work out the expected trajectories for likely populations under likely conditions. Even the simplest models turn out to be disappointingly messy; hints of the difficulties emerge as soon as one considers three alleles at a locus.

Yet the subject is essential to the appraisal of evolutionary explanations and to the serious assessment of sociobiological claims. Hypotheses about the evolution of a trait depend on two assumptions. It must be supposed that the account given is compatible with theoretical population genetics. It must also be assumed that rival explanations, for which population genetics allows, have not been overlooked. Both assumptions may merit examination.

The story so far has omitted a number of important points that figure prominently in the elaboration of theoretical population genetics. We have looked only at the most elementary case, a single locus with two alleles. Complications ramify when we consider the problem of multiple loci. Here, considerations of linkage must be addressed. Moreover, genes at different loci may interact with one another: the phenotype associated with a pair of alleles at one locus may depend on the alleles present at some other locus (or loci). It is not hard to appreciate how such factors disturb the simple accounting used in our derivation of the equation for changes in gene frequency.

Another source of trouble lies in the assumption that we can derive descriptions of long-term change by iterating the effects found in a single generation. This assumption will be unwarranted whenever the organisms under study cannot be conceived as falling into discrete, non-overlapping generations and also whenever the fitnesses of allelic combinations do not remain constant from generation to generation. One common violation of the assumption of constant fitness occurs when selection is frequency-dependent—that is, when the fitness of an allelic combination varies with the allele frequencies in the population. A classic case of frequency-dependent selection occurs in fruit flies: males with certain arrangements of genes are at an advantage when their genotypes are rare. A little reflection should convince us that frequency dependence is likely to be prevalent whenever we are concerned with animal behavior. The advantages that accrue from a tendency to exhibit a particular type of behavior are likely to depend crucially on what the animals all around are doing. Unfortunately, as we shall discover later, the course of evolution under frequency-dependent selection belies some of our naive intuitions about the maximization of fitness.

Another important omission from my account so far is the neglect of stochastic factors (chance) in evolution. In deriving the Hardy-Weinberg law, I imagined an infinite population of organisms (or, at least, a population so large that it could be treated as if it were infinite). Why was this supposition needed? The answer is that I made tacit use of the law of large numbers. I assumed that I could move back and forth between probabilities and actual relative frequencies. However, actual populations of organisms are not infinite—and they may even be quite small. Imagine that we have a relatively small population, at maximal size for its available habitat, and that we want to understand the evolution of this population with respect to a single locus. There are N organisms in the parental generation, and they will produce N offspring. Suppose that the allelic combinations AA, Aa, and aa are all equally fit. If the frequency of A in the parental generation is p, will we expect the frequency of A in the next generation to be p? That is possible, but not likely. Just as we do not expect to get exactly ten heads when we toss a fair coin twenty times, we should not expect that the actual frequency of A among the zygotes will be the probability of finding A in a randomly chosen parental gamete.

Coin-tossing analogies are useful. Suppose that we were to repeat the experiment of tossing a fair coin exactly twenty times. Suppose we were to run this experiment again and again. We would obtain a distribution of values for the frequency of heads. The mean (average) of all these values would be close to ten, and it would be closer to ten the greater the number of repetitions of the experiment. Similarly, if we conceive of a large number of populations, each of size N, then if all the populations begin with a frequency of A equal to p, the next generations will show a distribution of values for the frequency of A, and the mean of these values will be p. Further, just as there will be occasional trials in which we obtain all heads or all tails, so there will be occasional populations in which either A alone is present or a alone is present. Such populations are said to be fixed for A or fixed for a. Fixation for either allele is more likely if N is small. Even if N is not small, there may be a reasonable chance of fixing A if p is close to one (or of fixing a, if p is close to zero). We can foresee that a probabilistic theory of evolution under selection can take account of the disturbing effects of the whims of organic fortune.

A quick look under the hood of the evolutionary machine should convince us that the connections are not always made quite as we would have expected them to be. There are more things in the workings of evolution than are dreamt of in the naive scenario of an advantageous allele sweeping to victory.

Fitness

The concept of fitness is important both to informal presentations of evolutionary theory and to the mathematical formulations of the last section. But what exactly is fitness? Why do evolutionary theorists spend so much time talking about it? Critics of evolutionary theory often revel in the fact that biologists and philosophers have offered some remarkably maladroit answers. In a popular caricature Darwin's evolutionary theory is simply the "principle of the survival of the fittest." Conjoining the caricature with the suggestion—which has some basis in the biological literature—that biologists take the fittest organisms to be those that leave the greatest number of descendants brings jubilation to the opponents of evolutionary theory. Critics may triumphantly proclaim that evolutionary theory has been reduced to the thesis that those who leave the greatest number of descendants leave the greatest number of descendants, that this is a tautology, and that the theory of evolution is therefore unscientific.

I shall postpone to the next section the task of explaining why the popular caricature misrepresents evolutionary theory. For now, it is important to clarify the concept of fitness, a concept that will loom large in our investigations of the claims of sociobiology. We can begin with an important distinction (see Mills and Beatty 1979 for insightful presentation). Sometimes biologists talk of individual organisms as being fit (as I did in the last paragraph). Sometimes they talk of combinations of alleles or phenotypic properties as being fit (as in the discussions of the last section). Let us start with the fitness of organisms.

Plainly, fitness has something to do with survival and reproduction. Just as obviously, the fitness of an organism cannot be identified with the actual number of offspring that it produces. Imagine the scenario of the nursery song:

> Two little monkeys bouncing on the bed;
> One falls off and bumps its head.

Suppose the monkeys have the same genes, that the bump is fatal, and that the surviving monkey goes on to a long life and many progeny. Because the inequality in actual reproductive prowess is simply a matter of fortune, it would be counterintuitive to declare that one monkey has greater fitness than the other. Nevertheless, it is fairly clear how to refine the idea that fitness is reproductive success. Given the genotype and the environment, there is a number of offspring that each of the monkeys could have been expected to produce. That number is the value of the (absolute) fitness of each monkey. The fallen monkey failed to produce that number of offspring, a failure that testifies to lack of luck, not lack of fitness.

The idea of an expected number of offspring may seem unfamiliar, but it is easily understood by comparison with commonplace examples. If I toss a fair coin ten times, there is an expected number of heads—five. I do not obtain exactly five heads each time I make the trial: sometimes I get six, sometimes three, very occasionally ten, and so forth. The expected number is just a probabilistically weighted average of the possible values. If p_n is the probability of obtaining exactly n heads, then the expected number is

$$\sum_{1}^{10} p_n n \quad (= p_1 + 2p_2 + 3p_3 + \ldots + 10p_{10}).$$

The fitness of an allelic pair (or of a phenotypic property) can be defined in terms of the fitnesses of organisms. The fitness of an allelic pair (relative to an environment and a population) is the average of the fitnesses of the organisms in the population that carry that pair of alleles. Similarly, the fitness of a phenotypic property (relative to an environment and a population) is the average of the fitnesses of the organisms in the population that have that property. (See Mills and Beatty 1979; it is possible to refine these definitions in various ways, but except as noted below, it will not be necessary to do so for the purposes of this inquiry.)

We can now begin to understand the role of the concept of fitness in evolutionary theory (and so obtain, as a side benefit, a simple response to the detractors). When we understand the presence of a trait in terms of a history of selection, we identify a property of the organisms we are studying and a benefit that the property confers on them. (Thus a student of the early evolution of birds may propose that the avian precursors found it advantageous to have feathers because of greater efficiency in regulating their body temperature.) The heart of our story is that the property we have picked out enhances fitness: it raises the expected number of descendants above the value that it would otherwise have had. Evolutionary explanations are not simple reiterations of the claim that ancestral organisms with the property were fitter than those without it. A genuine explanation must identify the grounds of the fitness, showing how the presence of the property that is singled out gave the organisms with that property a greater expected number of offspring. We point to an enhanced ability of moths to camouflage themselves, of sticklebacks to attract mates, of flowers to attract insects, of birdlike reptiles to improve their thermoregulation, or to any of a myriad of other properties. The determinants of fitness are highly diverse. The concept of fitness has no direct role in the explanation of the particular characteristics of particular organisms. What then is the use of it?

Reference to fitness is helpful in theorizing about the evolutionary process. When we want to understand the ways in which populations will evolve over time under the influence of particular conditions, we need to abstract from the diverse ways in which organisms can gain advantages. (See Sober 1984, 48–49, for concise discussion.) We are concerned with the general properties of fitness, conceived as a measure of the propensity to leave offspring. Our concern is met by introducing the symbols w_{AA} and so forth into the equations of mathematical population genetics. We frame theorems that apply to any situation in which a particular collection of fitness relationships are present—whether those fitness relationships derive from superior foraging ability, resistance to toxins in the environment, ability to attract mates, or whatever.

Moreover, once we have the general concept of fitness, it is possible to consider the ways in which fitness might be enhanced in a broad range of cases. If a species is confronted with a changing environment, will its members achieve greater fitness if they are assorted into a number of distinct types, each suited to one of the environmental conditions, or will they do better if they are not particularly well suited to any of the environments but able to cope with all of them? Under what conditions will the fittest combination of genes fail to become fixed in a population? These important theoretical questions about the evolutionary process can be raised (and answered, at least in part) once we abstract from the diversity of properties that enhance the ability to leave descendants, and introduce the notion of the expected number of offspring—the concept of fitness.

As there are uses, so there are temptations. Once having acquired the concept of fitness, it is easy to think that the canonical problem of reconstructing the history of life is to appreciate the ways in which the characteristics of organisms promoted their fitness. However, we should not forget that an explanation in terms of natural selection is ultimately committed to the existence of an underlying process in which gene frequencies change. Awareness of the complexities of population genetics should forestall the simple assumption that, once we have shown how a trait enhances fitness, we are done. It is easy to forget, in the flush of enthusiasm, that the genetics of evolution do not match single alleles with single characters, such that the alleles corresponding to advantageous properties march inexorably to fixation.

I shall return to this theme later, in the context of pop sociobiological proposals. For the moment I want to expose a different snare. Short-term reproductive success, enshrined in our definition of fitness, need not be the key to long-term evolutionary change. As J. M. Thoday pointed out (1953), organisms that are remarkable at producing offspring may not leave any descendants many generations hence. When our interest is in the long-term evolution of a lineage, we must make sure that fitness, as defined, does not differ from the propensity to leave descendants in later generations.

For a well-known illustration of the problem, we may turn to one of Fisher's many insights. Imagine a population of animals with a hereditary tendency to produce more male than female young. Within this population there arises a variant that has a propensity to produce female offspring. The variant animals will have a greater ability to transmit their genes, even though they may produce no more young than their rivals whose litters are male-biased. For in a population with a preponderance of males, sons have to compete for mates, while daughters are virtually assured of mating. As a result, the variant animals will have a greater expected number of grandoffspring than their rivals. Here we have an example of a delayed fitness effect: the expected number of offspring for both types of animal is the same, but, if we understand fitness in terms of the propensity to leave descendants in later generations, the variants have greater fitness. (For Fisher's beautiful treatment of the issue, see Fisher 1930, chapter VI; 1958, 158–160.)

The example reveals that the official definition of fitness as expected number of offspring will not always answer to the needs of evolutionary theory. It also reminds us that it is easy to take short cuts with fitness. We see that a characteristic increases life expectancy—and claim that organisms with the characteristic are fitter. We show that an animal that behaves in a particular way will copulate more—and we claim that the animal is fitter. We demonstrate that animals that follow a certain pattern of behavior will leave more offspring—and we conclude that those animals are fitter. All the inferences deserve scrutiny, even the last. Other things being equal, the more children we have, the more likely are our genes to find their way into remote generations. But sometimes other things are not equal.

What Is Evolutionary Theory?

Our quick tour of some highlights of evolutionary biology prepares us for a look at the methodological issues. To appraise sociobiology in general and pop sociobiology in particular, we need a clear view of the structure of evolutionary theory and of the ways in which parts of evolutionary theory are tested, confirmed, and sometimes rejected. We can begin with a look at the character of sociobiology's presumptive parent.

As I noted at the beginning of this chapter, Darwin's original theory of evolution is not readily identified as a small set of axioms and their logical consequences. The fundamental principles advanced in the *Origin* were common knowledge, and the passage from those principles to the principle of natural selection is deductively trivial. Yet, with good reason, Darwin's proposals caused a revolution in biological science.

How? I propose to adopt a perspective on Darwin's theory (or, as we shall see later, on Darwin's theories) and on the versions of evolutionary theory that descend from Darwin, that is different from the one in vogue among scientists, philosophers, and historians. Instead of thinking that the content of any theory worthy of the name can be identified with a small set of axioms—Newton's laws, Maxwell's equations, and so forth—I suggest that some theories are best approached by focusing on the questions they address and the patterns of reasoning they offer for answering those questions. (I have defended this general perspective and applied it to a number of examples, including evolutionary theory, in various places; see Kitcher 1981, 1982a, 1984, 1985.) The approach avoids trivializing Darwin's achievement, and it also does justice to the most striking feature of the *Origin:* Darwin's multiplication of examples to demonstrate the importance of uncontroversial principles whose significance nobody had previously appreciated.

Because Darwinian evolutionary theory is something that comes in a large number of versions, it will help to begin with what all the versions have in common. Darwinian evolutionary theory is a three-part affair. It focuses on a collection of questions, it supplies strategies for addressing these questions, and it offers some general views about the evolutionary process. The components are not independent. In the light of general ideas about the process of evolution, scientists come to adopt certain preferred strategies for answering evolutionary questions and to accept certain questions as legitimate.

The kinds of questions addressed by evolutionary theory overlap those that Darwin brought within the domain of science. Evolutionary theorists continue to wrestle with the topics of why organisms live where they do, why they share morphological and physiological characteristics, why the fossil record shows the features it does, why plants and animals possess the structures and behavior patterns they do. More exactly, all versions of contemporary evolutionary theory accept the legitimacy of at least some of these investigations. As we shall see, there are grades of allegiance to Darwin, and it is possible to hold that some of the inquiries that the master began were misconceived.

Evolutionary questions are answered by using strategies that descend from those introduced by Darwin. To explain the distributions of organisms, we relate histories of descent with modification. To account for the presence of apparently adaptive characteristics, we add to a history of descent with modification a hypothesis about the causes of the evolutionary changes that we envisage. It will be helpful to have a name for these structured narratives. Let us call them "Darwinian histories."

The kinds of Darwinian histories that are put forward will depend on general ideas about the process of evolution. Prominent among these will be the collection of views about the genetics of evolutionary change. Population genetics consists in some general ideas about what kinds of genetic systems are important in evolution and what kinds of forces impinge on evolving lineages. These general ideas serve as the basis for models of the evolutionary process, obtained by working out the mathematical analysis of the allegedly important situations, just as we did in the simple examples considered earlier. For any acceptable Darwinian history it will be supposed that there is a genetic model that underlies it. For any legitimate evolutionary question it will be assumed that the genetic models do not allow for numerous alternative Darwinian histories, each compatible with all the evidence we could ever obtain. Good evolutionary questions should not be effectively unanswerable. Thus the general ideas of the population geneticist constrain the activity of providing evolutionary explanations.

The third component of evolutionary theory divides neatly into two parts. There are claims that certain types of situations occur (more or less frequently) in the course of evolution. Then there are detailed mathematical analyses of what happens when these situations arise. The division is illustrated by our treatment of population genetics. Evolutionary theorists may propose that there are common situations in which two alleles occur at a locus and in which the heterozygote proves to be superior. They also derive a result to the effect that, in any situation of heterozygote superiority, then, discounting stochastic factors, the population will reach a polymorphic equilibrium (a state in which selection maintains more than one type of organism in the population). The proposal is a biological claim about the character of evolution. The precise result is a piece of mathematics.

So much for evolutionary theory in general. Let us now consider the differences among particular versions. Many kinds of evolutionary theory have their origin in the *Origin*. The differences result from divergent conceptions of Darwinian history, reflecting varying degrees of involvement with such themes as adaptation and evolutionary gradualism.

There is a minimal notion of Darwinian history, one that embodies the fewest assumptions about the tempo and mode of evolutionary change. On this conception a Darwinian history for a group of organisms, between two times, with respect to a family of properties, consists in the specification of how the group exhibits the properties in the family between the two times. For example, we might trace the growth in brain size along a hominid lineage by specifying the distributions of sizes during an interval of time; or we might describe the way in which the range of a species of deer has contracted with time, relating this to changes in the habitat or in the presence of predators.

This minimal conception can be used to answer some biological questions. To understand the present (disconnected) distribution of the pangolins—who now live only in Southeast Asia and Southern Africa—we might simply relate the history of a group that was once continuously distributed, describing the way in which the poor pangolins in the middle declined from generation to generation. To explain the similarities in skull morphology between dogs and bears, we might describe the common ancestor and the successive modifications along two independent lineages that have led to the contemporary animals. Even the minimal conception has its biological uses.

Cautious Darwinians believe that we are never in a position to give more than minimal Darwinian histories. They hold general views about the character of the evolutionary process. They may even think that natural selection has been an important agent of evolutionary change. However, they refrain from trying to identify the workings of selection (and of other evolutionary agents) in the history of life.

For cautious Darwinians the controversy about sociobiology is soon ended. To corroborate hypotheses about the causes of the evolution of behavior is impossible—no more difficult than what usually passes in evolutionary theory, but impossible nonetheless. Ambitious Darwinians will not achieve so easy a resolution. For them the question of adaptation is a serious question. They hold that it is possible to give a justified account of the evolutionary forces that have shaped some of the present characteristics of plants and animals. There is no bar in principle to the practice of giving accounts of this kind when the characteristics are behavioral and the animals are human.

A stronger conception of Darwinian history involves us in specifying not only the changes that take place along an evolving lineage but also the causes of these changes. Those who appreciate the possibility of such histories are not necessarily exclusive devotees of natural selection. One type of ambitious history takes the form of identifying the benefits that possession of a particular property conferred on its past bearers. So we claim that butterflies that mimic unpalatable forms gained—and continue to gain—an advantage by fooling their predators. But this should not be our only paradigm. Darwin himself appreciated the "correlation and balance" of characteristics. He knew that "cats with blue eyes are invariably deaf," that "pigeons with feathered feet have skin between their outer toes" (1859, 12), and he predicted that discovery of "the laws of correlation and balance" would be important to evolutionary theory. Modern genetics and embryology advance the project Darwin envisaged. We know that there can be cases of pleiotropy (in which one gene affects many traits), of linkage (in which genes are bound together in groups that challenge the disruptive powers of evolution), of allometry (in which change in the size of one structure induces change, at a different rate, in the size of another structure). Knowing this, we can see that ambitious Darwinian histories may take any of a number of forms. Sometimes it may be appropriate to identify an ensemble of properties, to show that that ensemble brought benefits to ancestral organisms, and to claim that a particular property is present just because it belongs to the ensemble. We can break the spell of naive selective thinking, which assumes that for each trait there is a corresponding advantage.

Sibling rivalry promotes the life of the Darwinian family. Ambitious Darwinians can bicker with one another about the merits of various kinds of Darwinian histories, and they can join hands to point to the narrow-mindedness of their cautious colleagues. There are other kinds of family disagreements, too. Committed gradualists propose that reconstructions of the history of life should reveal the accumulation of slight modifications. Others believe that the Darwinian histories should show (geological) moments of brief excitement in the middle of long stretches of constancy. Contemporary evolutionary theory appears, hydra-headed, in a number of variations.

For our purposes only one type of internal dispute is relevant. Should evolutionary theorists take up the Darwinian task of explaining the presence of the characteristics of organisms? Or should they adopt the easy agnosticism of the cautious Darwinians? If they are entitled to be more ambitious, how are they so justified? How do we obtain evidence for explanations of the evolutionary causes of the characteristics of organisms? If ambitious Darwinism is subject to methodological rules, do the pop sociobiologists play by those rules?

A Red Herring

Unfortunately, we must begin by clearing out of the way an influential mistake. Critics of evolutionary theory are sometimes inclined to try a quick line of indictment: evolutionary theory, they claim, is methodologically bad science, perhaps even "pseudoscience." Ironically, evolutionary theorists who oppose pop sociobiology often make similar remarks. Our first task should be to advance beyond the sloganeering to the serious issues.

There are slogans for all seasons, and for our time the magic word has been "falsifiable." Thanks to the influence of Sir Karl Popper, many scientists are convinced that there is an important distinction between science and pseudoscience, that real science is falsifiable while pseudoscience is unfalsifiable. Challenged by critics who dispute the status of evolutionary theory, prominent scientists rise to the bait and give much energy to the task of specifying statements that, if true, would falsify evolutionary theory. As if purposely to confuse an already obscure situation, however, some of those who insist on the methodological purity of evolutionary theory lambaste sociobiology as unfalsifiable and thus condemn it as pseudoscience. There results the undeniable impression that a double standard has been applied.

Early critics of Wilson's *Sociobiology* charged that "the entire theory is so constructed that *no tests are possible*" (Sociobiology Study Group, in Caplan 1978, 287; italics in original). The criticism lives on in recent discussions (see, for example, Leeds and Dusek 1983, ix, 75–79). Pop sociobiologists have responded vigorously by emphasizing the "predictions" furnished by their theory (Lumsden and Wilson 1983a, 38–41; Alexander 1979, 156ff.). The upshot is thorough confusion. Proponents of pop sociobiology and many of their critics seem to agree on the falsifiability of evolutionary theory (compare Gould 1981 with Alexander 1979, 7–8, 19–22). The debate seems to degenerate into a series of blunt assertions, with one side claiming that (pop) sociobiology and evolutionary theory are fundamentally different and the other maintaining that they are methodologically on a par.

The villain of the piece is the notion of falsifiability. Unless we can dispose of the red herring that there is a clear criterion—falsifiability—for demarcating science from pseudoscience, then the methodological issues surrounding sociobiology will be swathed in permanent obscurity. Opponents will insist on an irrelevant standard. Champions will labor to produce an irrelevant defense.

In one obvious sense, evolutionary theory, sociobiology, and all major scientific theories are unfalsifiable. In another equally straightforward sense, any favored doctrine can be formulated in a way that makes it falsifiable. For if theories are taken to be small sets of general laws (Newtonian dynamics consists of the three laws of motion, electromagnetic theory comprises Maxwell's equations, and so forth), then the theories have no observational consequences. There is no statement whose truth or falsity can be determined by observation and whose falsity is incompatible with the theories (construed in the suggested way). To relate the theory to observationally determinable statements, it is necessary to use auxiliary assumptions (often a very large number of them); and if the observations turn out badly, the blame can always be laid at the door of these extra hypotheses. On the other hand, if the commitment to falsifiability requires only that a theory be part of some system that, taken as a whole, yields observational consequences, then the condition is toothless. Whatever statement we choose, we can generate any "predictions" we fancy. If our cherished doctrine is the hypothesis that "it's Love, Love that makes the world go round" and we are eager to predict that elephants eat peanuts, then all we need to do is make sure that our system contains the statement "If Love makes the world go round, then elephants eat peanuts" and the desired "prediction" will be forthcoming. The moral is obvious. The road to a sophisticated understanding of the methodological issues surrounding evolutionary theory and sociobiology does not lie through the simple notion of falsifiability that typically figures in the debates. (See Kitcher 1982a, 35–44, for further details.)

To appreciate the need for a different methodological perspective, let us briefly consider the failure of struggles to show that evolutionary theory is falsifiable. A much-quoted passage from the *Origin* inspires defenders of evolutionary theory to claim that Darwin himself recognized the importance of falsifiability:

> Darwin's statement far antedated methodologies proposed and made prominent by contemporary philosophers of science, some of whom have doubted the validity of evolution as a scientific theory because of what they see as the absence of suitable falsifying propositions or operations. Darwin said the following: "If it

> could be demonstrated that any complex organ existed, which could not possibly have been formed by numerous, successive slight modifications, my theory would absolutely break down." This challenge is one of several which showed that Darwin was trying to postulate ways in which his theory could be falsified. (Alexander 1979, 11–12; see also 19–20)

Not only is Alexander too quick in crediting Darwin with methodological insight, but the passage cited would not touch the central objection pressed by the most acute of the early reviewers of the *Origin*. To be sure, we can announce that a doctrine is "falsifiable" by pointing out that a particular statement is incompatible with it—perhaps our chosen statement will simply be the denial of the doctrine—but this procedure should cut little ice with critics unless the favored statement is one whose truth or falsity can be ascertained by relatively direct observation. Anyone seriously worried about the falsifiability of evolutionary theory will regard as grotesque the suggestion that we might "demonstrate" the existence of a complex organ that "could not possibly have been formed by numerous, successive slight modifications." How could we *demonstrate* any such thing?

The precise point made by Darwin's detractors was that evolutionary theory had ready-made strategies for accommodating any possible observations of animals, plants, and rock strata. Fleeming Jenkin catalogued the resources he took to be available to the committed evolutionist:

> He can invent trains of ancestors of whose existence there is no evidence; he can marshal hosts of equally imaginary foes; he can call up continents, floods, and peculiar atmospheres, he can dry up oceans, split islands, and parcel out eternity at will; surely with these advantages he must be a dull fellow if he cannot scheme some series of animals and circumstances explaining our assumed difficulty quite naturally. (Jenkin 1867, 319; see also Hull 1974, 144, 264)

The force of Jenkin's indictment and the ineptness of the reply that Alexander ascribes to Darwin can readily be seen if we turn our attention to a different case. Creationists stand little chance of establishing the scientific respectability of their approach by announcing that if it could be demonstrated that the universe could not possibly be less than 10,000 years old, their theory would collapse. Our reasons for denying credit to the Creationists would parallel Jenkin's: they can induce cataclysms of whose existence there is no evidence;

they can modify decay rates at will and invoke unknown modifications of geological states; they can alter the speed of light, create light waves that emanate from the interstellar spaces, and call up the operations of a supernatural being; surely with these advantages they must be dull fellows if they cannot overcome any apparent evidence for an ancient universe.

Those who follow Alexander in finding easy ways to establish the methodological legitimacy of evolutionary theory will find it equally simple to dismiss doubt when the subject under scrutiny is pop sociobiology. Critics who allow that evolutionary theory could be "falsified" in direct, readily describable fashion discover that their concessions come back to haunt them. Zealots rightly claim that there are equally direct "falsifications" of pop sociobiology and so turn back critical challenges. We should reject the slogan that generates the dispute. Employing the falsifiability criterion to settle methodological issues in either the theory of evolution or pop sociobiology is about as profitable as trying to perform delicate surgery with a rusty kitchen knife.

Nevertheless, there is something right about the Popperian intuition. Deliberately protecting pet ideas from the threat of criticism is bad science—and, in general, bad thinking. As Fleeming Jenkin saw, theories that come equipped with tactics of self-protection should be viewed with suspicion. The appeal to falsifiability starts from these sensible points, but it fails to articulate them in a sensible way. We can do better by applying the picture of evolutionary theory developed here.

Jenkin's Legacy

Jenkin argued that Darwin's theory of evolution received no genuine support from the observations that Darwin assembled because the theory ran no genuine risks of defeat. To understand the methodological status of evolutionary theory, we need to do two things. First, we must show how a particular sequence of observational findings would have made it rational to abandon the theory. Second, we must examine the ways in which the actual evidence has confirmed evolutionary theory and attempt to understand which parts of the theory receive the strongest support. This section will take up the first task.

Our conception of Darwinian evolutionary theory as a three-part affair enables us to focus the issue. No methodological scruples should attach to population genetics. Some of its claims are matters of mathematics. The remainder are hypotheses to the effect that particular kinds of genetic systems and constellations of forces occur—some are infrequent, some common—in the history of life. These claims are tested in straightforward ways through the investigation of laboratory and natural populations.

The methodological difficulties lie in attempts to answer questions about groups of organisms by relating Darwinian histories for those groups. Skeptics worry that favored Darwinian histories can be maintained come what may and that there will always be alternative Darwinian histories available to ease the evolutionist's burden, should things start to prove difficult. Both kinds of anxiety are unfounded.

Individual Darwinian histories are rationally abandoned when observational findings continually generate puzzles that resist solution. Imagine that evolutionary biologists propose to explain the distribution of the tenrecs—a primitive group of insectivorous mammals, confined for the most part to the island of Madagascar. They allege that there was an ancestral population living on the African mainland, that Madagascar was once sufficiently close for the insectivores to reach it (perhaps the colonization even took place before Madagascar became detached from the continent), that the animals that later eliminated the mainland descendants of the primitive insectivores arose only after Madagascar was too far from the continent for predators or competitors to reach it, that the tenrecs are all related through descent from the primitive mainland insectivores. Observations, supplemented with independently testable hypotheses from geology and physiology, could undermine the proposal.

Consider the possibilities. Fossils might appear in the wrong places at the wrong times. Or the fossil record might be blank, and blank in ways that would be improbable if the Darwinian history were true. By comparing the fossil records for different groups of animals and performing mineralogical experiments, we can estimate the probability that the absent fossils should be AWOL (see Meehl 1984; Raup and Stanley 1978, chapter 1). Hypotheses about animal dispersal can be tested by trying to disperse the relevant animals, or, at least, animals that are physiologically similar. Hypotheses about the past positions of Madagascar and Africa can be tested by using results from geology that are independent of evolutionary theory. We can scrutinize claims about animal competition by arranging the right conditions, looking to see if our hypothetical predators would indeed have eliminated our hypothetical victims. We can investigate the anatomical, physiological, and biochemical relationships among the contemporary tenrecs.

Crucial "falsifications" are hard to come by. An enthusiastic author of the account of tenrec distribution can wave away the finding that the fossils are so rare that the probability of such colossal absence is extremely low ("This seems to be one of those unlucky cases, but maybe we need to look a bit harder"); can treat the inability of comparable contemporary mammals to replicate the aquatic feats of their alleged ancestors as a temporarily unexplained puzzle ("Perhaps they reached Madagascar in ways that we haven't yet thought of"); can dismiss the ability of contemporary tenrecs to live in harmony with the kinds of animals that allegedly eliminated their mainland ancestors ("Maybe the island tenrecs evolved a protective behavior that the mainland mammals did not have"); can blame the findings of diversity in anatomical and biochemical properties on a poor choice of characteristics ("Those traits are highly labile, those molecules are subject to variation"). Each hypothetical negative test will be met by a retreat. The imagined author admits that there are puzzles, that the account cannot be elaborated in the most straightforward way, and that there is no *specific* alternative development of it. The story is progressively weakened: explanations are replaced by hopeful hints about the possibility of explanations. A single result does not compel abandonment of the Darwinian history. Sometimes it is reasonable to insist that it is worth looking for revised auxiliary assumptions and that, for the time being, the findings must be seen as unresolved puzzles. Yet when puzzles mount and remain unresolved, it is reasonable to give up the Darwinian history in favor of something different.

A negative test result is an observation statement (that is, a statement whose truth or falsity can be ascertained in ways that do not depend on assumptions from evolutionary theory) that is logically incompatible with the historical claims and auxiliary assumptions tacitly or explicitly accepted prior to the test. Darwinian histories are rationally abandoned when puzzles multiply at a faster rate than explanatory successes. (For similar methodological ideas, presented in a more general context, see Lakatos 1969 and Laudan 1977.) The sensible core of the idea that the individual accounts put forward by evolutionary theorists are falsifiable is the thesis that certain combinations of observational findings could generate a sufficient number of puzzles to make it unreasonable to persist in defending the accounts in question.

We are half done. What remains is to see if evolutionary theory is so flexible as to be able to shrug off the failures of particular evolutionary explanations. Skeptics worry that Darwinian histories are like pennies from heaven. If they should be lost, there will be others to replace them. In extremis, true believers may resort to the bare assertion that there is some (unknown) Darwinian history that will overcome the difficulties, hailing the task of finding this history as an "interesting research project."

Two points dispose of the charge. First, there are global constraints on the collection of Darwinian histories. Pleas of bad luck in the face of absent fossils look suspect when we are considering a single Darwinian history. They would take on a ludicrous appearance if we were to find that the entire fossil record was marked by missing ancestors vastly in excess of the estimates from our theories of the fossilization process. Too much bad luck begins to look like carelessness. Even more obviously, our Darwinian histories must form a coherent collection. If we suppose that a group of animals was present at a particular place at a particular time and that it had an impact on another species, then we cannot ignore these hypotheses when we turn our attention to a different group. We must achieve a consistent collection of claims about times and places of origin, characteristics, abilities, interactions, and so forth.

The second, and more important, point is that evolutionary theory *as a whole* might encounter the same predicament that we envisaged for a particular Darwinian history. Sometimes, when Darwinian attempts at explanation go awry, no substitute Darwinian history is available and the evolutionary theorist is driven to a confession of ignorance. Imagine that we have proposed the history sketched above for the case of the tenrecs, but that we face recurrent difficulties in explaining the initial colonization. Experiments with contemporary mammals convince us that our hypothetical pioneering insectivores would have been unable to swim the distance, that they are too large to be carried by birds, incapable of clinging sufficiently tightly to pieces of driftwood to survive the rough seas, and so forth. Considerations of morphology—the very structural similarities that we take as indications of recent common ancestry in numerous other instances—do not leave us the option of supposing that the mainland and insular forms are unrelated. Geological findings eliminate any possibility of a land bridge that would span the sea channel. We are forced to confess ignorance, to admit that we have no Darwinian history to explain the observed distribution of the animals.

Confessions of ignorance do not come gratis. Just as it would become unreasonable to maintain a Darwinian history in the face of mounting puzzles and few successes, so it would become unreasonable to continue to defend evolutionary theory in the face of an increasing number of instances in which no Darwinian history was available, unless the increase was offset by a greater number of new successes. A puzzle for evolutionary theory is a question that is admitted as legitimate, for which there is no available Darwinian history. Puzzles arise when the observational findings force us to abandon every Darwinian history that we can think of. If puzzles for evolutionary theory were to arise with great frequency, resisting determined efforts at solution, and if there were no compensating instances in which evolutionary theory was applied to answer new questions, then it would be rational to abandon evolutionary theory and seek a different approach to the history of life.

A sad, imaginary, destiny for evolutionary theory would have it collapsing in just the ways that Ptolemaic astronomy succumbed in the sixteenth and seventeenth centuries. Ptolemaic astronomers were committed to the enterprise of explaining the orbits of the planets by identifying those orbits as combinations of circular motions, and there were explicit constraints on how the circles were to be combined. Comparison of a detailed proposal about the orbit of a planet with the observed motions tested the skill of the author of the proposal, not the theory that was being applied (see Kuhn 1970, 36–40). Ultimately Ptolemaic astronomy itself was called into question as the result of persistent failure by the best astronomical minds to resolve the puzzles that had arisen. Ptolemaic astronomers were compelled to insist again and again that there was some combination of allowed motions that would account exactly for the planetary orbits, until at last some students of the planets were rationally moved to cry "Enough!" Darwinian evolutionary theory was potentially vulnerable to a similar predicament—and a similar fate.

The Happy Ending

Understanding how evolutionary theory might have failed helps us to understand how it can and does succeed. Individual Darwinian histories have won support, time after time, by making claims about the existence and characteristics of past and present organisms that can be checked observationally. The hypothesis that early mammals reached Australia by way of Antarctica was recently confirmed through the discovery of mammalian fossils in the Antarctic. Wilson and two colleagues (F. M. Carpenter and W. L. Brown) used a study of the comparative morphology of living ants to propose a hypothesis about the structure of a primitive ancestor. Dramatic support was obtained when researchers discovered fossils in amber, fossils that corresponded very closely to the hypothetical ancestor. There are literally thousands of similar cases.

The blanket question, How is evolutionary theory tested and confirmed? is best broken into parts. We can start with the confirmation of the central claim of the theory, the claim that it is possible to address major families of biological questions by constructing Darwinian histories for groups of organisms. The claim is supported by performing the trick again and again. It receives especially strong support when we manage to produce histories in situations where the background constraints seem to defy our ingenuity—charting the route of the marsupials to Australia, finding the form of the ancestral ant—and where the articulation of the history leads us to some highly surprising observational discoveries.

It is important to see that the central claim can be confirmed even when we have no basis for accepting any particular history about a group of organisms. If evolutionary theorists are challenged to find a Darwinian history in some particularly puzzling case, if they produce two such histories, each of which generates some surprising new discovery, then the credit of the general theory is increased, despite the fact that neither rival can claim exclusive right to our credence.

Consider next the confirmation of our general views about the evolutionary process. Hypotheses asserting the existence of various kinds of genetic systems with particular relations among the fitnesses are verified in the laboratory and in the field when researchers find populations with those systems and those fitness relations. To confirm the hypothesis that there are single-locus genetic systems with heterozygote superiority, we can point to the case of sickle-cell anemia. Claims about the *common* existence of certain types of genetic systems require sampling across populations. Claims about the relative weakness of mutation as an evolutionary force turn on our measurement of mutation rates in a wide variety of populations. There is nothing mysterious here.

Hypotheses of the kinds considered in the last paragraph are conjoined with theorems about what occurs in hypothetical populations with specified genetic systems, fitness relations, rates of mutation, and so forth. The theorems, as my description of them implies, depend on the application of the laws of probability to the description of the hypothetical situation (possibly with the aid of some very general, and independently confirmed, principles of cell biology). There is nothing very mysterious here either.

The crucial issue in the confirmation of evolutionary theory concerns the confirmation of particular Darwinian histories. The main strategy in justifying particular scenarios for the history of life is one that is well known to practicing biologists, though not much emphasized by recent philosophers of science. Darwinian histories win their way to the top through the elimination of rivals. Given our commitment to the central claim of evolutionary theory, given our justified general views about the character of the evolutionary process, given well-supported constraints from independent fields (physiology, geology, and so forth), there is, in effect, a *space* in which potential Darwinian histories can be located. In ideal cases we can actually locate the hypotheses, list the rivals, and proceed to acquire evidence that will eliminate all but one of them. So, for example, in approaching a biogeographical problem, we may start from information about the relations among the species under study, their present dispersal powers, the previous positions of the continents, and so forth. Using the constraints, we may be able to draw up a finite list of possible dispersal routes. (Example: the marsupials could only have reached Australia from Antarctica or from the Indonesian archipelago.) A search for fossils may disclose them in regions unique to one of the dispersal routes. So we achieve evidence for a particular Darwinian history.

Dobzhansky's investigation into the relations among populations of *Drosophila pseudoobscura* and *Drosophila persimilis* provides an illustration of the same procedure in the context of a different type of problem. Dobzhansky and his co-workers discovered that the order of the genes in particular chromosomal regions varies among populations. They were able to formulate rival hypotheses about evolutionary relationships, using the principle that if two arrangements cannot be obtained from one another by a single inversion of a chromosome segment, then they are related through an intermediate arrangement. (For example, if the order of the genes in one population is *ABCDEFGHI* and the order in another population is *AEHGFBCDI*, we hypothesize an intermediate population whose gene arrangement is *AEDCBFGHI*.) On this basis Dobzhansky was led to look for previously unsuspected arrangements. By discovering some of them, he was able to eliminate some evolutionary scenarios and confirm others (see Dobzhansky 1970, 129–131).

The method of confirmation used in these cases and in a host of other evolutionary examples is *contextual* elimination of rival hypotheses. It depends on our having constraints that enable us to construct the space of possible Darwinian histories. It is possible that, given some observational findings, we shall be led to believe that none of the available hypotheses was correct and thus to reconsider our original formulation of the constraints. It is possible that general evolutionary theory itself will falter and collapse. But while that theory is justified by its long record of successes, while we have independently justified hypotheses about the physiology of the organisms with which we are concerned, while we have well-supported views about certain aspects of their evolutionary history, all such information provides the context within which our current efforts at finding a unique hypothesis are conducted.

I have illustrated the common strategy of confirming evolutionary accounts by considering minimal Darwinian histories. That is no accident. Minimal Darwinian histories are more readily confirmed than their ambitious cousins. The reason is easily found. Ambitious histories frequently agree on their hypotheses about descent with modification but do not allow any means of discriminating the rival suggestions about the causes of the modifications. While the evidence for a single view of the pattern of relationships among the carnivores may be compelling, there may be numerous possibilities for explaining the causal processes that underlie that pattern.

The point may lead to despair or to cautious Darwinism. However, it is premature to suppose that there are no means for distinguishing rival hypotheses about the causes of evolutionary change. Cautious Darwinism is a bit too cautious. Sometimes we can gain evidence for the conclusion that a particular property in a particular group has become prevalent through selection, and we can identify the basis of selection, the benefit that the property brings to the organisms that possess it. It will help to begin with what is commonly hailed as the best case, so that we can have not only a clear view of how the confirmation of ambitious histories is possible but also some conception of what may be lacking elsewhere.

A classic demonstration of the impact of selection on a natural population is Kettlewell's series of investigations of industrial melanism in British moths (Kettlewell 1973). Prior to the industrial revolution the speckled form of the peppered moth, *Biston betularia,* was common in Britain. With increasing urbanization moths with black coloration, melanic variants, became ever more conspicuous near industrial centers. Speckled moths continued to predominate in relatively unpolluted parts of rural England. The change in frequency of the melanic trait in populations near cities is easily explained. Pollutants kill lichens that normally grow on tree trunks. Lichen-covered trunks form a protective background for the speckled form, which is nearly invisible to both humans and, apparently, predatory birds when it is resting against them. When the lichens are killed, the surface of the tree trunk is uniformly dark, and the speckled form is much more conspicuous. The melanic form is more conspicuous than the speckled form against a lichen-covered trunk, but it is better camouflaged against a uniformly dark trunk. In all cases superior camouflage confers greater life expectancy because of the diminished risk of predation.

The explanation accounts for the observed distribution of forms of the peppered moth, but it is not hard to construct rival stories. Perhaps the larvae of the speckled form are less viable in industrial areas because of the presence of pollutants (which the melanic larvae are able to resist). Perhaps speckled forms have a greater variation in fertility, the variation depending on the density of food items, which are more readily available in rural areas. To be justified in believing that one has identified the properties that were selected and the benefits that they conferred, one must have reasons for believing that any alternative effects are slight in comparison with the advantage (or advantages) that are heralded as the focus of selection. In other words, one must eliminate alternative Darwinian histories.

The elimination may be accomplished either by appealing to background theoretical considerations or by performing experiments. We might invoke our knowledge of the physiology and development of the moths to resist the suggestion that males of the speckled form are less energetic at mating in the vicinity of cities. From what we know about physiology and development, there is no basis for any such distinction, and the hypothesis is excluded from the space of possibilities. In other cases rival accounts are eliminated by making the kinds of careful tests for which Kettlewell is famous. He observed the rates at which birds picked the two kinds of moths off various kinds of trees. He measured the viabilities of larvae under different conditions. He compared the fertilities of the two kinds of moths when they were reared in different environments. The results showed clearly that superiority in camouflage is an important part of the selection process.

Ambitious Darwinian histories identify the causes of evolutionary changes. Those that appeal to natural selection, henceforth *selectionist* histories, identify a property (or properties) of the organisms under study and the benefit (or benefits) that the property confers on its bearer. If we are to draw general morals from Kettlewell's achievement, it will be important to recognize that there are a number of properties that distinguish the melanic forms from the speckled moths. To name but three: the two forms differ in the allele present at a single locus, they differ in the amounts of melanin produced in certain tissues, and, most obviously, they differ in color. In this case it is the obvious difference that is the crucial one. (If Kettlewell had not been so careful, and if his conclusions about the importance of camouflage had been mistaken, we can easily imagine how he would have been accused of concentrating on "superficial differences.") In industrial areas there is selection for moths with dark coloration, and any moth with dark coloration (but similar to the speckled forms in other respects) would have increased fitness, even if its coloration were produced in some different way.

Moreover, even if we are right about the property that has been selected, there may be alternative benefits from possessing that property. Coloration may be relevant to success in evading predators; it may also be relevant to capturing prey or to increasing efficient temperature regulation.

Kettlewell and his colleagues were able to explore the possibilities of error by systematically investigating rival hypotheses about selected properties and resultant benefits. Detailed genetic, physiological, and developmental investigation was the key to making the investigation systematic. Kettlewell knew that melanic moths and speckled moths differ in a single allele, and he knew a great deal about the mechanisms through which that allele is expressed. As a result, he could draw up an inventory of possible properties and benefits. He could then eliminate many of the hypotheses that figured on the resultant list, gaining strong support for his own proposal that the selected property is dark coloration and the benefit conferred is reduced predation in polluted environments.

Kettlewell's study is a classic. As John Tyler Bonner admits, there are a "mere handful" of other "well-authenticated examples" (Bonner 1980, 187). For almost all traits of almost all animals, we lack the knowledge of genetic, physiological, and developmental differences that brings system to Kettlewell's work. The necessary information is very difficult to collect—a cause for much frustration among those who hope to understand the causes of present characteristics.

In the pursuit of ambitious Darwinian histories, three types of situations may arise. *Ideal* cases are those in which our background knowledge enables us to construct the space of possibilities and our observations and experiments allow us to eliminate all but one of the competing histories. Cases of *underdetermination* are those in which we can recognize two or more alternative histories that cannot be discriminated by the information available. (This may occur even when our background knowledge is inadequate to the construction of the space of possibilities.) Neither of these types of situations is methodologically problematic. Ideal cases are rightly prized, cases of underdetermination are seen as examples in which we must acknowledge at least temporary ignorance. It is the third kind of situation that causes trouble.

What should we say when we have one detailed Darwinian history—a history that may even have led us to some surprising results—but know that we are too ignorant of the ecology, genetics, physiology, and development of the organisms under study to articulate many of the possible rival accounts? Confronted with situations of this kind, skeptics will urge caution, enthusiasts will claim that the single history earns the right to our acceptance. Each party has something to say on its behalf.

Skeptics can point out that there are potential rival accounts that have not yet been elaborated in detail. The fact that they have not yet been articulated should not be held to their discredit—they are not intrinsically implausible—but to our ignorance of important biological facts about the organisms. We cannot parlay that ignorance into a device for obtaining knowledge of the preferred account, any more than we can justifiably assign candidates to positions of great responsibility, when we have done nothing to check their credentials, on the grounds that "we know nothing against them." Just as in the selection of office holders we know the kinds of information we need to make a defensible judgment, so too our general understanding of the evolutionary process enables us to appreciate the kind of evidence we need to assess the merits of the preferred Darwinian history. Lacking that evidence, we should suspend judgment.

The enthusiast replies. Skeptics set the standards of evidence too high. Invocation of bare possibilities cannot be allowed to stand in the way of positive results, garnered from experiments. Underdeveloped reminders to the effect that there may be alternative approaches are rightly not taken seriously as threatening the status of successful theories. Actual rivals must be considered, and their credentials evaluated; skeptical obscurantism, however, is the road to regressive science.

The dispute is delicate precisely because it lies between two cases in which we know just what to say. Ideal cases command our allegiance. Biologists who jump too soon in cases of underdetermination attract reproach. Enthusiasts clamor for including the tricky cases with the ideal; skeptics prefer to assimilate them to cases of underdetermination. Both achieve genuine insights. Both lapse into error. Skeptics are surely wrong to insist that amassing a corpus of surprising results that the ambitious history leads us to expect does nothing to increase our rational confidence in the history. Enthusiasts err equally in dismissing the possibilities of rival histories as "bare possibilites." They are unrealized because of our ignorance; and, as the skeptic notes, ignorance is not a key for unlocking the doors of knowledge.

Can we avoid the dilemma? I think so. The dispute presupposes that there should be some definite, all-purpose, epistemic appraisal of the ambitious Darwinian history. We are beguiled by the picture that the history should either be inscribed in golden letters in the Great Big Book of Truth or left to moulder in the penciled scribblings of a discarded notebook. There is no reason to believe that rational decision would consign the proposal to either fate. We can accept it for what it is, the best account so far developed, while appreciating the various ways in which it may later have to yield to rivals. We can even attempt to assess the likelihood of developing rival accounts by examining the extent of our ignorance about the underlying genetics, ecology, physiology, and so forth.

In the extreme cases we are clear: we know, or we know that we do not yet know. The intermediate case resists either appraisal. We have a working hypothesis, something that may turn out to be true, but our prudent acknowledgment of our own ignorance prevents us from treating it as established. In the context of everyday science, tentative acceptance of the hypothesis may be accompanied by an assessment of the extent of our ignorance, or, more positively, by efforts to remedy it and thus to achieve a more definitive verdict.

We began our methodological investigations of evolutionary theory with some questions. We can now give answers. There is no bar in principle to the Darwinian enterprise of explaining the presence of the characteristics of organisms. We are not compelled to become cautious Darwinians and view all sociobiology as a dramatic outgrowth of a disease prevalent in evolutionary practice. Ambitious Darwinian histories can sometimes be systematically tested and confirmed. Ambitious Darwinism is subject to methodological rules.

The large question remains. Do pop sociobiologists play by the rules? We cannot yet tell. It is already possible, however, to discern a potential source of trouble for pop sociobiology. If the methodologically tricky situations that I have described abound in the study of the evolution of social behavior (in particular, of human social behavior), then pop sociobiologists will have to take sides with the enthusiasts. For otherwise, the conclusions they draw from the theory of evolution will provide an insufficient foundation for the grand theory of human nature that they wish to erect.

The Persistence of the Unfit

There is a further type of evolutionary question that needs to be distinguished from the issues discussed in the last sections. Sometimes evolutionary biologists are not primarily concerned with discovering the *actual* Darwinian history that stands behind the presence of a trait, but rather with showing that it is *possible* that the trait should be the product of evolution. Inquiries of this kind are often undertaken when we discover characteristics that appear to reduce the fitness of their bearers. How can these characteristics be reconciled with the theory of evolution by natural selection? How did they first become established? Why does selection not eliminate them?

Darwin was troubled by a prominent example of the general difficulty, an example that has been central to much of the work in sociobiology. How, he wondered, could the existence of sterile workers among the social insects be reconciled with his theory? The problem is not to show how apparently unfit organisms—the sterile workers—might actually manage to reproduce: they do not pull off anything so miraculous. Darwin was puzzled by the presence of a propensity in the parents to produce sterile offspring, a propensity that is transmitted to the fertile offspring, accounting for the presence of sterile workers in each generation. The characteristic would seem to diminish the capacity of an insect to leave descendants: if a queen bee can produce a fixed number of eggs, then we anticipate that she will leave more descendants in subsequent generations if all the eggs develop into fully fertile insects than if some of them become sterile.

Darwin tried to show that appearances may be deceptive. Let us compare insects with a propensity to produce some sterile offspring with rivals who would seem to do better under a regime of selection. There is no reason to think that the imagined rivals will do better in the first generation. Presumably the total number of offspring will be the same whether all are fertile or some are sterile. We anticipate differences at the next stage. With all those extra reproductives in the first generation, it seems that the rivals will have more descendants in the second generation. Not necessarily, claimed Darwin. If the sterile sisters help, then perhaps a smaller number of reproductives will have a greater number of offspring than the larger number of reproductives in the rival hive who do not profit from the aid of workers (Darwin 1859, 238).

Darwin's suggestion solves one puzzle, the problem of how it is possible for selection to fashion (or maintain) a characteristic that initially appears to reduce fitness. Another, quite different, question is how the trait was actually fashioned (is actually maintained). Although the questions can be posed in the same form of words, it is crucial to recognize the difference between them. Answering the first requires only that one provide a possible scenario, compatible with what is known. To address the second, we must follow the rules elaborated in the last section.

It would have been premature for Darwin to maintain that he had identified the crucial factor in the preservation of sterile insects. There are too many alternatives. Perhaps the sterile workers do not increase the reproductive output of their fertile siblings but enable their mother to produce further fertile offspring. Perhaps the propensity to produce some sterile offspring is connected, by the laws of "correlation and balance," with highly beneficial properties. As we shall see, there are further possibilities.

Cynics might think that the problem of evolutionary possibility could easily be sidestepped. Besieged Darwinians might always turn to invocations of "blind chance and historical accidents." In fact, it is not correct to think that this strategy will always meet the difficulty. When we find types of behavior that have arisen independently in several groups of organisms, it strains credulity to suppose that they were formed and maintained as the result of a number of accidents, all independent, but all tending to the same lucky state. (One of the challenging things about the social insects is the presence of sterile workers in unrelated groups.) We are in the position of Tom Stoppard's Guildenstern, confronted with a run of ninety-two heads. It can happen, of course, but when it does, the rational response is to take a closer look at the coin—or, in our case, to consider a different type of explanation.

Even when we allow that selection is only one agent of evolutionary change, there are genuine questions about the possibility of Darwinian histories for traits that appear to reduce fitness. Answering those questions is considerably easier than achieving justified accounts of the actual evolutionary processes that have led to the presence of the traits. When the questions (and the answers) are confused, we may be misled into thinking that we know more than we do.

Sociobiological curiosity sometimes starts with Darwin's kind of perplexity: can the theory of evolution reconcile itself to the persistence of the apparently unfit? As we shall see in the next chapter, today's students of the evolution of social behavior can pride themselves on having made great advances on this general problem. But pop sociobiology is enticed beyond the puzzles about possibility. Pop sociobiologists hope to trace the actual evolution of patterns of animal behavior and to apply their favored explanatory techniques to the understanding of human behavior. We should make sure that the move from questions of evolutionary possibility to questions of actual history does not short-cut the systematic process of investigation that ambitious Darwinian histories need.

We should also recognize that the emphasis on human behavior departs from Darwin's predicament in another way. The problem of evolutionary possibility is primarily a puzzle about organisms that cannot deliberate and plan, that cannot adjust their actions to their perceived interests. Those who read Dickens's account of the death of Sydney Carton do not typically exclaim, "How odd that this character should act so as apparently to reduce his fitness!" We appreciate the possibility that people may choose to behave so as to achieve nonevolutionary ends, to do what they conceive as far better things than they have ever done. There is no puzzle about human beings that initially strikes us as comparable to Darwin's worry about the social insects. When people apparently reduce their own fitness, as, most dramatically, in cases of self-sacrifice, there is no immediate threat to the theory of evolution.

Pop sociobiologists claim that we must learn to reevaluate the stock examples. They claim to have followed a road that leads from Darwin's original theory and Darwin's original problems to radical conclusions about human nature. But the road seems to have two sharp bends. It veers from evolutionary possibility to evolutionary actuality; and organisms that seemed to pose a threat to Darwin's theory of evolution, precisely because their actions could not be viewed as subordinating fitness to nonevolutionary interests, give way to the organism we identify as the being that chooses its own ends. Perhaps we are compelled by reason to travel the route that pop sociobiologists map out for us. Yet, as we set out, we should ponder the possibility that we may occasionally find ourselves lost; and we should be prepared to think critically about the confident claims of our guides that Darwin's finger points clearly in a particular direction.

Chapter 3

A Beginner's Guide to Life, Sex, and Fitness

Counting Kin

Pop sociobiology's claim to fame rests on the possibility of transcending the unsystematic approach of traditional ethology. Champions of the "new synthesis" can improve on the efforts of their predecessors, not only because they can draw on better observations of animal behavior, genuine facts from the field, but because they have an arsenal of gleaming techniques, clever refinements of evolutionary theory that are beautifully adapted for overcoming the problems of social behavior. My aim in this chapter is to review the techniques and so to prepare the way for a clear conception of sociobiology.

Prominent among the theoretical advances is an idea developed by William Hamilton, often known as the "theory of kin selection" or the "theory of inclusive fitness." Some critics have assumed that all there is to sociobiology is the conjunction of an updated ethology with Hamilton's ingenious idea (Sahlins 1976; Hinde 1982, 152). They are wrong. Sociobiologists can point to a variety of theoretical ideas—the notion of inclusive fitness, the application of game theory to the understanding of fitness relations, the use of optimality models to analyze fitness. Yet there is a basis for the critics' assumption. Many sociobiological discussions give prominence to the concept of inclusive fitness, and they do so because Hamilton's insight provides an extremely valuable extension of the neo-Darwinian concept of fitness. Without supposing that theoretical sociobiology begins and ends with Hamilton, we can reasonably start by trying to understand inclusive fitness.

First, some stage setting. Hamilton's proposal figures in a debate that divided biologists in the 1950s and 1960s, the quarrel about "units of selection." The debate was tangential to the controversy with which we are concerned, although most sociobiological discussions contain ritual allusions to it. (The original "group selection" controversy inspired the development of some theoretical machinery with which sociobiologists have gone diligently to work. The applications are largely independent of the provocative theses about units of selection currently debated by, for example, Richard Dawkins [1976, 1982] and David Sloan Wilson [1980, 1983].)

In 1962, V. C. Wynne-Edwards attempted to solve some of the problems posed by the existence of patterns of behavior that appear to reduce the fitness of those who exhibit them. He suggested that the behaviors are maintained because of the benefits that accrue to the group to which the animals belong. So far there was nothing new. Biologists had long been appealing, in a general way, to "actions for the good of the group (or the species)," and the theme would continue to be a favorite with ethologists and students of human nature (Lorenz 1966, 27ff.; Morris 1967, 80 ff.; Eibl-Eibesfeldt 1971, 93, 105). But Wynne-Edwards put forward the first explicit and detailed proposal for how group advantages might be selected even at a cost to individuals. Study of social arrangements among birds convinced him that some birds act in ways that detract from their fitness. He tried to show precisely how their behavior contributes to the welfare of the group to which they belong, how such groups are stable over time, and how groups with particular advantageous traits are able to maintain themselves while rival groups succumb.

Although the idea is an attractive one, it immediately encountered criticism. In a deservedly influential book George Williams reviewed the range of cases to which group selectionist explanations had been applied. (Ironically, although Wynne-Edwards had given the most systematic treatment of group-selectionist ideas, Williams's primary focus was on the earlier tradition, exemplified in the work of W. C. Allee and others.) He concluded that, with one exception—the persistence of an allele in populations of mice that renders males sterile when they have two copies of it—alternative accounts in terms of the selection of individuals were always available and preferable. Williams's guiding principle was that considerations of parsimony ought to induce us to avoid group-selectionist explanations when we can (Williams 1966, 5, 262). His extensive study convinced him that appeals to group-level advantages and group selection were "gratuitous and unnecessary."

Complementary objections were urged by John Maynard Smith. Where Williams had pondered the necessity of invoking group selection, Maynard Smith worried about the sufficiency of doing so. Maynard Smith's basic intuition was simple and powerful. If groups containing members who act against their own reproductive interests are to prosper because of the elimination of rival groups, then the group-level competition must go forward sufficiently rapidly to ensure that the groups with the alleged advantages cannot be subverted from within. Imagine that a species comes in two kinds, the Goodies and the Nasties. Goodies forgo some opportunities for reproduction in order to benefit the group to which they belong. Nasties look out only for themselves. A group of Goodies may ultimately eliminate a group of Nasties if they compete for resources in the same region. However, if it takes many generations for the competition to be resolved, then migrant Nasties may infect the group of Goodies and, through ordinary individual selection, sweep to prevalence within that group. Maynard Smith worked out this intuition precisely (see Maynard Smith 1964 and, for clarifications, 1976a). His discussion does not show that group selection, as Wynne-Edwards envisaged it, is impossible. Rather, the conditions under which the selection of groups can be a significant evolutionary force are stringent. (Attempts to specify these conditions precisely were later made in Boorman and Levitt 1973, Gilpin 1975, and D. S. Wilson 1980 and 1983; E. O. Wilson refers to some of this work, in a somewhat idiosyncratic way, in his 1975a, 106ff.; a lucid recent discussion is part II of Sober 1984.)

The work of Williams and Maynard Smith was complemented by that of Hamilton. In 1964 Hamilton published a pair of papers under the title "The Genetical Evolution of Social Behavior." The first of these offered a mathematical formulation of an idea that, according to folklore, had originally been touted by J. B. S. Haldane in a conversation in a British pub. (Williams had also arrived at similar ideas; see Williams and Williams 1957.) Asked if he would be prepared, on evolutionary grounds, ever to sacrifice his life for another, Haldane is supposed to have grabbed a beer mat and a pencil and, after a few quick calculations, to have declared that he would willingly lay down his life if he could save more than two brothers, four half-brothers, or eight first cousins. (Hamilton [1964a, 42] repeats these apocryphal words.) After working out the details, Hamilton devoted his second paper to indicating how "the model described in the previous paper can be used to support general biological principles of social evolution" (1964b, 44).

Let us start with Haldane's reasoning. Suppose we have some sexually reproducing diploid organisms. Consider two full siblings; select a particular locus. One of the siblings has a particular allele at this locus. What is the probability that the other sibling shares this allele through transmission from the same parent? Clearly 0.5: the parent who passed on the allele to the first sibling transmitted to the second either the allele in question or the allele on the homologous chromosome. Now imagine that life is hard and that either sibling, acting alone, can be expected to produce just one offspring. If one of them gives up reproduction in the interests of aiding the other, then the expected number of offspring produced by the beneficiary will be three. Potential helpers are faced with two possible scenarios—they can reproduce or they can help siblings. From the standpoint of transmitting their genes into the next generation, which will be more successful? (More exactly, if we suppose that there is a locus at which alleles for helping kin or for not helping kin may occur, then which allele is likely to achieve greater representation in the next generation?)

Organisms that choose to go it alone, neither helping nor being helped by siblings, pass on half their genes to their single offspring. Self-sacrificing helpers fail to transmit any of their genes directly. However, for any one of their genes, there is a probability of one quarter that, in each production of an offspring by the sibling, that gene will be passed on. (This probability is the product of the probability that the gene is present in the sibling and the probability that, if present, it is transmitted.) Since the sibling produces three offspring, there are three independent occasions on which the gene has a one-quarter chance of transmission. The expected frequency of the gene in the next generation is three quarters. For there is a probability of $(1/4)^3$ that 3 copies will be present, a probability of $3(1/4)^2(3/4)$ that 2 copies will be present, and a probability of $3(1/4)(3/4)^2$ that 1 copy will be present. The expected value is thus

$$3(1/64) + 2(9/64) + 1(27/64) = 3/4.$$

Since the same calculation goes for any other locus in the helping organism, it is not hard to see that the expected representation of the organism's genes in the next generation is just as good if it helps its sibling.

Why isn't it better? After all, loners achieve representation for only half their genes. So it might seem on first reflection—but first reflection needs tutoring. We have forgotten to add the genetic representation that the loner would have achieved anyway in virtue of the sibling's production of a single offspring. The expected frequency of a gene in the next generation is 1/2 (through the single offspring) plus 1/4 (through the sibling's offspring). So the strategies are equally successful. Alternatively, we may summarize the situation by noting that a helper trades one offspring for the sibling's two extra offspring. The change in expected genetic representation through the trade is zero. This second perspective is useful in thinking about the Haldane-Hamilton insight. Recognizing that animals may obtain genetic representation through their relatives, we must appreciate the importance not of the absolute extent of the relatives' reproduction but of the difference that the helper animals' presence and action make to the reproduction of relatives.

Hamilton showed how to generalize the simple calculation I have given. His first goal was to define a notion of inclusive fitness that would accommodate the idea that our genes are spread into subsequent generations by our relatives. It is easy to go astray here. (Many of Hamilton's successors have gone astray; see Grafen 1982 and 1984 for reviews, and, for critical discussion of Hamilton's own proposals, Michod 1982.) Let us begin with the organism's expected number of offspring. We now modify this in two ways. First, we discount aid to the organism from relatives and identify the organism's expected number of offspring in a situation in which it did not receive this aid. Second, we consider all the effects that the organism has on the expected number of offspring of relatives, discounting these effects by the probabilities of sharing genes—the *coefficients of relatedness* (or *relationship*). Weighted in this way, the changes in the expected number of offspring are added to the organism's own expected number of offspring.

Puzzles about apparent reduction of fitness may admit resolution once we adopt Hamilton's insight. It is easy to see this qualitatively. If there is a genuine adverse effect on personal fitness that is offset by the effect on kin, Hamilton's concept can save the day, showing us that there is no threat to the workings of selection. More exactly, let us suppose that the personal fitness (or individual fitness, or classical fitness) of an organism is the expected number of the organism's adult offspring. We now "strip" the fitness by discounting aid from relatives. Assume that the "stripped" personal fitness of the organism is (at least approximately) the organism's personal fitness. The inclusive-fitness effect of the organism is the increase in the expected number of offspring of relatives that can be attributed to the organism's presence and behavior, weighted by the coefficients of relatedness. So, if we have an ordering of the n relatives of an organism in a population, and if the organism's effect on the expected number of offspring, w_i, of the ith relative is δw_i, then the inclusive-fitness effect is $\Sigma r_i \delta w_i$, where r_i, the coefficient of relatedness of the ith relative with the original organism, is the probability that, for a particular allele chosen at random from the organism's genome, the ith relative will share a copy of that allele by descent. The inclusive fitness is just the sum of the personal fitness and the inclusive-fitness effect. The inclusive fitness of an allelic pair, a combination of genes, or a phenotypic property is computed by averaging over the bearers of the alleles or of the property.

Outlines of solutions to puzzles about apparent reduction in fitness can now be seen more clearly. One type of puzzle arises when we have two kinds of organisms in a population who behave like the Goodies and Nasties of the example in chapter 2. The personal fitness of the Goodies is less than that of the Nasties. Yet the Goodies survive and multiply. How can this be? Enter the Hamiltonian insight. If the inclusive-fitness effect for the Goodies offsets their deficit in personal fitness compared to the Nasties, then Goodness will increase and Nastiness decline.

The conclusion that selection acts on differences in inclusive fitness, presupposed in our outline problem solution and in Haldane's tavern calculations, is motivated by intuitive ideas about the spreading of genes. Hamilton made those ideas precise. Analysis of changes in gene frequencies in a single generation is more complicated when we consider inclusive fitness rather than personal fitness, but Hamilton was able to emulate classical results. Just as Fisher and Wright had been able to show that populations—under certain conditions, including the absence of stochastic factors and of frequency-dependent selection—move to local maxima of mean fitness (intuitively, selection works to make the population as fit as it can be), Hamilton was able to demonstrate that, under similar conditions, populations go to local maxima of mean inclusive fitness (Hamilton 1964a, 32). A corollary is that allelic pairs (or larger combinations) having an inclusive fitness that is higher than the population mean will be (at least temporarily) favored by selection. That is, if a prevalent phenotype has greater inclusive fitness than some rival phenotype, and if there are genetic differences between the organisms bearing the two phenotypes, then there is no real mystery about why the phenotype persists in the population. We have a basis for the triumph of the Goodies.

Though the mathematical derivation of the result about maximization of mean inclusive fitness is tricky, most sociobiological applications employ a simple corollary. Provided that we remember the conditions under which Hamilton's analogs of the Fisher-Wright theorems were derived, and provided we use the concept of inclusive fitness correctly, we can gain insight into the persistence of characteristics that appear to reduce fitness by comparing inclusive-fitness effects with deficits in personal fitness. So, if we have a situation in which one organism, Donna, aids another organism, Benny, then we can dissolve the puzzle of the persistence of Donna's behavior under selection if we are able to apply a simple inequality. As long as we can show that the extra costs incurred by Donna (the loss in expected number of offspring that results from the performance of the behavior) are less than the gains in inclusive-fitness effect (the gain in expected number of Benny's offspring weighted by the degree of relatedness of Benny to Donna), then there will be no mystery about the persistence of behavior like that which Donna displays. Hamilton's inequality states that a form of behavior can be maintained under selection, provided

$$C < rB \qquad \text{or} \qquad B/C > 1/r,$$

where C is the cost of the behavior (in reduced personal fitness), B is the benefit brought by the behavior in terms of inclusive-fitness effect, and r is the coefficient of relationship. The inequality underlies Haldane's calculations, and hordes of sociobiological arguments of the last decades. We should remember, however, that the justification for the inequality lies in Hamilton's proof that a population under selection will move to a local maximum of mean inclusive fitness. Thus the inequality is inapplicable when the conditions of the theorem are not fulfilled and when the inclusive fitness cannot be regarded as the sum of personal fitness and an inclusive-fitness effect. (We should also bear in mind a point that Hamilton makes explicitly: apparently fitness-enhancing behavior may be selected against because of harmful effects on relatives; here, an addition to personal fitness would be swamped by a negative inclusive-fitness effect.)

My discussion has approached Hamilton's insight in terms of the concept of inclusive fitness, and I have avoided reference to "kin selection." Despite the popularity of talk of kin selection in the sociobiological literature, it seems salutary to continue my practice. Those who yearn to use the term can think of it as covering cases in which reduction in personal fitness is outweighed by an inclusive-fitness effect. However, this may all too easily generate the thought that episodes in which organisms act so as apparently to reduce their fitness fall into one of two distinct types: cases of kin selection and cases in which some other analysis of the organism's behavior is needed (to uncover the hidden advantages). To think in this way is to miss Hamilton's fundamental insight. Evolutionary change is change in gene frequencies in populations, and, for a population under individual selection, the direction and magnitude of the change will be determined by the relations among the inclusive fitnesses of the organisms in the population. When a phenotype that appears to reduce fitness persists in a population, we may sometimes be able to resolve the puzzle by considering the inclusive fitness of the phenotype, without pursuing any extensive analysis of the factors that contribute to inclusive fitness. But even when the switch to thinking in terms of inclusive fitness does not dispel all our wonderment about the persistence of the phenotype, considerations of inclusive fitness may still be relevant. In some cases, to be sure, our understanding of the situation will turn on recognizing the role of fitness-promoting factors that enhance personal fitness in unobvious ways, and in such cases the evolutionary change may be treated as if it were dependent on the relations among the classical fitnesses. (There will be many phenotypes for which inclusive fitness is either identical with or only trivially different from classical fitness.) In others, however, the problem must be approached by considering rival strategies for promoting inclusive fitness; and when this occurs, disputes about whether the maintenance of a phenotype is due to kin selection can only be sterile.

Where Sisterhood Is Powerful

Enough of hypothetical cases. Skeptics may legitimately wonder if the introduction of the concept of inclusive fitness is simply a pretty piece of applied mathematics—or, if they have read Hamilton's original paper with its sensitivity to the intricacies of the problem of computing expected changes in gene frequency, a not-so-pretty piece of applied mathematics. Where is the biology?

Hamilton mentions one possible biological application at the end of his mathematical paper: he suggests that the apparent reduction in personal fitness caused when a bird gives an alarm call to signal the presence of a predator may be compensated if the call alerts a relative. However, the prize example of the utility of the notion of inclusive fitness, and one of the high points of sociobiology, is the study of the behavior of social insects.

Darwin found the social insects deeply puzzling. Colonies of ants, wasps, bees, and termites are often notable for the cooperation among their inhabitants and for the many ways in which the insects seem to act against their individual fitness. Ant workers do not forage primarily for themselves, but frequently regurgitate food for the consumption of others; they lay trophic eggs to be fed to larvae; they sacrifice their own lives defending a colony in which they have no direct descendants. How have the complex systems of the social insects evolved? Hamilton noted a peculiarity of the *Hymenoptera*—the order that includes ants, bees, and wasps and in which sociality has evolved on eleven separate occasions. He used the distinctive feature for an application of the concept of inclusive fitness. (Sociality in the termites, which do not share the peculiarity, has to be explained in a different way.)

The oddity concerns sex determination. Ants, bees, and wasps, unlike most other insects, have a haplodiploid system of sex determination. Males develop from unfertilized eggs containing half the number of chromosomes found in the cells of the mother (more exactly, in the somatic cells of the mother, those cells that are not gametes). The number of chromosomes in the unfertilized eggs is the haploid number for the species. Females develop from fertilized eggs and are therefore diploid. A consequence of this system of sex determination is that there are some unusual coefficients of relationship.

Consider two individuals, O_1 and O_2. The coefficient of relationship of O_1 to O_2 is the probability that a gene found at a locus in O_2 is present in O_1 by virtue of descent. The coefficient of relationship between two sisters is easily computed. Consider any locus in one of the sisters. Pick one of the alleles present at this locus. That allele could have been transmitted to the organism we are considering in one of two ways: either it came from the father or it came from the mother. If it came from the father, then, since the father's genes are passed on to all offspring (paternal gametes are genetically identical to all other paternal cells), the other sister must share the gene. If it came from the mother, there is a probability of one half that the gene was also passed on to the other sister. (The maternal gamete that became the sister's zygote had half the maternal genes.)

Probability that sister has the gene
= [(Probability that gene came from father)
× (Probability that sister has gene, given that it came from father)]
+ [(Probability that gene came from mother)
× (Probability that sister has gene, given that it came from mother)].

Substituting the figures we have derived, we learn that the coefficient of relationship between two sisters is 3/4.

Here are some other coefficients of relationship for a haplodiploid mating system. For daughter to mother, $r = 1/2$. For son to mother, $r = 1/2$. For brother to brother, $r = 1/2$. For brother to sister, $r = 1/4$. (If the gene in sister came from father, the probability is zero that brother has it by virtue of descent; if it came from mother, there is a probability of 1/2 that brother has it by virtue of descent.) For sister to brother, $r = 1/2$. (The gene in brother came from mother, and there is a probability of 1/2 that mother transmitted the same gene to sister.) For mother to son, $r = 1$. (For any gene in a male there is a probability of 1 that the gene is present in his mother.) Plainly, there are asymmetries here. Because of the asymmetries the definition of the last paragraph is more than an exercise in pedantry.

How do the curious facts about the *Hymenoptera* enter into an account of the evolution of their sociality? In our initial discussion of the existence of sterile workers, we looked at the situation from the perspective of the queen. Darwin suggests, in effect, that the queen may maximize her expected genetic representation in subsequent generations if she has a propensity to produce some offspring who do not reproduce, but who aid their mother in raising fertile sisters. But why should the intended helpers carry out the plan? It is not enough that the workers be sterile, itself hardly an advantage to the mother, but that they *work*, and, indeed, work toward the production of fertile sisters. Here is Hamilton's explanation. Imagine that evolution has already shaped the species so that females produce offspring over a long period of time. If a mother is still at her reproductive work when the first of her daughters becomes mature, then we can compare two options for the daughter. She can mate, construct a nest, and begin a reproductive career of her own. Or she can stay at home and help mother by raising more sisters. Because females are more closely related to their sisters than they are to their daughters (by 3/4 as opposed to 1/2), the expected representation of their genes in future generations is increased by their pursuing the second strategy. We can articulate the idea as a possible explanation of both the origin and the maintenance of the propensity to produce some sterile offspring.

Imagine that some daughters inherit a propensity to stay at home and help mother. The genes of any such daughter, in particular the gene(s) underlying the cooperative behavior, would be expected to spread. Subsequently there would be selection for any system of redeploying resources that would otherwise be wasted in developing the reproductive systems of the helpers. (The system could be flexible, allowing any offspring to develop into either a reproductive adult or a sterile worker, depending on the environment during development; or there could be genetic differences among fertile and sterile forms. The social *Hymenoptera* follow the former pattern.) The insect colony might thus come to contain some members who, after a certain stage of their development, are incapable of reproducing. Moreover, the propensity to produce some offspring that become sterile workers would be maintained by selection because insects lacking this propensity would have daughter helpers who were less efficient. So Darwin's puzzle about the existence of sterile workers among the social insects—the problem of *how it is possible* for worker sterility to persist in the face of selection—is resolved.

Hamilton's account can easily be extended to illuminate other puzzling facts about the *Hymenoptera*. Among ants, bees, and wasps the workers are invariably female. Males contribute nothing. Hamilton's explanation turns on the values of the coefficients of relationship. Daughters are exactly as closely related to males as siblings are: in both cases the coefficient of relationship is 1/2. Hamilton concludes that "the favorable situation for the evolution of worker-like instincts cannot ever apply to males" (1964b, 63). But this is only the beginning of an explanation. To explain the lazy parasitism of the drone, we have to show that noncooperation is favored by selection. Wilson attempts to articulate Hamilton's idea, arguing that the best way for a hymenopteran male to spread his genes is for him to prepare to be as vigorous as possible at mating time. Lazing about the colony and cadging what he can from his sisters are the drone's strategies for enhancing his fitness. (There are occasional instances in which drones rise above their stereotype. See Hölldobler 1966 and, for brief discussion, Oster and Wilson 1978, 128.)

In going to the ant, we become wiser about the ways of evolution. Hamilton's insight proves its worth in showing how characteristics that seem doomed for elimination might be able to become prevalent. It would be wrong, however, to think of the evidence we have reviewed as providing strong grounds for conclusions about the actual course of evolution. It would be equally wrong to think of it as confirming Hamilton's detailed mathematical results about changes in gene frequencies and evolutionary trajectories, or even the simple corollary that promotion of the reproduction of a relative will be favored if $B/C > 1/r$.

Confirming the hypothesis that the account we have sketched tells the story of the evolution of worker sterility in the social *Hymenoptera* would require more than we have so far achieved. Alternative suggestions would have to be elaborated and explored. (See, for example, the ideas presented in Seger 1983, Brockmann 1984, and Bull 1979.) Hamilton's mathematics, by contrast, stands in no need of the data from the social insects. Like the central claims of classical population genetics, Hamilton's theorems are grounded in the laws of probability and in the basic concepts and principles of genetics. Like Darwin's argument for the principle of natural selection, Hamilton's defense of the claim that, under specified conditions (random mating, frequency-independent selection), populations will move to local maxima of mean *inclusive* fitness is compelling once we have achieved the crucial insight. To echo T. H. Huxley, "How extremely stupid not to have thought of that!" (F. Darwin 1888, II, 197).

Hymenopteran haplodiploidy should have its due. Recognizing the unusual coefficients of relationship and applying Hamilton's ideas, we can learn something important. Is it possible for worker sterility to develop and be maintained under natural selection? That was Darwin's question. Hamilton answered it by finding a way to give a resounding "Yes."

Evolutionary Gamesmanship

Appeal to inclusive fitness is only one among several approaches that can be tried when we discover an organism that seems to behave in ways that reduce its fitness. Another approach is to introduce some ideas from the theory of games. Suppose we regard a population as a collection of "players," pursuing strategies from a given set, and try to analyze the expected payoffs (in terms of future genetic representation) for the available strategies. Perhaps we can then show that a particular strategy (or distribution of strategies), once achieved, is likely to persist in the population because those who deviate from it will be less fit than their orthodox rivals. The idea is an attractive one, and it has been developed with considerable sophistication by Maynard Smith. (I hasten to note that the idea can also be applied in cases where there is no appearance of reduction in fitness.)

Simple examples provide the best introduction. I shall begin with a case that is not only simple but, in certain respects, "ridiculously naive" (Maynard Smith 1982a, 5). Imagine that we have some animals who compete for certain divisible resources—prize fruit trees, places in the sun, or whatever. When two of them meet near a resource, both would like to obtain it, preferably entire. Suppose, with the abandon born of familiarity with population genetics, that the population of animals is infinite and that it reproduces asexually. (We make the last assumption to underwrite the idea that like organisms will beget like progeny. Behavior that inevitably results from a given genotype will be transmitted from parent to offspring.) The pairwise contests between animals are always symmetrical. Neither contestant is bigger, quicker, wiser, needier, than the other. Finally, there are two genetically distinct types in the population. *Hawks* always fight over the resource, escalating until they are injured or until their opponent gives way. *Doves* invariably begin with conventional displays, retreating at once if the opponent starts to fight seriously (see Maynard Smith 1982a, 12).

We need to measure the outcomes of various sorts of contests. Continuing our simple tale, suppose that all conflicts involve a resource whose value, in units of fitness, is V. This means that an animal able to gain the resource without cost will increase its expected number of offspring by V. It does not mean that animals who fail to get a resource have zero fitness (see Maynard Smith 1982a, 11). When encounters escalate into fights, losers suffer a cost C, also measured in units of fitness. Doves suffer no cost in any encounter.

Consider the possible meetings. When two Hawks meet, there is a fight. The winner takes the entire resource and the loser limps away injured. Because the situation is symmetrical, any Hawk can expect to win half of its encounters with other Hawks. So the expected change in fitness to a Hawk is $\frac{1}{2}(V - C)$. When Dove meets Dove, all is peace and sharing. The expected change in fitness to a Dove due to a dispute with another Dove is $\frac{1}{2}V$. Finally, when Hawk meets Dove, Hawk appropriates the entire resource, and Dove runs away empty handed. The expected change to Hawk is V, that to Dove is 0. We can present our results as a payoff matrix.

	Hawk	Dove
Hawk	$\frac{1}{2}(V - C)$	V
Dove	0	$\frac{1}{2}V$

The entries in the table represent the change in the fitness of animals in the left-hand column when they meet animals in the top row.

Now for some definitions. A strategy is a way of behaving. More exactly, it is a function that maps a context onto a particular piece of behavior. In our example the strategies are simple. Hawks fight resolutely in all contexts. Doves never fight; they run away if challenged. There are obviously more subtle strategies. A mixed strategy is a strategy with probabilistic assignments to other strategies. *Indecisives* play Hawk with probability 1/2 and play Dove with probability 1/2. More generally, we can envisage all the strategies of the following form: play Hawk with probability p, play Dove with probability $(1 - p)$. Finally, we can imagine context-dependent strategies. *Discretion*, playing Hawk against all and only animals smaller than you, may be the better part of valor.

A strategy is said to be an *evolutionarily stable strategy* (or ESS) if it is uninvadable. More precisely, J is an ESS only if, in a population whose members all play J, no mutant strategy could invade under selection (see Maynard Smith 1982a, 10). Invasion occurs when a mutant strategy is able to maintain itself within a population; in other words, the strategy is not immediately eliminated by selection. A strategy is an ESS only relative to a set of potential rival strategies. For almost any game that we might consider, a little imagination will suggest strategies (some of them implausible) that would be able to invade a population whose members played any of the strategies explicitly considered. *Clever Hawk*, the strategy of attacking exactly those animals that can be defeated without injury to the attacker, can clearly invade a population playing either Hawk or Dove (or any mixture of Hawk and Dove). Moreover, the idea of an ESS should not be confused with that of a globally optimal strategy. Mutants playing a rival strategy will be at a disadvantage in a population of animals playing an ESS. Even if J is an ESS, we can suppose, quite consistently, that if a particular mutant strategy K became prevalent, all the animals would receive a greater payoff by playing K than they do by playing J. Indeed, there are many games with multiple ESS's, some of which are better—in the sense of bringing greater payoffs—than others.

Back to our simple example. If $V > C$, then Hawk is an ESS. Mutants playing Dove would be rare, would mostly encounter Hawks, and would have a smaller payoff from contests than members of the savage majority: 0 as opposed to $\frac{1}{2}(V - C)$. Dove is not an ESS, because mutants playing Hawk would have a field day, meeting Doves virtually wherever they went. The mutants would gain double payoff at each encounter. However, it is probably more realistic—as well as more interesting—to assume that the costs of injury are much higher than the benefits obtained from the contested resource. The plump banana is not worth a broken arm, and those who run away may live to find uncontested resources elsewhere.

If $V < C$, then Hawk is no longer an ESS. A mutant Dove would encounter Hawks, and its expected payoff from contests would be 0. Not striking, but better than the expected payoff—$\frac{1}{2}(V - C)$—to the Hawks battling bloodily around. (Recall that the payoff is an increase in fitness. The result should not be interpreted as indicating that Doves have a fitness of zero and Hawks have negative fitness.)

Generality and precision can easily be introduced (see technical discussion A). The general formulation supports our qualitative assessment of the Hawk-Dove game. It also helps us to go further. Consider mixtures of Hawk and Dove, all strategies of the following form: play Hawk with probability P, play Dove with probability $(1 - P)$. Is there a strategy of this form that is an ESS in the interesting case where $V < C$? Yes. If we have a population of animals, *all of whom* play Hawk with probability V/C and Dove with probability $(1 - V/C)$, no mutant playing a different mixture of Hawk and Dove can invade by selection (see technical discussion A). This result should not be confused with a claim about the stability of a polymorphic *state*, in which V/C of the population play Hawk and the rest Dove. So far, we have considered whether J is an ESS relative to the set of all mixtures of Hawk and Dove. It is a quite different question to focus on a population whose members can only follow one of the pure strategies and to ask if there is a stable polymorphism. However, that question is worth asking.

Imagine a population of Hawks and Doves with frequencies p and

Technical Discussion A

Let I be an evolutionarily stable strategy (ESS) with respect to rival strategies $J_1, \ldots J_n$ (or with respect to an infinite collection of strategies). Then we shall require that the fitness of animals playing I must be greater than the fitness of any mutant organism playing one of the rival strategies. Assuming that the mutant organism has the same basic fitness (the fitness unaugmented by the spoils of contests) as any of the I-playing animals it seeks to displace, the fitnesses can be written as

$$W(I) = W_0 + (1 - p) \cdot E(I,I) + p \cdot E(I,J_i),$$
$$W(J_i) = W_0 + (1 - p) \cdot E(J_i,I) + p \cdot E(J_i,J_i),$$

where W_0 is the basic fitness, p is the frequency of a mutant playing J_i, and $E(K,L)$ is the expected payoff to an organism playing K in an encounter with an organism playing L. (Notice that we also suppose that only one mutant strategy tries to invade at a time. Matters are much more complicated if there are several mutants that arise at once. The conditions given by Maynard Smith only guarantee stability under multiple invasion if all mixed strategies from the set of rival strategies are included in the set of rival strategies. Maynard Smith makes this point somewhat obliquely [1982a, appendix D].) The value of p is very small (J_i players are *rare* mutants). I will be an ESS if, for any rival strategy J, $W(I) > W(J)$. So, for each i, we require that

$$(1 - p)\,[E(I,I) - E(J_i,I)] + p[E(I,J_i) - E(J_i,J_i)] > 0.$$

Because p is close to zero, the inequality will hold if $E(I,I)$—recall our argument that if Doves are rare, they encounter mostly Hawks, so that we can compare the fortunes of Hawks and Doves by considering the payoffs to Hawks from Hawk-Hawk contests and to Doves from Hawk-Dove contests. The inequality will also hold if $E(I,I) = E(J_i,I)$ and $E(I,J_i) > E(J_i,J_i)$. I can maintain itself if mutants "draw" when playing against I but do badly against one another.

Now let us consider the possibility of mixed strategies for the Hawk-Dove game. First we establish a simple result. Let J be the mixed strategy for some value of P that is neither zero nor one. Suppose first that $E(\text{Hawk},J) < E(J,J)$. By definition of J,

$$\begin{aligned} E(J,J) &= P \cdot E(\text{Hawk},J) + (1 - P) \cdot E(\text{Dove},J) \\ &< P \cdot E(J,J) + (1 - P) \cdot E(\text{Dove},J). \end{aligned}$$

Hence, $E(J,J) < E(\text{Dove},J)$. So J is not an ESS. Thus, we have shown that if $E(\text{Hawk},J) < E(J,J)$, then J is not an ESS. So, if J *is* an ESS, $E(\text{Hawk},J) \geqslant E(J,J)$. But if J is an ESS, $E(J,J) \geqslant E(\text{Hawk},J)$. So, if J is an ESS, $E(J,J) = E(\text{Hawk},J)$. In parallel fashion we can show that if J is an ESS, $E(\text{Dove},J) = E(J,J)$. So, if J is an ESS,

$$E(\text{Hawk},J) = E(\text{Dove},J) = E(J,J). \tag{A}$$

(A more general result, proved by D. Bishop and C. Cannings, 1978, is derived in appendix C of Maynard Smith 1982a.) I should point out explicitly that (A) is not a sufficient condition for J to be an ESS. A simple counterexample is the case in which all parties receive the same payoffs from all encounters.

Using (A), we find that any ESS must satisfy the condition

$$\begin{aligned} E(\text{Hawk},J) &= P \cdot \tfrac{1}{2}(V - C) + (1 - P)V = P \cdot 0 + (1 - P) \cdot \tfrac{1}{2}V \\ &= E(\text{Dove},J). \end{aligned}$$

So, *if* there is an ESS, the value of P must be V/C. (Notice that P has to be between zero and one, so that the assumption that $V < C$ is crucial in allowing a sensible interpretation.) It is not hard to check that the strategy is an ESS.

$(1 - p)$ respectively. The fitnesses of Hawks and Doves can be written as

$$\begin{aligned} W(\text{Hawk}) &= W_0 + p \cdot E(\text{Hawk},\text{Hawk}) \\ &\quad + (1 - p) \cdot E(\text{Hawk},\text{Dove}), \\ W(\text{Dove}) &= W_0 + p \cdot E(\text{Dove},\text{Hawk}) \\ &\quad + (1 - p) \cdot E(\text{Dove},\text{Dove}). \end{aligned}$$

(Reason: the fitness of a Hawk $W(\text{Hawk})$ is the basic fitness W_0 plus the increases that come from encounters with others; p of the encounters are with other Hawks, $(1 - p)$ with Doves. $E(\text{Hawk},\text{Hawk})$ is the expected payoff to Hawk from an encounter with another Hawk.) At equilibrium the fitnesses will be equal. So, if $\hat{p}$ is the equilibrium frequency of Hawk, then

$$\hat{p} \cdot \tfrac{1}{2}(V - C) + (1 - \hat{p}) \cdot V = (1 - \hat{p}) \cdot \tfrac{1}{2}V,$$

which yields the result that $\hat{p} = V/C$. So there is a stable polymorphism when V/C of the population play Hawk and the rest play Dove.

So why should we distinguish between evolutionarily stable strategies and evolutionarily stable distributions of strategies? There are two reasons for caution. First, the similarity of the algebra (see technical discussion A) should not blind us to the fact that the conclusions drawn are quite distinct. In one case, we have one collection of strategies and we want to determine which (if any) of the strategies in the set would resist invasion by any other strategy in the set. In the other case, we have a different set of strategies and we want to know if there is some equilibrium distribution of these strategies among the members of the population. Second, there is no general formal equivalence between the two kinds of questions. Given more than two pure strategies, we can have a stable mixed strategy without any stable polymorphism, and vice versa (see Maynard Smith 1982a, appendix D).

We have come a long way with a very simple game. How can we extend and refine our approach? One assumption that has so far been crucial is that "like begets like." If Hawks had a propensity to produce some Doves as offspring, then all bets would be off. The frequencies of players could change, even given a "stable" distribution. This possibility was blocked by the drastic step of insisting on asexual reproduction. Given an equilibrium distribution at which the fitnesses are equal, the frequencies in the parental generation will be the frequencies of the offspring, and because the offspring are genetically identical to the parents, the distribution will be preserved. But is the requirement necessary? In particular, can we apply the game-theoretic approach to sexually reproducing, diploid organisms—animals like us?

There is no great problem if the ESS can be produced by a homozygote. Under these circumstances the progeny of animals in a population playing the ESS will also play the ESS. Complications enter if several loci are involved or if the ESS is produced by a heterozygote. (It might seem that multiple loci are not bothersome. After all, if there are five relevant loci and the ESS is produced by a combination of homozygotes at each locus, what can go wrong? However, trouble can develop if the ESS can be produced by different homozygous combinations—if, for example, all that matters is that any three of the five loci be filled by a particular type of homozygote. And this is only the tip of the iceberg. If the loci are linked (that is, if they are close together on the same chromosome) or if they interact, then the messy details of population genetics intrude on the sanitized game-theoretic analysis.)

Consider a modification of the Hawk-Dove game, played in an infinite, random-mating population of sexual organisms. Assume that $V = 1$ and $C = 1/2$, so that the ESS (Indecisive) is Hawk with probability 1/2, Dove with probability 1/2. We start the population with individuals who are all heterozygous at the *A* locus. *AA* animals play Hawk, *aa* animals play Dove, and the *Aa* heterozygotes are Indecisives. Is the strategy of the *Aa* individuals an ESS? No. Both pure strategies invade in the first generation and are maintained within it, as a consequence of the sexual reproduction of the individuals who play Indecisive. Here, we *can* identify a stable polymorphism ($\frac{1}{4}AA$, $\frac{1}{2}Aa$, $\frac{1}{4}aa$), but there is no ESS.

The moral is that the existence of an ESS depends not only on the payoffs and the available strategies but on the genetic basis underlying those strategies. I shall conclude my discussion of evolutionary gamesmanship by introducing a different kind of complication. What happens when we relax the requirement of perfect symmetry?

One important type of animal interaction involves attempts by an intruder to wrest a resource (for example, a territory) away from its erstwhile owner. All kinds of asymmetries enter into this situation. The owner may have an enhanced ability to defend, based on knowledge of the territory—intruders may be lured into bogs and thickets, for example. The value of the territory may be greater to the owner than to the intruder (more soberly, loss of the territory may cause the owner a greater loss of fitness than the gain in fitness that would accrue to a successful usurper). Let us ignore these asymmetries and concentrate on the simple fact of ownership. We augment the set of strategies by including a third pure strategy, *Bourgeois*. Bourgeois individuals act like Hawks when they own a resource and like Doves when they confront another owner. The payoff matrix for the Hawk-Dove-Bourgeois game (assuming that Bourgeois are owners half the time) is as follows:

	Hawk	Dove	Bourgeois
Hawk	$\frac{1}{2}(V - C)$	V	$\frac{3}{4}V - \frac{1}{4}C$
Dove	0	$\frac{1}{2}V$	$\frac{1}{4}V$
Bourgeois	$\frac{1}{4}(V - C)$	$\frac{3}{4}V$	$\frac{1}{2}V$

Bourgeois is an ESS against Hawk and Dove. This is not surprising. In a population of Bourgeois the expected payoff for a mutant Dove will be $\frac{1}{4}V$ (as opposed to $\frac{1}{2}V$ for the Bourgeois majority); and the expected payoff for a mutant Hawk will be $\frac{3}{4}V - \frac{1}{2}C$, which differs from the Bourgeois payoff by $\frac{1}{2}(V - C)$—a negative quantity, since we are continuing to suppose that $V < C$. We can show that a tendency to fight in defense of territory and to retreat before another animal's defense of its territory would be maintained by selection (if the alternative strategies are Hawk and Dove, and if there are no genetic complications).

Interestingly, there is an alternative strategy that would also be an ESS against Hawk and Dove. *Proletarians* play Hawk when intruding and Dove when they are in possession. A moment's reflection reveals that the payoff matrix for the Hawk-Dove-Proletarian game is the same as that for the Hawk-Dove-Bourgeois game. What happens when we include both alternatives?

	Hawk	Dove	Bourgeois	Proletarian
Hawk	$\frac{1}{2}(V - C)$	V	$\frac{3}{4}V - \frac{1}{4}C$	$\frac{3}{4}V - \frac{1}{4}C$
Dove	0	$\frac{1}{2}V$	$\frac{1}{4}V$	$\frac{1}{4}V$
Bourgeois	$\frac{1}{4}(V - C)$	$\frac{3}{4}V$	$\frac{1}{2}V$	$\frac{3}{4}V - \frac{1}{4}C$
Proletarian	$\frac{1}{4}(V - C)$	$\frac{3}{4}V$	$\frac{3}{4}V - \frac{1}{4}C$	$\frac{1}{2}V$

Here there are two ESS's. In a population of Bourgeois none of the other strategies can invade—the payoff to any mutant from an encounter with Bourgeois is less than the payoff to Bourgeois from a Bourgeois-Bourgeois contest. Exactly the same holds for Proletarian. Hence, either strategy can be maintained by selection. (Maynard Smith [1982a, 101–103] shows that this conclusion also holds when the value of the resource is greater to the owner; he offers an interesting discussion of "paradoxical" ESS's. There are complications when we consider the possibility that animals make mistakes in assessing their positions.) Plainly, an ESS may not be "better" than the available alternatives. There may be cases in which a population cries out for an evolutionary accident, a catastrophe that will free it from the tyranny of the inferior behavior with which it has been saddled.

The Natural Bourgeoisie

As in the case of Hamilton's insight, it is tempting to think that the mathematical analysis is more elegant than biologically useful. Here, as before, we should not succumb to temptation. Even simple game-theoretic analyses can afford some understanding of animal behavior.

Animals often seem to hold back when aggression would appear to pay. Victorious animals sometimes emulate the "thumbs up" response of Roman emperors and fail to finish off their vanquished rivals. Sometimes they do not fight at all, even though there is a valuable resource that they could easily wrest from the owner. How is such magnanimity compatible with selection?

The eminent primatologist Hans Kummer performed a famous experiment to determine the extent to which fighting ability is relevant to the "owning" of females by male hamadryas baboons. (As sociobiologists constantly remind us, hamadryas society is remarkable for the severity of male aggression and the burdens imposed on the invariably submissive females. Hamadryas females are herded by the males to whose "harems" they "belong," and they are chased, bullied, and bitten when they stray too far from their lord and master.) Kummer found that when males are removed from their troop, "their" females are taken over by other males. When a previous "owner" is reintroduced, there will be a fight, and fighting ability dictates whether the old "owner" or the new "possesses" the female. Kummer noted that there is a puzzle here: if the new "owner" wins, why did he not fight for the female before? (See Kummer 1971, 101, 104.)

One answer would be to apply the analysis of the Hawk-Dove-Bourgeois game. Suppose that hamadryas males are programmed to play Bourgeois, and the only available rival strategies are Hawk and Dove. Then we can understand the behavior of the males who respect "ownership" by other males. They are playing Bourgeois. We can also understand why Bourgeois persists. It is an ESS with respect to the available rivals. So we have an explanation of how it is *possible* that the observed behavior, despite the initial appearance of reducing fitness, should be maintained in the population.

The fact that we have explained the possibility of maintaining the observed behavior under selection does not mean that we have isolated what is actually going on. Kummer attempted to test his hypothesis that "a male does not *claim* a female if she already belongs to another male" (1971, 104) by trying to screen out the obvious factors, other than ownership, that might have introduced an asymmetry into the situation. Removing baboons from familiar terrain and from their troops, Kummer placed a female in an enclosure with one male while another male watched from a cage. After fifteen minutes of separation the onlooker was allowed into the enclosure. In conformity to Kummer's hypothesis, he failed to fight the other male and apparently fidgeted about uncomfortably. Nevertheless, the observed behavior might have resulted from recognition of superior fighting ability. So Kummer reversed the roles. The previous onlooker now became the proud "owner" of a new female; the previous "owner" became a dispossessed onlooker. As before, the onlooker refused to fight (1971, 105).

Kummer's experiments provide some support for the hypothesis that hamadryas baboons play Bourgeois when the "contested resource" is a female. One obvious question is why the rampant aggression that these animals sometimes display (for example, in cases where there are two "claimants" for a female, a previous "owner" and a new "owner") is inhibited in this context. How is it possible for Bourgeois behavior to be maintained in the face of selection? We envisage an evolutionary scenario. We imagine uninhibited baboons, or, more exactly, uninhibited baboons who are good fighters, and we think of them as being more successful in leaving descendants than their Bourgeois rivals. Game-theoretic analysis reveals that our imagination may deceive us. If the costs of losing a fight outweigh the increase in fitness derived from gaining any one female, then it is possible for Bourgeois to resist invasion from Hawk. We discover that what we conceived as a more successful rival behavior does not promote fitness.

I should note that subsequent work by Kummer and his colleagues—notably Bachmann and Kummer (1980)—showed that the actual forces maintaining Bourgeois behavior in hamadryas baboons are more complicated. Female preferences play an important role in male attempts at takeover. Yet even in the absence of this experimental research it should be clear that there is a logical gap between explaining the possibility of preserving the Bourgeois behavior and locating the actual factors that have been at work.

Hamadryas baboons are not the only possible members of the natural bourgeoisie. Speckled wood butterflies may also respect rights of residence. In a series of experiments N. B. Davies showed that competitions for territories among male butterflies were resolved in favor of the owners. (Davies notes that "a very tatty male" once ousted an intruder who was "a perfect male in mint condition" [1978, 145].) Like Kummer, Davies was careful to show that he could reverse the outcome by reversing the roles. Moreover, by tricking the butterflies into "thinking that they were both the rightful owners," he could bring about long and mutually damaging battles. (See Davies 1978 and, for a brief account, Krebs and Davies 1981, 104.)

So game theory has its evolutionary uses. It can help us to see that our intuitive ideas about maximizing fitness may be wide of the mark. As a result we can understand how it is possible for apparently fitness-reducing behavior to be maintained. Another class of Darwinian worries dissolves under analysis.

"If I Can Help Somebody . . ."

Evolutionary theory does not seem to respect the old spiritual. Those who help somebody as they walk along seem prime candidates for those whose living has been in vain—provided that the somebody is not a relative and that our standard of futility is that of reduced genetic representation. Can we avoid the gloomy conclusion that aid to unrelated animals will be punished by the evolutionary process? Can we allow for broad cooperation?

The problem was tackled in a seminal article by Robert Trivers

(1971). Trivers imagined a population of animals repeatedly faced with situations in which the fitness of one individual would be greatly enhanced if another animal dispensed aid at small cost to itself. The classic illustration is highly implausible. Individuals constantly find themselves in danger of drowning in ponds, in the presence of onlookers on the bank, who can save them at negligible risk. Suppose that the population divides into two types, Saints and Exploiters. Saints leap in to help, without regard for the identity of the endangered animal. Exploiters are happy to be saved but never so much as get their own toes wet in aid of others. It is not difficult to show that the gloomy Darwinian conclusion holds: Sainthood is eliminated. (Strictly, the conclusion depends on genetic assumptions. If we assume that the behavior is controlled by a single locus, that one homozygote always produces Saints, that the other homozygote produces Exploiters, and that one allele is dominant, then the reasoning goes through without trouble.) However, as Trivers showed, Sainthood can be maintained if saints are a little less high minded, dispensing greater benefits to those who are inclined to reciprocate than to those who are not (1971, 192).

Suppose that two animals interact on two occasions. The first time, one animal has the opportunity to gain G units of fitness if the second acts so as to decrease its fitness by C. On the second occasion the roles are reversed. There are two strategies, *Trusting* and *Cheat*. Trusting animals invariably give aid, irrespective of the track record of the animal with whom they are interacting. Cheats invariably fail to help. The payoff matrix is as follows:

	Trusting	Cheat
Trusting	$G - C$	$-C$
Cheat	G	0

We assume that $G > C$. Given this assumption, the game is a familiar one. As Trivers noted, it is Prisoner's Dilemma. In the classic form, two prisoners are in separate cells. If both confess, they will each receive a two-year sentence; if one confesses and the other does not, then the one who confesses will receive a sentence of ten years and the one who does not will go free; if neither confesses, both will receive sentences of eight years.

What should our imaginary animals do? If each pair of animals only engages in a single pair of interactions, it is easy to see that there is only one ESS. Given any mixed strategy with a nonzero probability of Trusting, Cheat will invade. Pure Cheat is uninvadable by any rival strategy, pure or mixed. However, matters change if we suppose that there are repeated pairwise interactions and that the animals have some ability to remember the ways in which particular "partners" have behaved in the past.

Robert Axelrod has investigated the problem of iterated Prisoner's Dilemma. He invited a number of game theorists to submit computer programs for playing the game repeatedly against the same opponent: each played against virtually all the others, although each interaction was pairwise. (One important condition on these programs was that they should not be based on knowledge of the total number of interactions. Rather, the programs are to be conceived as providing strategies for a sequence of unknown length of Prisoner's Dilemma games against the same opponent.) Axelrod held a tournament in which the programs played against one another. Each program played two hundred games against each other program. Deciding the winner by computing total payoff (which would be total increase in fitness in the evolutionary context, and which is thus the quantity that seems to be relevant for thinking about the evolution of cooperation), Axelrod found that a relatively simple program, TIT FOR TAT, submitted by Anatol Rapaport, was victorious. TIT FOR TAT is very straightforward. In the first game against any opponent it plays Trusting; thereafter it does exactly what its opponent did on the previous interaction.

TIT FOR TAT is an ESS with respect to the other strategies proposed, and it has been claimed that it is an ESS with respect to any rival strategies that do not exploit knowledge of the length of the sequence of interactions. (A program that knows the number of interactions can invade TIT FOR TAT by playing Trusting for all interactions except the last and playing Cheat on the last interaction. It can be invaded by a program that Cheats one stage earlier. So it goes until the population collapses into Cheats.) TIT FOR TAT does not quite conform to the definition of ESS given by Maynard Smith, however, for the expected payoff to Trusting against TIT FOR TAT is the same as that to TIT FOR TAT against itself. Similarly, the payoff to Trusting against Trusting is the same as that to TIT FOR TAT against Trusting. So, in principle, Trusting animals could invade a population playing TIT FOR TAT.

We can show that in a population playing TIT FOR TAT, no mutant strategy can do better. (The argument is from Axelrod 1981; it is elaborated in Axelrod and Hamilton 1981 and in appendix K of Maynard Smith 1982a. See technical discussion B for details.) Gloomy Darwinian expectations are confounded. Cheats are not necessarily able to prosper and invade under selection; if the conditions are right, cooperative interactions among unrelated organisms can be maintained against subversion by what initially appear as more profitable

Technical Discussion B

Suppose that the payoff matrix for a game of Prisoner's Dilemma is

	Trusting (T)	Cheat (C)
Trusting (T)	P	Q
Cheat (C)	R	S

where $R > P > S = Q$. We also suppose that $2P > R + Q$. Players engage in a sequence of games with each other. After each game between two players the probability that the same two players will play again is w.

What are the alternatives to TIT FOR TAT? TIT FOR TAT bases its present action on what occurred in the previous game. If a strategy I plays against TIT FOR TAT, a play of T at any stage will effectively start the sequence over again. Hence, if I produced the best response at the beginning, then it should produce the same response following any occasion on which it plays T. So if I begins by playing T, then it should play T on all subsequent occasions. But suppose I begins by playing C. Then at the next stage it can either play T or play C. If it plays T at the second stage, then by the previous argument it should play C at the third stage (since C was the best response to the initial situation). Now, because T was the best response at the second stage after an initial play of C, and because TIT FOR TAT only remembers one stage back, it follows that at the fourth stage I ought to play T again. Plainly, we can iterate the argument: if I begins CT, then I should play $CTCTCTCT$. . . . Finally, I can begin with CC. But if C is the best response after an initial play of C, then C is the best third move for I, because, given TIT FOR TAT's tactic of basing present action on the behavior at the immediately prior stage, the situation at the third stage is exactly the same as the situation at the second stage. Again, we can iterate the argument: if I begins CC, then I should play $CCCCCCCCCC$. . . .

What we have shown is that if some strategy can do better against TIT FOR TAT, then one of three strategies—Trusting, Cheat, and Alternate—can do better. (Alternate plays $CTCTCTCT$ . . . .) We can now compute the expected payoffs in games where the opponent is TIT FOR TAT.

TIT FOR TAT: $P + wP + w^2P + \ldots = P/(1 - w)$.
Trusting: $P + wP + w^2P + \ldots = P/(1 - w)$.

Alternate:

$$R + wQ + w^2R + w^3Q + \ldots$$
$$= R(1 + w^2 + \ldots) + wQ(1 + w^2 + \ldots)$$
$$= (R + wQ)/(1 - w^2).$$

Cheat:

$$R + wS + w^2S + \ldots = (R + wS)/(1 - w).$$

Obviously, TIT FOR TAT and Trusting do equally well, and TIT FOR TAT will do at least as well as the others, provided

$$P/(1 - w) > (R + wQ)/(1 - w^2) \text{ and}$$
$$P/(1 - w) > (R + wS)/(1 - w);$$

that is,

$$w > (R - P)/(P - Q) \text{ and } w > (R - P)/(R - S).$$

The condition that $R > P > S > Q$ ensures that $(R - P)/(R - S) < 1$, and the condition that $2P > R + Q$ ensures that $(R - P)/(P - Q) < 1$. So it is possible for both conditions to be met. (Recall that w is a probability, so $w < 1$.) Hence, as long as w is sufficiently large, TIT FOR TAT will do at least as well as any mutant rival that attempts to invade a population of TIT FOR TAT.

rival strategies. (The attentive reader will recognize that the door has not been completely bolted against Cheats. If Trusting is able to drift to fixation in a population originally composed of animals playing TIT FOR TAT, then Cheat will subsequently be able to invade, so long as Cheats bide their time.)

Axelrod's argument does not provide an answer to the different question of how cooperative arrangements might have become fixed under selection. Given an appropriate assortment of strategies in the initial population, TIT FOR TAT can establish itself. But, unfortunately, Pure Cheat is an ESS. Hence, if we are to suppose that cooperative arrangements could have arisen from a population of Cheats, more work will be required. (Axelrod and Hamilton [1981] argue that TIT FOR TAT might secure a foothold in a population of Cheats originally divided into clusters of relatives.)

Another potential Darwinian puzzle seems less vexing. Thanks to Axelrod and Hamilton, we can begin to see how cooperation among nonrelatives might survive the disruptive workings of selection. Moreover, we can appreciate what selection might favor, the forms that animal cooperation may be expected to take.

Consorts and Coalitions

Cynics will wonder whether the labor is to any purpose. Are there cases in which unrelated nonhumans cooperate? One major difficulty in finding good examples (a difficulty that provokes some sociobiologists to hymn the praises of inclusive fitness) is that prominent systems of mutual aid among animals involve individuals who may well be relatives. Symbioses among members of different species can be found: some large fish allow small cleaners to enter their mouths and devour parasites, forgoing the opportunity to terminate the cleaning by terminating the cleaners. (Dawkins 1976, 201–202, cites this as the prime example of "reciprocal altruism"; the example is one among many offered by Trivers—see 1971, 196–203.) However, the very reason that enables us to discount the hypothesis that the behavior is maintained through aiding kin (and thereby maximizing inclusive fitness) also tells against assimilating the case to the kind of game-theoretic analysis considered earlier. Cleaners and their hosts do not compete for genetic representation in later generations of the same population. By contrast, when unrelated animals of the same species dispense aid to one another, they are apparently acting doubly against their own reproductive interests. Not only do they take on the costs in fitness associated with helpful behavior, but they also promote the genetic representation of a competitor.

Yet there is at least one case in which nonhumans do seem to cooperate with unrelated members of their species. Unlike their unfortunate hamadryas relatives, female olive baboons are not herded as the "property" of males. When an olive baboon is in estrus, she typically consorts with a single male, and the consortship may last for several days. Craig Packer found that coalitions of adult males sometimes formed to attack other single males, especially when there was a chance of displacing a male from a consortship (Packer 1977). One male, wishing to challenge a consortship, would solicit the aid of another. If the coalition was successful, the soliciting male formed a new consortship, while the helping male ran the risk of injury from a fight with the previous consort. Packer discovered evidence of reciprocation: males tended to solicit the aid of those whom they had aided in the past. "Favorite partners" for various males could be identified. Favorite partners typically returned the solicitation more than they solicited other males.

Trivers's discussion of reciprocal altruism, or the more systematic approach pursued by Axelrod and Hamilton, can help us understand how it is possible for selection to maintain cooperation among male olive baboons. If the baboons play TIT FOR TAT (or some strategy that converges on TIT FOR TAT), then the correlations discovered by Packer *might* result. There is reason for caution here, because the costs and benefits are not constant from one occasion to another; a solicited male might fail to join a coalition because the risks of doing so were too great. Moreover, we would expect the formation of coalitions to be determined not only by the past actions of the soliciting male but also by those of the male who is to be the object of the attack. Baboons may be reluctant to take on the local heavyweight champion or a bosom buddy.

Even if we waive these doubts, there is a deeper worry. When we considered Kummer's hamadryas baboons, we applied the results of the Hawk-Dove-Bourgeois game by supposing that the value of "gaining" a female was less than the costs of injury in a fight. (If $V > C$, then Hawk beats Bourgeois.) In applying the Trivers-Axelrod-Hamilton approach to Packer's olive baboons, we need to suppose that the value of a consortship is greater than the costs of a fight. If that were not so, then it would always pay to Cheat. Can the suppositions be reconciled?

Let C_h and C_o be the costs of fights to hamadryas and olive baboons respectively; let V_h and V_o be the respective values of access to the female. Because hamadryas "harems" are forever, it seems that $V_h > V_o$. Hence, the problem of consistency seems worse than before, for consistency requires that $V_h < C_h$ and $C_o < V_o$. Champions of the analyses will respond that the fights that occur in hamadryas baboons typically lead to great damage, while those in olive baboons frequently involve no more than much chasing around. As a result, C_h is very large and C_o is small, so that both inequalities can be satisfied.

We might originally have wondered how it is possible for selection to maintain cooperative action among olive baboon males. We are now led to wonder how it is possible for selection to maintain a strategy of limited aggression in displaced consorts. Could a tendency to inflict severe punishment establish itself in the baboon population, thereby destroying the features on which the persistence of cooperation rests? Two points should be noted. Our analyses of animal behavior, achieved piecemeal, must fit together: we cannot make assumptions in one case that are denied elsewhere. Moreover, as the analysis unfolds, we may uncover constraints that themselves call out for explanation; in trying to give an account of the constraints, we may change dramatically the shape of the analysis.

If all that is at stake is a question of evolutionary possibility, the second issue is not threatening. We may invoke game theory to explain how the behavior of hamadryas and olive baboon males is possible by showing that there are values of V_h, V_o, C_h, and C_o that would satisfy the crucial inequalities and that these values conform to the intuitive assessments of field naturalists. If we want to describe the actual evolutionary history, then we shall have to explain the prior fixing of costs and benefits. Constraints cannot simply be invoked to fit our preferred scenarios. But I am getting ahead of my story.

Perfect Designs

The last sections have looked at situations in which animals might adjust their behavior to the behavior of the animals around them. The theory of games is useful in understanding cases where the effect of a form of behavior on fitness is highly dependent on what other animals do. Not all situations are of this type. Sometimes the effect on fitness of a particular piece of behavior is determined by factors that are largely independent of the actions of others. One popular approach to these cases is the use of optimality models. Faced with an organism that seems to be acting so as to reduce its fitness, we try to show that, given the constraints to which the organism is subject and the range of strategies available, it is doing the best it can. Under analysis the organism is seen to be maximizing its fitness. The challenge for the biologist is to find the right analytic perspective.

Sometimes the analysis is conducted qualitatively. Consider an example provided by George Williams. We imagine a species that breeds once a year. Members of the species grow throughout their lives, and fecundity increases with size. Production of offspring is directly proportional to fecundity. All the offspring has to do is release its gametes at the right moment, and it is guaranteed that a fixed proportion of them will turn into zygotes. Prudent organisms postpone reproduction. Eager organisms go to it as soon as they can. Which will prove fitter? It depends. Prudent organisms run the risk that an untimely end might prevent them from reproducing at all. Eager organisms slow down their growth rate and thus produce fewer gametes later on. Relative fitness is determined by the annual survival probability and the relation between the growth rates (Williams 1966, 173).

Williams's idea could easily be elaborated as a formal model. We would consider the expected number of offspring produced over a lifetime, computing this for both strategies in terms of the survival probabilities and the growth rates. Obtaining values for the relevant parameters, we could then show how a strategy that initially appears to reduce fitness is actually the fitter of the two. (Williams suggests that it is possible to understand the variety of reproductive schedules in fish by applying a model of this type. See 1966, 174ff., and, for elaboration, Charlesworth 1980.)

The approach can be pursued in a very large number of cases. To construct an optimality model, it is necessary to start with some measure of the fitness of the organisms under study. This measure will be taken to correspond to the expected number of descendants in later generations. Quite frequently, simply isolating an appropriate measure of fitness can solve the problem of how an apparently fitness-reducing behavior survives selection. If we wonder how Prudent organisms can be maintained, we can dispel our perplexity by focusing not on the number of offspring produced in a single season but on the output of a lifetime.

The model proceeds by claiming that the organisms are subject to particular constraints and that they have a specified set of alternatives available to them. Given these constraints, we compute the value of the chosen measure of fitness for the alternative forms of behavior. If we find that the apparently fitness-reducing form of the behavior maximizes the measure subject to the constraints, our work is done. We have revealed how it is possible for the behavior to be maintained by selection, by showing that there never was a genuine worry in the first place.

One of the most thorough and sensitive uses of optimality models is the work of George Oster and E. O. Wilson on the division of labor in colonies of social insects. Their approach can be illustrated by one of their least technical discussions, which, as in most of the cases they consider, is directed at analyzing the conditions under which the fitness of an insect *colony* would be maximized. After choosing as a measure of colony fitness the net rate of production of reproductives during the colony lifetime, Oster and Wilson look for a schedule of producing workers and fertile organisms that will maximize this rate. Intuitively, the question is whether the colony can do better by raising some fertile individuals from the beginning or whether it is best to start with sterile workers who can then be expected to help with the rearing of a fertile brood (Oster and Wilson 1978, 50ff.).

The problem is simplest in cases where the insects have an annual life cycle. We suppose that the old queens die each fall and that only the new queens survive the winter to found new colonies in the spring. Reproductive males have their moment of glory in the fall, and then perish along with their mothers and their worker sisters. The colony is assumed to have two "castes"—workers, who forage for food and who feed and care for the queen's offspring, and reproductives (the queen, her fertile daughters, and the necessary drones). How should the colony schedule production of these castes so as to maximize fitness? (If the colony is founded by a single queen, then we can reformulate the question in terms of the inclusive fitness of the queen mother or in terms of the inclusive fitness of her worker daughters. For simplicity's sake, however, I shall follow Oster and Wilson in their preferred formulation.)

Suppose that at time t there are $W(t)$ workers in the colony. The resources accumulated within the colony will obviously depend on the size of the work force. So let us write the resources returned to the colony per unit time as $W(t)R(t)$, where $R(t)$ is the rate of return per worker. (If the workers do a good job, R is high.) Some of these resources will be used to make further workers. The number of new workers made per unit time will depend on the resources available, the fraction of the resources that is put into making workers, and the amount of resources needed to make a worker. Suppose that $u(t)$ is the *scheduling* function, whose value is the fraction of colony resources devoted at time t to the production of workers. Finally, let b be a constant, measuring the rate at which resources are converted into new workers (b is inversely proportional to the amount of resources required for making a single worker). Then, at time t, new workers are being produced at a rate $buRW$. Old workers die, of course; at time t the number of workers in the colony is being depleted by an amount that is the product of the number of workers and the death rate. So the equation for the change in W over time is

$$\mathrm{d}W/\mathrm{d}t = buRW - mW,$$

where m is the worker mortality rate. Similarly, we can write the equation for the change in Q, the number of reproductives in the colony, as

$$\mathrm{d}Q/\mathrm{d}t = bc(1 - u)RW - nQ,$$

where bc is the rate at which resources are converted into reproductives (c is the worker/reproductive weight ratio) and n is the mortality rate for reproductives.

Now we can state the optimization problem. We want to choose the function u so as to maximize $Q(T)$, the number of reproductives alive at T, the end of the breeding season. But there are constraints. $Q(0) = 0$, for there are initially no descendant reproductives—that is, reproductives who will survive the next winter; $W(t) \geqslant 0$, and $Q(t) \geqslant 0$ for all t, since we cannot allow the colony to boost its production of one caste through having a negative number of the other; finally, $0 \leqslant u(t) \leqslant 1$ for all t, because we cannot assign more than 100 percent of the resources to the production of either caste. We also assume that b, c, m, and n are constants.

Obviously, which schedule is optimal will depend on the nature of the return function. Oster and Wilson argue that for a wide variety of return functions the optimal strategy is qualitatively the same. In the first phase of colony life the entire effort is directed toward making workers. No reproductives are produced. At a critical time (which will depend on the character of the return function) the colony switches, and in the second phase of its existence it produces nothing but reproductives (daughter queens and males). (See Oster and Wilson 1978, 54–56, 71–73; Macevicz and Oster 1976, 265–282.) This is an interesting result about production of offspring in social insect colonies. If we had considered the possibility that an insect colony (or a founding queen) might achieve greater reproductive success by producing some reproductives from the beginning, or by continuing to build up the worker force even after it had started to raise reproductives, then the analysis reveals how these alternative strategies might have been defeated by selection.

The general technique of constructing optimization models promises to reach farther than the dissolving of particular puzzles about forms of behavior. Once we have shown that a pattern of behavior is optimal, it is tempting to think that we have isolated the actual evolutionary forces that maintain the behavior (perhaps also those that shaped it in the first place). Yet, as I have repeatedly emphasized, there is a gap between demonstrating an evolutionary possibility and justifying hypotheses about the causes of evolutionary changes. Later we shall consider whether the disclosure of an apparently perfect design automatically bridges the gap.

Integrating Insights

Critics of sociobiology are often impressed by the fact that sociobiologists use a number of different techniques in explaining animal behavior. The Sociobiology Study Group lambasted Wilson (1975a) on the grounds that the "new synthesis" had fashioned tools to guard itself against any conceivable failure:

> The trouble with the whole system is that nothing is explained because everything is explained. If individuals are selfish, that is explained by simple individual selection. If, on the contrary, they are altruistic, it is kin selection or reciprocal altruism. If sexual identities are unambiguously heterosexual, individual fertility is increased. If, however, homosexuality is common, it is the result of kin selection. Sociobiologists give us no example that might conceivably contradict their scheme of perfect adaptation. (Caplan 1978, 288)

The last sentence suggests an important point in its identification of the sociobiological commitment to seeing adaptation everywhere. But the main line of criticism is misconceived. Sociobiologists are not in the business of making an all-purpose tool kit from which instruments may be selected to meet the needs of each occasion. Their claim is—or ought to be—that the techniques reviewed in this chapter can be integrated. By using them in conjunction, we would achieve a more refined understanding of the fitness of various pieces of behavior. In different examples different approaches receive emphasis, and they do so because those approaches yield results that would be unaffected (or not appreciably altered) by a more elaborate analysis deploying all the techniques.

Casual sociobiologists might fancy the approach attributed by the Sociobiology Study Group. Where it became difficult to understand behavior as maximizing individual fitness, our hypothetical dilettantes might grasp the notion of inclusive fitness as a deus ex machina. That would be to pervert Hamilton's insight. Hamilton extended classical evolutionary theory by recognizing that the classical notion of fitness ought to be replaced by the notion of inclusive fitness if our aim is to chart the evolution of a population under individual selection. Selection always acts on differences in inclusive fitness. Classical evolutionary theory—including classical population genetics—works well enough in many cases, however, and it does so because in these cases inclusive fitness is equal to (or insignificantly different from) classical fitness. (Recall that the inclusive fitness of an organism does not add to the classical fitness fractions of the fitnesses of relatives. It augments a "stripped" classical fitness with fractions of the increase in fitness of relatives that results from the actions of the organism. In many cases the stripping effect and the augmentation will both be zero; in many other cases they will cancel.) Sociobiologists, and indeed all evolutionary theorists, can ignore considerations of inclusive fitness when the classical fitness is not significantly different—and when they do so, they can reasonably be challenged to show that the difference is insignificant.

Analogous remarks apply to the other ideas we have reviewed. In some examples we can apply the notions of game theory, without worrying about inclusive fitness or sophisticated models of optimization. The simple encounters in which nonrelatives contest a resource of immediate value provide obvious illustrations. In other contexts no game-theoretic apparatus is required to show that a particular type of behavior contributes to inclusive fitness. Some behaviors provide evident benefits to kin. Yet there are clearly situations in which the approaches need to be integrated. Sometimes there will be competition among relatives for the same valued resource. Sometimes the expected payoffs of various behavioral strategies can only be computed by engaging in a prior optimization analysis to determine the lifetime success of organisms following different reproductive schedules. Birds sometimes have the "option" of competing for territory and trying for immediate reproduction, or postponing competition and reproduction to the next breeding season. In cases like these, models of conflict and ecological optimality models are both relevant.

Careful sociobiologists will attempt, in all cases, to isolate the major factors affecting the inclusive fitnesses of the organisms under study, and may even envisage a vast theoretical analysis in which all relationships are taken into account, all sensitivities to the behavior of other members of the population considered, and all long-term contributions to reproductive success carefully computed. They are sometimes lucky. The vast analysis can sometimes be replaced with an account that identifies one important variable and uses just one of the techniques we have considered. What is crucial to the success of the replacement is the insignificance of the factors that are left out. Sociobiologists should not be chided for some general methodological error, but they may be guilty of omitting, in particular cases, considerations that, if incorporated, would subvert their favored conclusions.

One interesting example that requires the integration of techniques is Trivers's analysis of parent-offspring conflict (1974). The interests of parents and progeny, construed in terms of fitness, often coincide. An offspring's success in leaving descendants redounds to the credit of parents. But the comforting harmony of the relationship can be broken. Even though it would contribute to the fitness of a parent to promote the welfare—and thus the fitness—of one of its offspring, the value of such contributions is not unlimited. From the parent's point of view it will typically be a poor strategy to lavish on one of its progeny resources that would produce a greater number of expected descendants (in either the first or the second generation) if they were used to help another offspring. The potential target of the nurturing may see things differently. Not only may its expected number of offspring be increased if the contested resources are channeled in its direction, but its inclusive fitness may be greater as well. In this difference of perspective lies a conflict between the strategies of parents and those of their offspring.

Trivers's example illustrates the need for marrying optimality analyses with Hamilton's ideas about inclusive fitness (see technical discussion C). Both mother and young face problems of optimization. In each case the optimum is computed by appeal to considerations of

Technical Discussion C

Imagine a mammalian mother who has just produced her first young. Suppose that the increase in fitness to the offspring if it nurses through a time period of length t is $F(t)$ and that the decrease in the mother's expected number of offspring if the firstborn nurses for a time period of length t is $D(t)$. Assume that any further siblings produced would be full siblings of the firstborn. From the perspective of the mother the expected change in genetic representation in the second generation produced by allowing the firstborn to nurse through an interval of length $t + \delta t$ instead of an interval of length t is

$$[F(t + \delta t) - F(t)] - w_t[D(t + \delta t) - D(t)],$$

where w_t is the fitness of an offspring that nurses for a period of length t. From the perspective of the firstborn the expected change in genetic representation in the same generation, the generation of its offspring, is

$$[F(t + \delta t) - F(t)] - \tfrac{1}{2}w_t[D(t + \delta t) - D(t)],$$

since the lost siblings are expected to share 50 percent of its genes (in virtue of descent). Conflict will occur if these expressions have different signs: the change from a nursing period t to a period $t + \delta t$ will contribute to the offspring's first-generation fitness, while detracting from the mother's second-generation fitness.

Trivers (1974) suggests that there will be a critical interval during which weaning conflicts are likely to occur. The interval will be bounded by the mother's preferred time for terminating nursing and the offspring's preferred time. These times are given by

$$F(T_m) - F(0) = \int_0^{T_m} w_t \, dD(t)/dt \, dt,$$

$$F(T_o) - F(0) = \int_0^{T_o} \tfrac{1}{2}w_t \, dD(t)/dt \, dt.$$

T_m is the optimal time for the mother to wean the firstborn; T_o is the optimal time for the young to cease nursing. (My formulation of the problem and the solution departs from Trivers in avoiding the notion of parental investment. For a critique of Trivers's use of this notion and a clear account of the main insight, see Maynard Smith 1976b.)

inclusive fitness. The computations might serve as the basis for a further game-theoretic analysis, in which we could consider various parental and offspring strategies: offspring might be viewed as having the options of *Brat* (exhibit regressive behavior in attempts to exploit mother) or *Angel*; mothers might be seen as playing *Softie* or *Firm*. By elaborating the example, we could reveal the need for integrating the insights of the game-theoretic analysis with those of the other techniques. The moral is that careful sociobiologists ought to resist labeling the examples they study with such terms as "kin selection," "reciprocal altruism," and so forth. The main task is to explore the relative fitnesses of various forms of behavior; in doing so, one should take into account all the relevant factors. This "beginner's guide" has been an attempt to introduce the principal ways in which the analysis of fitness can be conducted.

Chapter 4
What Is Sociobiology?

Fields and Theories

In the beginning was the official definition. "Sociobiology is defined as the systematic study of the biological basis of all social behavior" (Wilson 1975a, 2; the phraseology recurs in Wilson 1978, 16, and Lumsden and Wilson 1983a, 23). Certainly there is a field of study here. Numerous investigators will perceive themselves as contributing something to "the systematic study of the biological basis of all social behavior." But from the beginning it was clear that official definitions run the risk of being without form and void. Wilson quickly assimilated his enterprise to the discipline of evolutionary biology. He singled out a question as the "central theoretical problem of sociobiology": "How can altruism, which by definition reduces personal fitness, possibly evolve by natural selection?" The reader did not have to wait long for an answer: "The answer is kinship: if the genes causing the altruism are shared by two organisms because of common descent, and if the altruistic act by one organism increases the joint contribution of those genes to the next generation, the propensity to altruism will spread through the gene pool" (1975a, 1–2). So what is sociobiology? The systematic study of the biological basis of all social behavior? A discipline that attempts to explain the evolution of altruism? A set of variations on Hamiltonian themes?

It is time to draw the threads together. In previous chapters I have tried to assemble the elements out of which a picture of sociobiology can be constructed. Any serious discussion of the topic must locate pop sociobiology in relation to studies of animal behavior and must distinguish various types of sociobiological endeavor. We are now ready to undertake these tasks.

Let us start with a pair of distinctions. We should separate sociobiology the *field* from sociobiology the *theory*. Fields of science, in my usage, are constituted by families of questions. Biogeography is a field, comprising questions about the geographical distribution of plants and animals. Biochemistry is another field, consisting of questions about the chemical interactions of biologically significant molecules. Not every collection of questions constitutes a field. The questions must form a family, that is, they must hold out the promise of a unified collection of answers. Lumping questions about taxes, taxis, and taxa does not create a field, the field of supertaxonomy. It simply makes a muddle.

Scientific fields offer invitations to the construction of theories. There was a field of astronomy (generated by questions about the orbits of the celestial bodies) long before there was an astronomical theory. Achievement of a theory may reconstitute a field, however, drawing boundaries around groups of questions that had previously seemed unrelated, distinguishing questions that had previously seemed to belong to the same family. Fields can separate (inorganic and organic chemistry). They can merge (electricity and magnetism). They can be abandoned as misconceived (astrology).

We can think about theories for fields by generalizing the account given for evolutionary theory. A theory for a field divides into three parts. The first part consists of the questions constitutive of the field. The second comprises some patterns of reasoning (problem-solving strategies, schematic solutions) that give the form of answers to questions. In the third part we find general claims about the entities or processes invoked in constructing answers to the questions.

Some rival versions of evolutionary theory can be regarded as addressing a common field. Those who take all Darwin's questions for their province agree on a field but may go on to quarrel about theory. Others restrict themselves to a subfield, rejecting as misconceived the efforts of their more ambitious colleagues. So, for example, cautious Darwinians change the field of evolutionary biology by abandoning any attempt to answer questions about the pervasiveness of characteristics. (See, for example, Eldredge and Cracraft 1980, Nelson and Platnick 1981.)

Sociobiology can be conceived either as field or as theory. More exactly, there are two fields, each of which might be called "sociobiology," and some candidate theories for each of these fields. The field of *broad* sociobiology corresponds to Wilson's official definition. It is the systematic study of the biological basis of all social behavior, including not only questions about the evolution of social behavior but questions about the mechanisms of social behavior, about the development of social behavior, about the genetics of social behavior, and perhaps even about the function of social behavior. (I say "perhaps" because it is far from clear that there are separate questions about function begging to be answered once all the problems about evolution, mechanism, development, and inheritance have been addressed. What else would be left to know?) Students of broad sociobiology may devote themselves with equal interest to questions about what happens in an animal when it engages in a particular kind of behavior or questions about the development of patterns of behavior in individual animals. Their vision need not be confined to questions about evolution. Broad sociobiology stands forth as approximately the field defined by Niko Tinbergen in his famous account of the four "whys" of behavioral biology (Tinbergen 1968, 79).

Narrow sociobiology is more selective. Its questions are evolutionary questions. So, in posing the question why animals engage in the forms of behavior that they do, narrow sociobiology construes the request as asking for a specification of the actual workings of evolution: How did the behavior originally evolve? How is it maintained? Evidently, any theory for the field of narrow sociobiology will bear an important relation to evolutionary theory. Its questions will be a subset of those addressed by ambitious versions of evolutionary theory. Moreover, since behavior that appears to reduce the individual fitness of the animal performing it is frequently encountered, and since such behavior challenges evolutionary theorists to find even a *possible* scenario, it is hardly surprising that such behavior should attract considerable attention. To hail the problem of "altruism" as the central theoretical problem of the field is surely overstatement. The field is equally concerned with the many cases of animal social behavior in which there is no appearance of fitness reduction. But it should not be hard to understand why the overstatement is made.

There is no single theory for the field of broad sociobiology. Biologists interested in this field draw on the resources of numerous disciplines, developing and mature. For example, in Hinde's conception of ethology (Hinde 1982) a theory of broad sociobiology is to be fashioned by integrating the findings of behavioral genetics, developmental psychology, neurophysiology, traditional ethology, and evolutionary biology. Here any "new synthesis" is, at best, a gleam in the fond parental eye, although there are partial achievements that encourage the hopeful. Much of the research reported by Wilson (in particular, 1975a, chapters 8–10) is concerned with questions of how various social phenomena actually work. (See also his beautiful descriptions of the mechanisms and life histories of insect colonies in Wilson 1971). Bernd Heinrich's monograph on bumblebees (1979) almost entirely forgoes evolutionary issues in favor of an attempt to understand precisely how bumblebees are able to do the things they do, to pinpoint the mechanisms of flight, of thermoregulation, or foraging behavior, of plant selection, and so forth. Work like this makes an important contribution to broad sociobiology. It does not touch the question Wilson hails as "the central theoretical problem."

If we are to hunt the "new synthesis," then we should look for a theory in narrow sociobiology. What are the prospects? First, any such theory will be concerned with a family of questions about the evolutionary origin and maintenance of social behavior. Internal disagreements can arise here, just as they erupt in evolutionary theory. Sociobiologists can bicker about the classification of behavior and about the level of generality to be sought. Not all sociobiologists find the question "Why are males generally more aggressive than females?" a theoretically appropriate one. Those who do envisage that there will be a pattern of evolutionary explanation, applicable with only slight modifications in the case of different species. Those who do not may reject the use of a blanket category of "aggressive behavior" or may suggest that a proper focus is on particular species. Some may hold that a cross-specific account of patterns of social behavior can only be assembled later—and then with caution. Still others will direct their research by posing comparative questions, by seeking evolutionary explanations of differences in social behavior across a group of related species.

Second, sociobiologists diverge in their preferred strategies for answering questions about the evolution of social behavior. For Richard Alexander, sociality is a symphony of nepotistic themes (Alexander 1979, 52): social behavior is usually to be explained by the recognition of hidden effects on kin that help to spread shared genes. Others, like Wilson, emphasize the importance of inclusive fitness but are also concerned to develop optimality models, allowing that sometimes the differences between maximization of individual fitness and maximization of inclusive fitness may be inconsequential. Still others, such as Maynard Smith, place great emphasis on game-theoretic analysis. All of these approaches to the questions of narrow sociobiology draw on the contemporary extensions of evolutionary theory that were reviewed in the last chapter. They differ by favoring different techniques.

The third part of any narrow sociobiological theory would consist of some general principles about the evolutionary processes that produce behavior. Here, the principle results are Hamilton's definition of inclusive fitness, his derivations of analogs of central results of classical population genetics, the introduction of the concept of ESS by Maynard Smith, and the delineation of optimality models by Wilson, Trivers, and others. This theoretical work can be viewed as refining the discussions of the genetics of the evolutionary process set forth in standard evolutionary theory and as enhancing our appreciation of the subtle ways in which fitness may be determined.

We seem to be within sight of a clear definition of sociobiology as theory, and of its relations to the general theory of evolution. Yet as we approach, the "new synthesis" turns out to be a mirage. There is no autonomous theory of the evolution of behavior. There is only the general theory of evolution.

Global Peace and Local Conflict

Nobody should deny that animal behavior, like animal morphology and animal physiology, is the product of evolution. Only the very cautious will suppose that it is never possible to answer questions about the evolution of animal behavior. Hence, the project of constructing ambitious Darwinian histories for behavioral characteristics of animals should not be rejected out of hand. Moreover, in implementing this project, we should appeal to the best available ideas about the evolutionary process for clues to the form that ambitious Darwinian histories should take. Thanks to Hamilton, Maynard Smith, and others, the last decades have seen great refinements in our understanding of fitness and of the ways in which fitness is promoted. In studying the evolution of behavior, we can put contemporary models to work: we can expect to explain some forms of behavior by showing that they maximize inclusive fitness, others by producing game-theoretic analyses, still others by devising optimality models. In other words, we shall study the evolution of behavior by using the best general ideas from contemporary evolutionary theory.

So there is no general theoretical dispute between the sociobiologists and those who accept evolutionary theory and do not limit themselves to cautious Darwinism. Theoretical sociobiology is irresistible simply because, insofar as there is a theory in narrow sociobiology, it is the general theory of evolution. The questions addressed by sociobiologists are a subfamily of those addressed by (noncautious) evolutionary theorists. The schemata accepted for answering those questions are the schemata provided by general evolutionary theory. Moreover, the general ideas about the evolutionary process, conceived as a process that brings forth animal behavior, are just the general ideas about the evolutionary process, period. True, sociobiologists may pride themselves on the fact that certain refinements of general ideas about evolution are the products of work on the evolution of behavior: Hamilton's concept of inclusive fitness is an outstanding example. Once assimilated, however, such insights lose sight of their ancestry. Like the concepts of heterozygote superiority and frequency-dependent selection, Hamilton's notion becomes part of the general framework within which our representations of the evolutionary process are fashioned.

If sociobiologists have any distinctive theory to offer, what might it be? Perhaps they propose to employ forms of reasoning that are not sanctioned by the schemata of general evolutionary theory? Unlikely. We should recall that the main champions of sociobiology have been proud to proclaim their allegiance to Darwin's banner, insisting that those who are not for them are against Darwin. More plausible is the view that sociobiology is distinctive because sociobiologists restrict themselves to certain preferred patterns of explanation in tracing the evolution of social behavior. But a moment's reflection should convince us that this would be a foolish claim. Just as it would be silly to contend that the morphology of organisms evolves solely in response to selection for defense against predators, so too it would be rash to declare that all animal social behavior involves selection for inclusive-fitness effects. Sober sociobiologists ought to expect that the explanations of social behavioral traits will be a motley—and that there will be appreciable diversity in the evolutionary maintenance even of those forms of behavior that appear to reduce fitness. (Moreover, they should be prepared for the possibility that some of the explanations may not involve selection in any direct way.) They should admit that there will be innumerable examples in which appeal to inclusive fitness does not differ significantly from appeal to individual fitness. Their general refrain should be that the theory of evolution provides outline answers to questions about the presence of behavioral traits and that the totality of answers will prove as diverse as the totality of evolutionary explanations generally.

Of course, just as there are important subtheories of the theory of evolution, questions that can be answered by articulating one style of evolutionary explanation, so we may expect that there will be subtheories of the theory of the evolution of social behavior that will address a family of questions by offering a single style of argument. Theories of mimicry and melanism find their counterparts in theories of the diversification of forms ("caste differentiation") among the social insects and in the theory of sex ratios. To recognize this is to say no more than that the field of animal behavior furnishes new subtheories for the general theory of evolution. It remains true that there is no general theory of the evolution of animal social behavior that is distinct from the general theory of evolution.

Exit the "new synthesis." Wilson launched no grand new theory. Proponents of sociobiology sometimes seem to insist that they have some distinctive general theory of the evolution of social behavior, which is grounded in contemporary evolutionary theory. They are half right. The theory is grounded in evolutionary theory. It is evolutionary theory. Viewed in this way, the credentials of sociobiology are impeccable.

Can we stop here, with a rousing endorsement of sociobiology? Surely not. While we may be content with the idea of addressing questions about the evolution of behavior, and while we may admire the schemata that the sociobiologists propose to employ, we may nevertheless protest the ways in which they construct instances of those schemata and the evidence they use to support the resultant explanations. On closer inspection, the sociobiology controversy is not a debate about the legitimacy of the concept of inclusive fitness or the mathematical results in game theory derived by Maynard Smith. The battles are fought around local issues of a general type: do the sociobiologists instantiate their preferred patterns of explanation so as to produce correct accounts of the evolution of animal behavior?

Critics of sociobiology are in the position of the car owner who goes to a local garage with hopes of finding mechanics who will fix some engine problems. There are all the necessary tools. They are in perfect working order. But when the mechanics descend, armed only with sledgehammers, announcing their resolve to tackle the troubles by these means and these alone, it is time to go elsewhere. Rational car owners do not stop to hurl maledictions at the hoist, the drills, the wrenches, or even the sledgehammers. These are all fine things that can be put to good use. But the present plan for employing them is not one to inspire confidence.

Cautious Darwinians will find the whole program of explaining the presence of behavioral traits wrong-headed—but no more awry than much of the research that passes in contemporary evolutionary biology. Others should not be trapped into accepting or rejecting the deliverances of sociobiology (or of evolutionary theory generally) as some monolithic whole. There are excellent instantiations of evolutionary schemata, including some in which the explanation generated accounts for the presence of a behavioral trait. Others are less impressive. As a result, investigation of the credentials of sociobiology must proceed piecemeal. Are differences in the behavior of males and females to be explained solely by reference to optimal male strategies for achieving copulations? Is the presence of homosexuality to be understood in terms of maximization of fitness through aid given to relatives? Is human aggression to be seen as the product of an evolutionary history in which relatively harmless animals failed to develop the inhibitions characteristic of more heavily armed species? When we pose these particular questions, the general theoretical machinery is not in dispute. The real issues concern what is to be done with particular styles of explanation.

We have reached a result that changes the shape of the sociobiology controversy. Narrow sociobiology proposes to use the techniques of contemporary evolutionary theory in charting the evolution of social behavior. The resultant accounts must be evaluated individually. Sociobiological enthusiasts would like us to think otherwise, to be swept along by a general commitment to general theory. Chagnon lays down the gauntlet:

> Professor John Maynard Smith asked, quite seriously, at the end of my comments: "Why are anthropologists interested in this theory for 'chaps' [humans]?" The question implies two kinds of answers. Either he thinks that it does not apply to man, in which case he would like to know why anthropologists are trying to apply it, or he thinks that it does not apply to *any* social behavior in any species, leaving himself in the unenviable position of having to explain why he goes about making theory that does not apply to real behaviour. (Chagnon 1982, 292)

Chagnon's argument is that there is a general theory of social behavior, to which Maynard Smith has contributed, and that only intellectual cowardice stands in the way of applying it to human social behavior. But the premise is wrong and the charge groundless. When we see that sociobiological explanations must be assessed piecemeal, we recognize that the assessments may be radically different in different cases. Some explanations—including some that focus on aspects of human behavior—may be extremely well supported. Others may fare less well.

Triumph for sociobiology, or, more exactly, triumph for evolutionary theory in the domain of social behavior, would consist in showing that the present range of schemata enables us to explain all cases of social behavior. There is plenty of room between triumph and complete failure. If we were to discover that some types of behavior could not be understood by applying our theoretical apparatus, we might decide that our evolutionary explanations had been focusing on the wrong factors, that a different style of evolutionary theorizing was needed. Yet whether or not evolutionary theory in its present form will triumph in the area of social behavior, whether or not further advances in our general understanding of the evolutionary process will be needed, the sociobiology controversy turns on issues that are much more limited. Narrow sociobiology is a legitimate part of evolutionary theory only if the explanations that its proponents offer are as well supported by the evidence as those in evolutionary theory generally. So we must turn again to the confirmation of Darwinian histories.

Equal Rights for Sociobiologists

Recall the charges of methodological corruption that critics have lavished on sociobiology. Allegedly, sociobiology has deserted the standards of good science to wander in the never-never land of unfalsifiable speculation. Its champions reply that sociobiology is as good as evolutionary theory generally and that their opponents employ a double standard. They are at least partly right. When the theory of narrow sociobiology is understood as the theory of evolution applied to social behavior, there is no surprise in the conclusion that both theories share the same methodological status.

Defenders of sociobiology sometimes point out that preferred methodological canons fit evolutionary theory rather badly, and they cheerily draw the moral that "as sociobiology is part of the evolutionary family, we ought not to judge it by standards more strict than we would apply to the rest of evolutionary theory" (Ruse 1979, 21). True enough. Before we acknowledge the equal rights of sociobiologists, however, we should consider that equal rights presuppose equal duties. We do not learn much about the merits of two baseball teams by finding that neither can translate a Sanskrit text. Similarly, failure to conform to inappropriate methodological standards is not an automatic guarantee of scientific legitimacy.

The crucial issue is whether sociobiologists can and do confirm their claims in the same way that workers in evolutionary biology support their Darwinian histories. We can set aside at the beginning the idea that there is something called "kinship theory" that is tested and confirmed. Although sociobiologists routinely talk in this way (Barash 1977, 88–93), they cannot intend that observation of a group of animals is used to assess the *truth* of Hamilton's claims about the dynamics of inclusive fitness or Maynard Smith's conclusion that Bourgeois is an ESS against Hawk and Dove. Like the principles of mathematical population genetics, such results are accepted because they follow from some definitions and some very well confirmed results about gene transmission (grounded quite independently of any evolutionary considerations). Hamilton saw that the classical approach to the understanding of evolution under individual selection omitted any effects that the bearers of phenotypes might have on their kin, and he saw that the omission might be serious. Maynard

Smith saw how to reinterpret some theorems of game theory. In neither case are observations of animal behavior relevant to the truth of the claims.

Empirical evidence is needed for supposing that the general results apply in particular cases. Have some birds evolved the behavior of giving alarm calls in the presence of predators because this behavior increases their inclusive fitness through alerting previously unsuspecting relatives? Are Kummer's hamadryas baboons playing Bourgeois? Are Packer's olive baboons playing TIT FOR TAT? These are questions that sociobiologists will have to answer by appealing to the kinds of observations that are gathered to support evolutionary hypotheses. They will have to avoid the pitfalls that wait for those attempting to support ambitious Darwinian histories.

Consider the tasks that an aspiring sociobiologist will have to complete. First it will be important to show that the phenomena are consistent with the explanation that is given. Suppose that the favored hypothesis claims that a form of behavior is present because of benefits conferred on relatives. Then it will be important to show that the fitness of relatives is indeed advanced. Suppose that the account identifies the animals under study as playing a particular strategy and specifies a collection of rival strategies against which the favored strategy is an ESS. Then it will be important to show that the behavior of the animals conforms to the preferred strategy and that there is no reason for believing there are rival strategies available to the animals, strategies that are not included in the list of alternatives. (Troubles arise when we have grounds for ascribing to the animals discriminatory abilities that would allow them to play more complicated strategies than those officially permitted—or when we hypothesize that some of their relatives actually play such strategies.) Establishing consistency may be no trifling matter. But consistency is not enough.

As we discovered in investigating the confirmation of ambitious Darwinian histories, success is achieved by eliminating rivals. Those who leap to embrace an evolutionary scenario without considering the credentials of obvious alternatives are rightly castigated by their colleagues. We also discovered a more tricky methodological situation. Sometimes a proper consciousness of our own ignorance may bring home to us the possibility of rival hypotheses that we are in no position presently to articulate. What we know about the physiology of the animals under study may simply be inadequate to the task of deciding whether a form of behavior genuinely diminishes the personal fitness of the animal that engages in it, whether it really enhances the fitness of a relative, whether certain alternative forms of behavior might have been available to the animal, and so forth. All too frequently we know too little about genetics, too little about development, and too little about the mechanisms of behavior to elaborate a convincing set of rival evolutionary hypotheses. Under these circumstances we are hardly justified in snatching at the first scenario that strikes our fancy.

Sociobiology inherits all the difficulties of evolutionary confirmation and generates a few of its own as well. Not only are the mechanisms and developmental processes underlying behavior often unknown, but we are usually ignorant about the appropriate classifications of animal behavior. Hinde frankly admits that "the problems of describing behaviour are considerable" (1982, 30; see 30–32). He cautions investigators against confusing causal and functional descriptions and against building in unwarranted assumptions. The dangers are easily appreciated. To classify birds as "giving alarm calls" already suggests a function for the behavior and an evolutionary scenario. Our taxonomy of behavior is not so firmly fixed that we can bring it to the enterprise of developing hypotheses about the evolution of patterns of behavior. That taxonomy itself begins to emerge from our evolutionary studies. It takes an evolutionist's eye to realize that baboons are playing TIT FOR TAT.

Once again the comparison with Kettlewell's investigation proves useful. Kettlewell began with a framework of concepts that he could employ in formulating hypotheses about the evolution of moths in industrial and rural Britain. Describing them as speckled or dark presupposed a scheme of classification, but not one that was likely to be challenged. Skeptics who mused about whether Kettlewell had described his moths in the proper way would be met with laughter or blank stares. Not so their cousins who question the description of behavior. If we are told that a butterfly is "defending its territory" or that baboons are "appeasing a dominant member of the troop," we should not waive the question of proper formulation. The categories we employ in reporting behavior are less well supported than those we bring to the study of morphology and physiology.

Instead of conceding that the properties of the animals have been correctly identified and inquiring whether there are pertinent aspects of mechanism and development that could shed light on the evolutionary history of these properties, we have to ask at the beginning whether our knowledge of the mechanisms of behavior is sufficient to give us confidence in the behavioral description. Kettlewell could focus on the property of coloration. He could consider the possibility that that trait is a side effect of the process that produces both it and a much more important characteristic; he could ponder possible alternative benefits. Those who engage in studies of the evolution of behavior need to do some work before they can reach the point from which Kettlewell began.

It would be wrong to think that the work is impossible. There are potential triumphs of sociobiological analysis. If we can come to know enough about behavioral mechanisms to arrive at a well-grounded classification of behavior, then we can begin to formulate and test rival evolutionary hypotheses. Triumphs are most likely to come in those areas where the behavior of animals can be specified in a relatively uncontroversial way—such areas as foraging, for example—and they are most likely to concern groups of animals whose behavior lends itself to causal and genetic analysis. They are most elusive just where pop sociobiologists would like most to achieve them, in the understanding of controversial aspects of human behavior.

Whether they have hopes for human sociobiology or are simply interested in some group of animals, sociobiologists generally concede that the enterprise confronts methodological difficulties. Wilson's remarks are exemplary:

> Paradoxically, the greatest snare in sociobiological reasoning is the ease with which it is conducted. Whereas the physical sciences deal with precise results that are difficult to explain, sociobiology has imprecise results that can be explained by many different schemes. Sociobiologists of the past have lost control by their failure to discriminate properly among the schemes. (1975a, 28)

Wilson hopes that the "new synthesis" can do better than the traditional ethology of Ardrey, Morris, Tiger, and Fox. As we shall see, when the difficulties loom, there is a temptation—to which Wilson sometimes, but not invariably, succumbs—to settle for lower standards. The lapses occur just where error is likely to work most harm. Speculation that would be rejected in the attempt to understand the behavior of ants flourishes freely when the animal under study is *Homo sapiens*. (For a prime example, see Wilson 1975a, 343.)

Many other students of animal behavior express their caution. Brian Bertram hopes that we may ultimately be able to measure the selective pressures that have shaped animal behavior, and that our understanding of the evolution of the social systems of animal species will eventually be "slightly less guesswork than it now is" (Bertram 1976, 180). Hans Kummer advises his readers that statements about the adaptiveness of traits require far more systematic testing than he has been able to achieve (1971, 16). A recent essay by Robin Dunbar is even more forthright:

> If we make a prediction based on some theoretical argument, but the data do not support the prediction, we may conclude either that the theory is wrong or that the behavioral measure was an inappropriate test, but we have no good grounds for preferring one explanation rather than the other. This situation will inevitably tend to encourage either rubrics of the kind "Accept the result if it fits your expectations; search for a 'better' behavioral measure if it does not" or a resort to *ad hoc* attempts to explain away anomalous results through the invocation of confounding variables. More often than not, this reflects a state of ignorance about the underlying biology, combined with a "super-Kuhnian" reluctance to dispense with the theory at all. (Dunbar 1982, 23)

All these comments reveal a sense of unease with current practice in understanding the evolution of animal behavior. By providing a picture of sociobiology and of the difficulties of confirming evolutionary hypotheses, I have tried to show what lies behind the complaints. The problems faced by the aspiring sociobiologist are just those of other evolutionary biologists. Writ large.

Rebuilding the Ladder

Understanding the difference between broad and narrow sociobiology is useful. Recognizing that sociobiology can be both field and theory enables us to sort out some confusions. Identifying the methodological problems that beset narrow sociobiologists helps us appreciate which sociobiological claims about the behavior of nonhuman animals need more support and which can stand proudly with the best of evolutionary theory. But the main concern of this book is with pop sociobiology. Where does pop sociobiology fit in?

There is a potential field, human sociobiology, whose questions are the questions of narrow sociobiology applied to *Homo sapiens*. At present we do not know how to pose those questions in a precise way, because we do not know which elements in human patterns of behavior have been fashioned through evolution. Pop sociobiology consists of attempts to organize the field and to answer the evolutionary questions so generated. Its practitioners assume that interesting and controversial aspects of human behavior are direct products of the evolutionary process, and they offer evolutionary scenarios to explain why we should expect those aspects to be present.

There have been protracted debates about human propensities for avoiding incest and for treating people of other races with suspicion and hostility. Pop sociobiology proposes to resolve such debates by offering evolutionary scenarios designed to show how natural selection might have been expected to favor hostility toward strangers and to frown on copulation with close relatives. Because of a purported link between an allegedly widespread pattern of human behavior and evolutionary expectations, pop sociobiologists announce large discoveries about human nature. The evolutionary explanations reveal us as we really are and thereby show that certain social arrangements are unalterable.

So conceived, pop sociobiology is a hybrid. It consists, in part, of a collection of proposals in narrow (human) sociobiology, proposals that must be evaluated according to the methodological canons that we have identified. It adds to this a method for connecting evolutionary explanations with dramatic claims about human nature. Evolutionary explanations, even evolutionary explanations of patterns of human behavior, are not in themselves the stuff of which headlines are made. Excitement is generated by focusing on particularly controversial aspects of human behavior and by drawing from the evolutionary explanations claims about the inevitability of human social institutions. Different pop sociobiologists have different ways of constructing the extra apparatus that is needed to hoist us from nature up to self-knowledge. For the moment, however, I shall concentrate on one type of pop sociobiology. It is time to try to rebuild Wilson's ladder.

Here is a new version of Wilson's ladder, one in which the weaknesses of our previous construction have been repaired; I take it to represent the argument that stands behind Wilson's provocative claims.

Wilson's Ladder (Revised Standard Version)

1. By using the standard methods for confirming evolutionary histories, we can confirm hypotheses to the effect that all members of a group *G* would maximize their fitness by exhibiting a form of behavior *B* in the typical environments encountered by members of *G*.

2. When we find *B* in (virtually) all members of *G*, we can conclude that *B* became prevalent and remains prevalent through natural selection, specifically through the contribution to fitness identified at step 1.

3. Because selection can only act where there are genetic differences, we can conclude that there are genetic differences between the current members of *G* and their ancestors (and any occasional recent deviants) who failed to exhibit *B*.

4. Because there are these genetic differences and because the

> behavior is adaptive, we can show that it will be difficult to modify the behavior by altering the social environment, in the sense that states in which *B* is absent are likely to be either precluded states or else states that are precluded given widely shared desiderata.

I attribute this style of argument to Wilson for four reasons. It reaches the kinds of conclusions at which he aims. It begins with the subject he takes as the source of insights that were unavailable to his predecessors—the analysis of fitness-maximizing behavior in present evolutionary theory. It avoids the positions from which Wilson has struggled to dissociate himself (the crude kinds of genetic determinism attributed by his critics). It proceeds by way of transitions that Wilson seems to approve (such as the passage from optimality analyses to conclusions about selection). If there is a line of argument to the early Wilson version of pop sociobiology, then this seems to be it.

Assume for the moment that the ladder is solid. Let us see how we could climb to dizzy heights of self-knowledge. We begin by identifying a group of organisms that includes some humans. This group might be a cluster of species containing *Homo sapiens*. It might be a cluster of subsets of species containing a subset of *Homo sapiens*. So, for example, *G* might comprise the primates; the family that includes the great apes, the gibbons, and ourselves; the set of all vertebrate males; or the set of lower-class individuals in humans and social termites. By appealing to the characteristics of *G* and of the environment in which members of *G* operate, we try to show that exhibiting *B* would maximize inclusive fitness. Perhaps we argue that a particular cooperative arrangement would be beneficial to the animals in *G*. Perhaps we give a game-theoretic analysis that identifies certain strategies as available to members of *G*, demonstrates that one of them is an ESS, and shows how following that strategy would have maximized inclusive fitness in a hypothetical animal population. So we stand on the first step.

Now, in looking at the animals in *G*, we discover that they do what our optimality analysis prescribes for them. Let us suppose for the moment that the fit is perfect. There is no need to wave hands in the direction of "confounding variables" (as Dunbar admits, this is sometimes a popular move). We conclude that the disposition to produce the form of behavior we have isolated is a genuine trait of the animals that has been shaped by natural selection. The second step brings us to the position of being able to talk confidently when using a particular behavioral description and of recognizing a history of selection for the form of behavior that we have uncovered.

Natural selection is powerless to operate unless there are genetic differences. If there were variation in behavior, but no *genetic* variation in behavior, then selection could not bring the group to its present position of manifesting the optimal behavior. Thus, in recognizing a history of selection, we also recognize that there are genetic differences between animals who exhibit *B* and other animals (hypothetical ancestors or recent deviants) who do not. If we want to talk casually, we can say that there are "genes for the behavior"—so long as all we mean by the remark is that there is an allele whose presence or absence against a common genetic background makes a difference to the manifestation of the behavior.

Step 4 is the most difficult. Even though we have struggled to a position at which we can claim that there are "genes for the behavior," the sense that we must give to this conclusion prevents us from making any automatic inferences about the fixity of the behavior. All we know is that, given a common background, there is an allele whose presence or absence makes a critical difference to the manifestation of the behavior in the typical environments encountered by members of *G*. To say this is not to suppose that the exhibition of *B* is part of the "nature" of members of *G*. We have no basis for conjectures about how members of *G* might be expected to act in different environments, including environments that are now quite common but that were not encountered during most of the evolutionary history of the group.

To see how simple inferences go awry, consider a well-known case. Most new parents learn about a disease that occasionally afflicts human beings, producing severe developmental problems if it is allowed to run its course. Babies are routinely tested for PKU. There is an allele that, on a common genetic background, makes a critical difference to the development of the infant in the normal environments encountered by our species. Fortunately, we can modify the environments. The developmental abnormalities result from an inability to metabolize a particular amino acid (phenylalanine), and infants can grow to full health and physical vigor if they are kept on a diet that does not contain this amino acid. So it is true that there is a "gene for PKU." Happily, it is false that the developmental pattern associated with this gene in typical environments is unalterable by changing the environment.

The top of Wilson's ladder needs some support. We can try to find the needed props by returning to the achievements of earlier steps. First, we focus our intended conclusion. Although we have so far labored to establish some general result about the members of an inclusive group *G* the name of the game is the fathoming of *human* nature. So let us derive from step 3 the conclusion that there is a "gene for the behavior" in humans. To show that the associated behavior is an expression of human nature, we need to argue that it is not something that would be varied if we redesigned our (physical, or, more probably, social) environment. We know from steps 1 and 2 that this gene has been selected. There are two lines of argument that might be used to parlay this knowledge into a result about the inflexibility of the behavior.

The first stresses the idea that adaptations are resilient to environmental change. If a form of behavior maximizes fitness across a wide range of physical and social environments, and if we can suppose that our hominid ancestors encountered this wide range of environments, then, it may be suggested, we should expect that small fluctuations in the environment will not modify the behavior. An environment that allows us to reach states in which the behavior is absent will have to be radically different from those for which selection has fashioned us. Ill-equipped by evolution for our new surroundings, we shall suffer much unhappiness. Environmental reform will go against the grain of human nature.

Let us revert to one of pop sociobiology's favorite examples. Assume that "genes for maternal care" have been selected in the course of human evolution. Pop sociobiologists may then argue that maternal devotion has maximized fitness across so large a range of human environments that there are probably buffer systems that prompt mothers to nourish and comfort their young even when conditions are far from propitious. If we attempt to create a social system in which mothers are encouraged not to spend time in child rearing, there will be genetically based feelings of resistance. There are costs in trying to disrupt the status quo.

The second line of argument assesses the available methods of social reform. Imagine that we want to reach a state in which a particular form of behavior—for example, hostility among nations—is absent. Suppose that we have discovered (by climbing the ladder) that there are "genes for aggression" in humans, and suppose that we are satisfied that such genes manifest themselves in hostility among nations. (Some readers may find these hypothetical attitudes hard to achieve. I sympathize. The point, however, is to understand what might follow, not to worry about the road so far traveled.) We want a peaceful world. We envisage certain kinds of social change that would create an environment in which the same genes no longer led to aggressive behavior. The proposal is not that feeding people on special diets would modify the behavior—as if consumption of spinach in our formative years would do the trick. (It is worth pointing out, however, that we would do much to diminish some hostilities if we could assure all people that they would be fed *something*.) Instead, we imagine educational programs, schemes of cooperation among nations, arms reduction, and so forth, as fostering a decrease in aggression. The imagined procedures would affect the ways in which we think about others. It is this emphasis on the cognitive that pop sociobiologists may seize on as fatal to the chances of success. Cognitive modifications, they may allege, are too shallow to affect our propensities for aggression.

Their argument begins with an emphasis on the efficiency of adaptations. Given that human fitness (or, perhaps, male human fitness) was maximized by a propensity for aggressive behavior, pop sociobiologists may challenge the idea that the simplest, or most efficient, way to construct a mechanism for the behavior in early hominids was to put our burgeoning cognitive abilities to work. Instead, they will suggest that the tendencies to aggression were fashioned in a system that was largely independent of our cognitive control—perhaps in the "hypothalamic-limbic complex" that Wilson loves to oppose to our "consciousness." An unwanted consequence is that the primitive drives remain with us, refine our perceptions of the world as we will.

A variation on the same theme is to insist that the propensity for certain forms of behavior—aggression, for example—was the product of selection in remote ancestors, whose behavioral mechanisms were far less complicated than our own, and to infer from this that the springs of our aggressive behavior are cognitively impenetrable. In either case the Wilsonian argument attains its goal, the conclusion about the fixity of an allegedly fitness-maximizing behavior, by suggesting that evolution had to short-cut our cognitive systems in the interest of maximizing fitness. (By contrast, some sociobiologists, such as Richard Alexander, seem to think of our cognitive systems as proximate mechanisms for fitness-maximizing behavior. As we shall see in chapter 9, Alexander's view of human nature differs from Wilson's in important respects, one of which is Alexander's emphasis on evolutionary understanding as the key to *modifying* our patterns of social behavior.)

Finally, should further buttressing be needed, those who hope to climb Wilson's ladder may appeal to the fixity of human behavior across different societies. "Behold the variety of cultures!" they exclaim. "In none of them is the behavior absent. How, then, can we hope to eliminate it by modifying our societal arrangements?" This appeal needs to be used with caution. As we saw in chapter 1, the societies actually developed do not constitute a sample that can be used in a systematic investigation of human behavior. Those who deny Wilson's claims about human nature can reply that all existing cultures share common features and that the removal of some of these features is crucial for attaining alternative forms of social arrangement. Furthermore, to suppose that the fixity of human behavior could be read off the anthropological record would make the lengthy detour through evolutionary theory quite unnecessary. Pop sociobiology is inspired by the hope that it can transcend the magical mystery tour of human cultures that has so often generated ideas about "human universals." If the anthropological data are used, they must be employed sparingly.

The final step of the ladder is negotiated with some artistic balancing. We are to lean a little on various props: here on the resilience of adaptations, there on the efficiency of evolution, just a little on the anthropological data. In this way the early Wilson program makes its final move. We are to draw the conclusion that states in which the forms of behavior are absent are either precluded states or states that are precluded with respect to widely shared desiderata. Either way, ideals of social justice are likely to lose.

Dismantling the Ladder

We are finally ready. We know what sociobiology is. We know how evolutionary claims are tested and confirmed. We know how pop sociobiology relates to sociobiology and how sociobiology relates to evolutionary theory. We have looked at the details of one form of pop sociobiology, the early Wilson program. Now we can ask the crucial questions. Is there good evidence for the fixity of forms of human behavior, specifically those forms of behavior that we might hope to change?

The answer, I think, is "No." In the next four chapters I shall argue that Wilson's ladder is rotten at every rung. Some of the conclusions that I draw will apply to other versions of pop sociobiology. However, they should not be construed as an indictment of all sociobiology. As I have been at some pains to emphasize, there is no general theoretical dispute between sociobiologists and other practitioners of evolutionary theory. It is eminently possible that some of the hypotheses now advanced about the evolution of animal behavior are well tested and well confirmed. Moreover, it is possible that we might some day achieve justified conclusions about the evolution of some aspects of human behavior. Although I shall try to expose the deficiencies of pop sociobiology by contrasting the claims of pop sociobiologists with the work of those who study the behavior of nonhuman animals, the defects lie in the method, not the matter. Wilson might have brought to his study of human behavior the same care and rigor that he has lavished on ants. But he did not.

We shall begin with the first step. Using the work of other sociobiologists as a foil, we shall examine some conclusions about fitness maximization in humans that are drawn by Wilson and his disciples. For it is here already that the rot begins.

Chapter 5
Tycho's Gift

A Moral from Mars

"To us Divine goodness has given a most diligent observer in Tycho Brahe, and it is therefore right that we should with a grateful mind make use of this gift to find the true celestial motions." So wrote Johann Kepler, in the midst of his struggle to compute the orbit of Mars. Kepler had reached a point at which almost all his fellow astronomers would have been satisfied. Using the Copernican system, he had worked out a combination of circular motions about a static sun, fitting the observed positions of Mars to within eight minutes of arc. Such accuracy was a major achievement. Kepler could have proclaimed a triumph for Copernicanism. Instead, he pressed on to more radical departures from traditional astronomy, moved by the desire to discover an orbit that would conform even more closely to the measurements of his former teacher and colleague, Tycho. His search led to the revolutionary proposal that the orbits of the planets are elliptical. What could easily have been dismissed as "reasonable error" proved to be the basis for a major change of perspective. Tycho's exquisite care was well known to Kepler, prodding him on toward a transformation of our vision of the solar system—one that Tycho himself would never have foreseen.

Astrology is bunk. Nevertheless Mars bears a moral for contemporary sociobiology. Students of animal behavior can comfortably congratulate themselves on finding matches between the behavior observed in the field and a favored story about the maximization of fitness. Failures of fit can be dismissed as minor exceptions, small errors whose eradication will not affect the basic picture. Kepler's determination to pursue the sources of inaccuracy should remind us of the possibility that the correction of what initially appears a small anomaly may be the source of a significant shift in vision.

Sociobiological research on nonhumans is a mixed bag. There is great latitude in the sizes of the groups for which Darwinian histories are sought. Some sociobiologists are concerned with particular species, such as hamadryas baboons or carpenter ants; others examine entire classes, focusing on aspects of the behavior of birds or mammals. There are all kinds of intermediate studies, whose targets are ungulates or cleaner fish, the social Hymenoptera or the social carnivores. The studies may be undertaken with the help of carefully formulated mathematical models. Or they may be carried out in verbal terms, with qualitative claims about advantages that accrue from particular forms of behavior. In some cases there are elaborate and detailed observations of the behavior of animals over several generations of a population. In other cases the observational base consists in a report of a single striking phenomenon. Sometimes the genetics of the animals studied is relatively well understood. In other instances inheritance, development, and behavioral mechanisms are all virtual mysteries.

It should not be surprising, then, that some biologists find sociobiology a field full of exciting ideas and studies that are as rigorous as any in evolutionary biology, while others dismiss the enterprise as methodologically misguided speculation. As we have seen, the issue comes down to the credentials of particular explanations. We must ask, has this explanation of this behavior in this group of animals been supported by the evidence in the ways that ambitious Darwinian histories lend themselves to support? Our methodological inquiries suggest that it is unlikely that the answer will be the skeptic's "Never!" or the aficionado's "Always!" A plausible answer is "Sometimes—and to varying degrees in different cases." My goal is to show that what is plausible is sometimes true.

We know in advance that there is a general methodological difficulty in obtaining strong confirmation for an ambitious Darwinian history when we are grossly ignorant of the genetics and mechanisms of the behavior under study. Even if we waive this difficulty, Darwinian histories may be presented with more or less precision, they may be defended by careful consideration of alternative approaches or they may not, they may rest on a body of exact observations about animals who have been the object of intense scrutiny or they may be supported by much more casual reports, they may dismiss exceptions as annoying irritants of no great significance or they may attempt to pursue the sources of inaccuracy.

A sympathetic student of recent work in sociobiology should weigh the merits and pardon the offenses. Even when sociobiologists offer accounts of the evolution of a behavior that do not conform to the high standards we rightly demand of accounts in evolutionary biology, their claims may nonetheless be taken as useful suggestions for further research. Ill-grounded conjectures are sometimes fruitful, and it would be counterproductive to impose a ban on speculation about the evolution of animal behavior. Danger looms only when we confuse ourselves into thinking that there is some undifferentiated mass of studies, all equally well supported, which goes by the general name "sociobiology." The stage is then set for the reporting of these findings as matters of "scientific fact"—that is, as if they were well-confirmed scientific hypotheses—in works that argue for the application of nonhuman sociobiology to humans. Pop sociobiology basks in the roseate glow of an optimistic vision of results about nonhumans.

Hence, even though the analysis to be undertaken in this chapter will attempt to differentiate a broad range of types of sociobiological study, I do not advocate nihilism or anything close to it. The challenge for the sociobiologist is to continue the imaginative, but critical, work that is found in the best studies. When this involves severe practical problems—for example, when it is not obvious how certain necessary tests can actually be carried out—then the correct response may be that the issue cannot currently be resolved. I do not suggest that the attempt to fathom the evolutionary history of animal behvior should be given up. Some behavioral traits have evolutionary histories, and it is worthwhile to attempt to fathom them. All that should be agreed is that we do not dupe ourselves with appealing anecdotes. Like Kepler—and like the most sensitive narrow sociobiologists—we should be wary of stopping too soon.

The Dung Fly's Dilemma

I shall begin with two studies that show narrow sociobiology at its best. The examples I have chosen are familiar textbook cases, and with good reason. They represent a class of examples in which careful observations are matched with refined theoretical analysis. Moreover, in neither case do the experts argue that they have arrived at the definitive account of the phenomena. Conclusions are advanced cautiously; and, especially in the second example, there have been persistent attempts to tackle unanswered questions and thereby to improve the evolutionary analysis.

The first example concerns the mating behavior of the dung fly. Female dung flies lay their eggs in patches of cow dung, the fresher the better. Males congregate around newly formed cow pats and await the arrival of the females. When a female lands, the males engage in strenuous competition to mate with her. Fights are intense and sometimes damage the females as well as the males.

Male dung flies face a dilemma. Even if they win the first round of battle, after copulating they are vulnerable to the efforts of newly arriving males who might copulate with the same female. Some of the female's eggs would then be fertilized by alien sperm, and the original male would gain a lesser number of offspring. How would we expect a properly selected dung fly to behave?

The British behavioral biologist Geoffrey Parker has studied the mating behavior of male dung flies and has advanced a number of analyses of various aspects of this behavior. I shall focus on his simple, but precise, model of male behavior after copulation (Parker 1978).

The first thing we need to know is what a male who has just copulated stands to lose. Parker resolved this issue with some ingenious experiments. The first experiment showed that if two males copulate in succession with the same female, the sperm of the second male will fertilize about 80 percent of the eggs. (Parker found a way to irradiate males, allowing the sperm to retain the capacity for fertilizing eggs but debarring normal zygote development. He then compared the numbers of zygotes that develop in two cases: first, cases in which an irradiated male copulates first and a normal male second; second, cases in which the order is reversed. In the former examples about 80 percent of the eggs develop; in the latter instances only 20 percent do.) The second experiment enabled Parker to determine the relation between the number of eggs fertilized and the time spent in copulation. The longer the time spent (up to about 100 minutes), the greater the number of eggs fertilized; but success comes more quickly at the beginning. (See figure 1 for the time dependence.)

Assume that a male's reproductive success is directly proportional to the number of eggs fertilized in his lifetime. Consider the situation

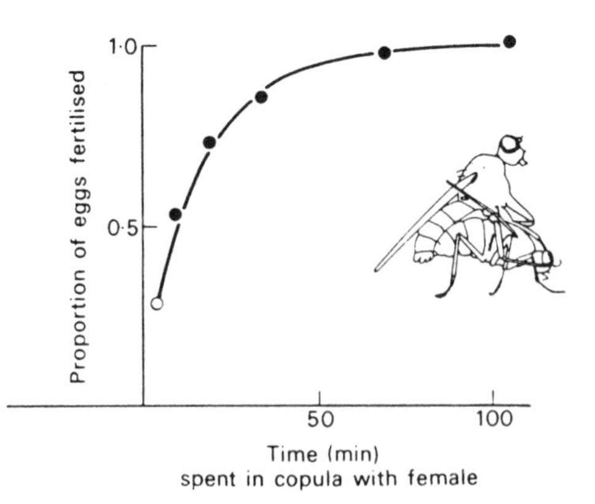

Figure 1
The dependence of fertilization of female eggs on male copulation time. (From Krebs and Davies 1981, who adapt Parker 1978)

from the perspective of the male. (Although this was once the common perspective in sociobiological analysis, there is now a rapidly growing literature on female choice and female sexual strategies. I shall go on to consider the situation from the perspective of the female dung fly.) We can divide the reproductive life of a male dung fly into three phases: there is time spent in searching for a female and in competing for access to her, there is time spent in copulating, and there is time spent in guarding after copulation to ensure that no interloper destroys the fruits of his labors. From the male's perspective—though not from the female's—the time spent in searching and guarding is fixed. Availability of mates is determined by the distribution of dung and by female propensities for arriving at patches of dung. The maximum time needed for guarding is fixed by the length of time between the end of copulation and the laying of the eggs.

What is the optimal copulation time? Parker suggests that males will maximize their reproductive success if the production of eggs per phase proceeds at the fastest possible rate. So, if t_c is the time spent in copulation, if t_{sg} is the sum of the times devoted to searching and guarding, and if the number of eggs fertilized in time t is $g(t)$, then t_c should be chosen so as to maximize $g(t_c)/(t_c + t_{sg})$. (Figure 2 gives a graphical solution to the optimization problem.)

Parker measured the mean value of the time spent in searching and guarding, using the result together with his determination of $g(t)$ to solve the optimization problem. He predicted that the optimum value

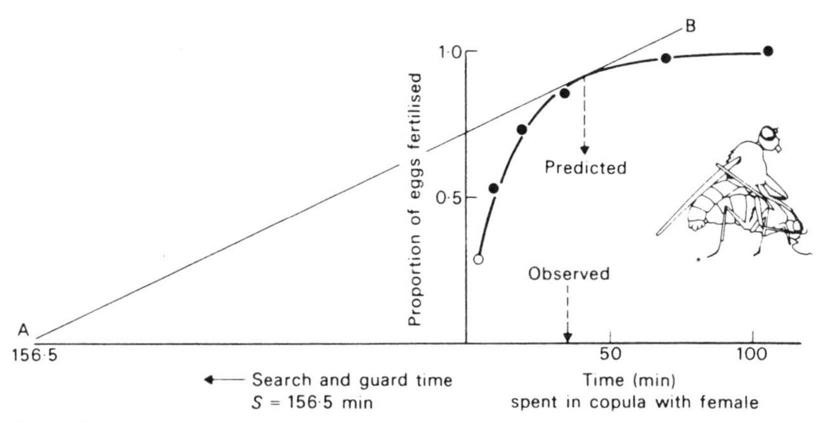

Figure 2
Graphical solution to the optimization problem. The optimum is found by maximizing the rate at which eggs are fertilized per unit time devoted to fertilization. In graphical terms this amouunts to finding the line of greatest slope that passes through $(-t_{s+g}, 0)$ and that meets the curve. This line is the tangent to the curve that passes through $(-t_{s+g}, 0)$. (From Krebs and Davies 1981, who adapt Parker 1978)

for copulation time is 41 minutes. There is an obvious test. We can go to the fields, examine the dung flies in action, and measure the average copulation time. If Parker's optimization analysis is to translate into a Darwinian history of the evolution of copulation time in the male dung fly, then we shall expect that selection has favored flies with a propensity to copulate for 41 minutes. What do we find?

"The average copulation time in the field is 36 min., quite close to the predicted 41 min."—so Krebs and Davies sum up the test results (1981, 61; see Parker 1978, 230–231). Certainly the two figures are of the same order of magnitude. (I take it that Krebs, Davies, and Parker would all have been convinced that they were pursuing the problem in the wrong way if the copulation time had proven to be 30 seconds or several hours.) But are they really any closer than we might have expected, given only Parker's graph of the rate at which eggs are fertilized? Is there any reason to think of selection as doing any more than operating against those who copulate either for very short times or for very long times, thus favoring copulation times between 20 and 100 minutes?

There are two responses that one might make. Enthusiasts will insist that the measured value is close enough to the predicted value to give us grounds for thinking that Parker's account is on the right lines—after all, the error is only about 12 percent, and who can hope for less, when dealing with the complicated behavior of complicated organisms? Skeptics may urge that that 12 percent is important, that it merits investigation, and that we are not justified in accepting this account of the evolution of male dung fly copulatory behavior until we have examined possible sources of error. We may find that the examination changes the entire picture of the situation.

Parker's result is certainly suggestive, and his skillful experiments are rightly admired. Questions arise only when we consider how the analysis might be extended to increase the accuracy of the prediction. Enthusiasts will hold that the actual copulation time is reduced because of some unrecognized constraint. Let us suppose them to be imbued with the Keplerian spirit. What can they say?

Here is one possibility. Perhaps males who copulate longer have less ability to resist newcomers, or diminished capacities for fighting their rivals when they fly off to fresh fields and cow pats new. Suppose we let $h(t_c)$ be the probability that a male can successfully guard a female through the period following copulation. Assume that $h(t_c)$ decreases as t_c increases. (The longer a male copulates, the less able he is to resist usurpers afterwards.) If dung flies displaced between copulation and laying forfeit 80 percent of the eggs they have fertilized, the copulation time should be chosen to maximize

$$\frac{g(t_c)\{h(t_c) + 0.2[1 - h(t_c)]\}}{(t_c + t_{sg})} = \frac{\text{expected number of eggs fertilized}}{\text{time spent}}.$$

Because h decreases with time, the effect is to reduce the optimal copulation time.

Here is another possibility. Perhaps we should question the assumption that reproductive success is directly proportional to the number of eggs fertilized. Ultimately, the dung fly's fitness is measured by the extent of its genetic representation in future generations. As a proximate measure, we might take the number of offspring who survive to maturity. This number need not be proportional to the number of eggs fertilized. Suppose that eggs that are fertilized late in copulation have a greater likelihood of being fertilized by sperm carrying alleles that diminish viability. Then the eggs that are fertilized earlier will be more "valuable"—in the sense of being more reliable vehicles for the transmission of genes to future generations—than those that are fertilized later. As before, the optimization problem will be to maximize a quantity differing from $g/(t_c + t_{sg})$ through scaling by a function that decreases with time. Again, the solution will be less than Parker's predicted optimum.

Enthusiasts have grounds for hope that the difference between Parker's predicted value and the value found in the field can be explained without altering the basic analysis. (For Parker's own proposal, see Parker 1978, 231.) Skeptics may suggest accounts that would introduce more substantial modifications. One challenge is to consider the genetic basis of male copulation timing, to use the classic techniques of behavior genetics (such as breeding experiments), and to see if there are developmentally correlated traits that interfere with direct selection for copulation time—or even involve costs and benefits so large that niceties of copulation time are trivial by comparison. This skeptical suggestion should be relatively familiar from earlier discussions. A second challenge is more novel.

We have been looking at the evolutionary situation solely from the perspective of the male. Females have been viewed as essentially passive, sources of fixed constraints on plastic male behavior. But that is not the only perspective. Female dung flies also face a dilemma.

A female dung fly does not face the difficulty of fighting off other females or of guarding her male. Her problem is to avoid injury in fights among competing males. A female also needs to ensure that her eggs are not fertilized by defective sperm. Hence, if extended copulation would put her at risk of damage in an attack from a newly arriving male or if the viabilities of eggs fertilized late in copulation are likely to be lower, the optimal copulation time for the female would be less than the apparent optimum of 100 minutes (the time it takes to fertilize all her eggs). Perhaps the length of copulation is not determined by the male at all.

How could this be? Males are bigger than females, and we might wonder whether even vigorous bucking could unseat a determined male. However, the resources of the female are easily underrated. Whatever proximate mechanisms serve to keep the male copulating can perhaps be subverted. So we can envisage an alternative to the simple picture that emerges from Parker's analysis. Instead of thinking that selection has favored an allele (or a combination of alleles), expressed in the male, that brings copulation time to the male optimum, we might suppose that female dung flies express the selected allele(s). These alleles cause a difference in their behavior, and the length of copulation is fixed at their optimum. Or we might consider the perspectives of both sexes, viewing each as attempting to bring copulation time to the appropriate optimum. Actual copulation time is determined by the outcome of a struggle.

Recall an important point from chapter 3. Sociobiological analyses of fitness attempt to simplify very complicated biological situations; in doing so, they presuppose that the factors that are neglected would make no difference to the conclusions. In the present case we can see that Parker's analysis is an idealization of a much more complicated game-theoretic treatment that would view copulation time as resulting from the pursuit of different behavioral strategies by males and females. There are certain things that females can do and certain things that males can do; given the general assumption that selection has worked to optimize, we shall expect the outcome to be determined by the ways in which different combinations of male and female strategies affect fitness. A full treatment would consider the difficulties posed for both sexes by male combat, the possibility of females adjusting their arrival times at cow pats, the possible confusions in sexual signals that a female might use, and a number of other variables. To identify reasonable strategies for males and females, we need, as usual, a much clearer view of the physiology of the dung flies and of the development of mating behavior. Only then shall we be in a position to see clearly whether Parker's elegant analysis is a legitimate idealization of a much more complex behavioral situation—or whether that five minutes, that 12-percent error, is really significant.

Nevertheless, recognition of the open-endedness of Parker's account should not dim our appreciation of what has been accomplished. Like many fine pieces of scientific work, Parker's model of dung fly guarding behavior structures our ignorance by identifying precise questions that we need to answer if the analysis is to be taken a step further. In this respect it is comparable to widely accepted studies of the evolution of morphological and physiological traits. Incompleteness is no vice—as long as it is not taken for final truth.

The Scrub Jay's Apprenticeship

Copulation is at least minimally social. However, sociobiology aspires to understand more complicated interactions. Let us see if we can retain Parker's rigor while tackling an instance of Wilson's "central theoretical problem of sociobiology."

Wilson presents an example that he hails as one of "the most recent and edifying studies of bird sociality" (1975a, 451–454). The work so honored is the explanation of helping behavior in the Florida scrub jay (see Woolfenden 1975; Woolfenden and Fitzpatrick 1978; Krebs and Davies 1981, 173–179; Emlen 1978; Emlen 1984). Florida scrub jays live in the sandy scrub of peninsular Florida. The jays pair for life and occupy a permanent territory. Many of the breeding pairs are assisted by other jays, who help feed the young and also defend the territory against rivals and predators (most prominently snakes). Why do the helpers pitch in in these ways? Wouldn't they do better to go off and reproduce themselves, instead of helping to rear the offspring of other jays? How did such "altruistic" behavior evolve?

One hypothesis is that the jays are maximizing their inclusive fitness by assisting relatives. Here is a simple version of the story. Birds who remain at home and help their parents forgo immediate reproduction. But perhaps they had little chance of establishing a territory of their own. Moreover, their lifetime expected contribution to the gene pool of the next generation will be augmented by the effects they produce in raising siblings. If the expected loss is smaller than the expected benefit, then helping at the parental nest will prove an optimal strategy. (There may also be effects on genetic representation in subsequent generations. If the population is growing, then the timing of reproduction is important. Birds who are able to rear a large number of young very quickly will achieve a greater number of grand offspring than birds who reproduce more slowly. If their children lend their aid to the family enterprise, then this may help all concerned. I shall henceforth ignore such complications.)

There are some obvious tests. Are the helpers indeed related to those they assist? What happens when territories are plentiful, so that the chances of founding a nest of one's own are relatively high? Is the reproductive success of nests with helpers greater than the success of nests without helpers? Is there a substantial loss in lifetime production of offspring when birds delay breeding for a year?

Woolfenden has systematically considered the first three issues. He has marked large numbers of jays and, by studying a population over a number of years, has identified the relationships between helpers and helped. Mostly, birds help their parents. Only in a single case was a jay observed assisting a pair unrelated to it.

Woolfenden also discovered that there are usually too few territories to accommodate all the potential breeders. As territories become available, birds who have previously been helpers occupy them and set up breeding on their own. So it appears that the probability of establishing a new territory is often low and that the expected gain to a jay who tries to reproduce directly is substantially less than the number of offspring expected from a first season of breeding.

Helping behavior contributes in two ways to the production of relatives—but they are not quite the ways we might have anticipated. Jays who rear young with help and those who manage alone provide the same amount of food for their nestlings. What alters is the pattern of provisioning. Unhelped birds have a harder time obtaining food for their young, and each parent has to spend more time in foraging. The toll is taken in the probability of survival from season to season. Jays who receive help have a greater annual survival rate (87 percent as opposed to 80 percent). Helpers ease the parental lot, extend the parents' lives, and are rewarded in the long run with a larger number of siblings. A second benefit is that the offspring of jays who have had help seem to be more viable. Even after the young leave the nest, those who come from nests with helpers have a greater chance of survival than those who do not.

Should we conclude that there is an inclusive-fitness effect due to helping behavior? Not necessarily. Maybe the birds who receive help are the more experienced breeders, and their success results from the experience, not from the help. Woolfenden tested carefully to eliminate this hypothesis. By making detailed observations, he could identify birds with comparable breeding experience, and he could compare the success of similarly experienced jays with and without help. Indeed, he analyzed the reproductive outputs of jay pairs in seasons when they received help and in seasons when they did not, and he was able to show clearly that receiving help correlated with offspring viability in these cases as well.

The tale now becomes a little tangled. Woolfenden and Fitzpatrick (1978) demonstrated that males who help gain an extra long-term benefit through increased probability of gaining a territory in later seasons. Success is likely to come in one of two ways: helpers may obtain territories through the deaths of the parents or, more commonly, through expansion of the parental territory. A gang of helpers is able to displace other jays from adjacent areas, and the extension of the parental territory can provide one of the helpers with enough room to establish a territory of his own.

The central hypothesis is that helping behavior increases the inclusive fitness of the helpers. The expected loss from forgoing reproduction is offset by the gains that accrue from helping—the increase in genetic representation through siblings and the greater probability of future occupancy of a territory. Two tests are relatively easy to conduct. Data on bird relationships, painstakingly compiled by Woolfenden, permit us to check the requirement that helpers be close relatives of those they assist. By removing the occupants of some territories, we can also show that helping birds are prepared to grab vacancies when they are available. It is slightly more difficult to test the claim that the helping behavior causes an increase in expected genetic representation. It is not enough to establish a correlation between helping and enhanced nest success. Woolfenden's approach to this problem applies a methodological principle beloved of philosophers and practitioners since John Stuart Mill. To show that *A* is a causal factor in the production of *B* in a system *C*, we hold constant as many properties of *C* as we can, allowing *A* to vary. We expect that if *A* is a genuine causal factor, then its presence will make a crucial difference to the presence of *B*. Woolfenden's analysis of jay pairs who receive help in some seasons and not in others is a clear illustration of Mill's Method of Presence and Absence.

The hypothesis passes three genuine tests and thus earns the right to serious attention. Nevertheless, it would be premature to suppose that we have the definitive account of the scrub jay behavior. In the decade since Wilson hailed Woolfenden's work, numerous sociobiologists have continued the investigation, refining the analysis and comparing the case of the scrub jays with other examples of "helping at the nest (or den)" (see Emlen 1984 for a lucid review). As the analysis has become more precise and new measurements have been achieved, hitherto unsuspected complications have begun to emerge.

Central to our story is the idea that birds who help at the old homestead will do better—by the criterion of genetic representation—than those who go out and scramble for their own space. So let us hypothesize two kinds of ancestral scrub jays. *Competitors* go out to seek fame, fortune, and genetic representation as soon as they are ready to breed. They devote themselves completely to searching and, very probably, fighting for areas in which to build and provision a nest. *Helpers* are prepared to grab empty territories when they come across them. Perhaps they make occasional surveys to assess the possibilities in the neighborhood. But if nothing turns up, they return to the parental nest and help to rear siblings.

To understand the relative fitnesses of the two strategies, we have to look at the entire reproductive lifetimes of the hypothetical jays. There are many important parameters. The reproductive success of a Competitor is a function of the following: its probability of finding a territory in the first season; its chances of surviving until the next season; the likelihood of its succeeding in finding a territory in the second season if it fails in the first; and the expected number of descendants produced from first-season occupancy of a territory, from second-season occupancy, and so forth. For Helper the determinants of reproductive success are even more complicated. Helpers may establish territories in the first season, may fail in the first season but establish territories by competition in the second season, may inherit a territory, or may die after the first season. The probabilities of all these occurrences, as well as the reproductive payoffs that accrue from them, must be taken into account before we can achieve a comparison.

When the details are elaborated (see technical discussion D), a number of points emerge. First, it should be clear that even more rigorous tests of Woolfenden's hypothesis are possible. The parameters actually measured are suggestive, but the analysis reveals that yet more detailed measurements are relevant. (As Emlen 1984 makes admirably clear, sociobiologists who are interested in cooperative breeding are actively involved both in formulating the crucial parameters and in trying to measure their values.) Second, it transpires that, if the parameters take on certain values, helping behavior may be favored because the probability for a helper of inheriting a territory or the probability for a competitor of increased mortality is relatively large. Under these conditions the relatedness of helpers and helped might be irrelevant. Perhaps the crucial factors are those of death and inheritance. If this were correct, then we might understand the tendency of birds to help at the *parental* nest simply as a disposition to undertake an apprenticeship in familiar surroundings. The one case in which the helper is unrelated to the breeders would no longer be seen as an anomaly—or even as a case of avian error.

A third point is that it turns out to be hard to determine whether there is a lifetime breeding loss for birds who help. In an important paper, Emlen attempted to evaluate the cost of helping parents instead of setting up a new breeding territory (Emlen 1978). He suggested that we compare the expected offspring production by new

Technical Discussion D: What the Scrub Jays Might *Be Doing*

I simplify quantitative treatment by making a number of assumptions. The population is assumed not to be growing, so that complications in the timing of reproduction are ignored. I also suppose that Helpers and Competitors achieve the same results in all seasons after the second. The latter supposition is unrealistic, but it has the virtue of making the analysis much clearer.

Let p_{HT1} = the probability that a bird with the Helper genotype obtains a territory in the first breeding season. In general, let p_{xy1} be the probability that a bird with genotype x obtains outcome y in season 1, and let p_{xy2} be the probability that a bird with genotype x obtains outcome y in season 2, given that it failed to obtain a territory in season 1. In accordance with this convention, I shall use the following definitions:

p_{CT1} = the probability that Competitor acquires a territory in season 1.

p_{CD1} = the probability that Competitor dies at the end of season 1.

p_{HD1} = the probability that Helper dies at the end of season 1.

p_{HT2} = the probability that Helper wins a new territory in season 2, given failure in season 1.

p_{HI2} = the probability that Helper inherits a territory in season 2, given failure in season 1.

p_{HH2} = the probability that Helper helps in season 2, given failure in season 1.

p_{CT2} = the probability that Competitor acquires a territory in season 2, given failure in season 1.

Let d_1 be the expected number of offspring of a first-season breeder who will survive to maturity. Let d_2 be the analogous number for a second-season breeder who bred in the first season, and let d_{12} be the analogous number for a second-season breeder who breeds for the first time in the second season. Finally, let b_1 and b_2 be the increases in the production of offspring equivalents that result from Helper's work at the nest in the first and second seasons, respectively. (The value b_i is computed as follows. First take the number of offspring produced by the parents over a lifetime, given help in season i; find the number that survive to maturity; now take the number of surviving offspring that would have been produced in the absence of Helper's contribution; form the difference; now multiply by the coefficient of relationship of Helper to the parents.)

For the first season the expected contributions to genetic representation are given as follows.

Helper: $p_{HT1}d_1 + (1 - p_{HT1})b_1$.
Competitor: $p_{CT1}d_1$.

After the first season life is complicated by the facts of death and inheritance. We may write the expected contributions to genetic representation from the efforts of the second season by considering all the possible scenarios.

Helper:

$$p_{HT1}(1 - p_{HD1})d_2 + (1 - p_{HT1})(1 - p_{HD1})(p_{HT2} + p_{HI2})d_{12} + (1 - p_{HT1})(1 - p_{HD1})p_{HH2}b_2.$$

Competitor:

$$p_{CT1}(1 - p_{CD1})d_2 + (1 - p_{CT1})(1 - p_{CD1})p_{CT2}d_{12}.$$

Using the assumption that later effects are equal for both strategies, we can state the condition under which Helper will have greater expected genetic representation in future generations:

$$\begin{aligned} & p_{HT1}d_1 + (1 - p_{HT1})b_1 + p_{HT1}(1 - p_{HD1})d_2 \\ & \quad + (1 - p_{HT1})(1 - p_{HD1})(p_{HT2} + p_{HI2})d_{12} \\ & \quad + (1 - p_{HT1})(1 - p_{HD1})p_{HH2}b_2 \\ & > p_{CT1}d_1 + p_{CT1}(1 - p_{CD1})d_2 \\ & \quad + (1 - p_{CT1})(1 - p_{CD1})p_{CT2}d_{12}. \end{aligned} \qquad \text{(A)}$$

What do we know about the terms that figure in this inequality? Previous studies already establish some relationships. Emlen's work (1978) tells us that b_1 and b_2 are both less than d_1. Woolfenden has evidence that experience in breeding pays. Hence, we can suppose that $d_1 < d_2$. The relationship between d_{12} and d_1 and d_2 is undetermined by the available studies. Given the benefits of experience, we can assume that $d_1 < d_{12} < d_2$. Since Competitors spend more time actively hunting out territories, $p_{HT1} \leqslant p_{CT1}$ and $p_{HT2} \leqslant p_{CT2}$.

This knowledge leaves open a wide range of possibilities. Suppose that p_{CD1} is large, p_{HD1} is small, and $p_{CT1} < 2p_{HT1}$. Then the inequality will almost certainly hold, whether or not b_1 and b_2 are significant. Competitors will be at a disadvantage because of their high mortality rate. Similarly, if both p_{CT1} and p_{HT1} are small, p_{CT2} relatively large, and $p_{HI2} + p_{HT2} > p_{CT2}$, then the inequality will hold. In this situation the crucial factor will be the probability of gaining a territory through inheritance.

Consider a simplification of (A). Assume p_{HT1} is very close to 0, p_{CT1} is very close to 1; let $p_{HT2} + p_{HI2}$ be 1, $p_{HH2} = p_{HD1} = p_{CD1} = 0$. We suppose, in effect, that all Competitors obtain territories at once and that all Helpers fail in the first season and succeed in the second. (A) reduces to

$$b_1 + d_{12} > d_1 + d_2. \qquad \text{(B)}$$

Since we know that $b_1 < d_1$ and that $d_{12} < d_2$, (B) is unsatisfiable. Result: Competitors should be favored if the assumptions just made are true.

Is that likely? Consider the possibility that Competitors do better the less of their own kind are around to compete with. Suppose p_{HT1} is independent of the frequency of Helpers, and that $p_{CT1} = f(q)$, where q is the frequency of Helpers. We can represent our imagined possibility by assuming that f increases with q, with $f(0) = p_{HT1}$, $f(1) = 1$. If we discount second-season effects, taking the fitnesses to be represented by the first-season contributions, then we shall expect the population to maintain a polymorphism. The equilibrium frequency of Helpers will be the solution of

$$f(q) = p_{HT1} + (1 - p_{HT1})(b_1/d_1).$$

Finally, we can easily relax the assumption that a Competitor in a population of Helpers would be assured of finding a territory. Suppose that f increases with q, with $f(0) = p_{HT1}$. But we do not demand that $f(1) = 1$. The condition for a polymorphic equilibrium is

$$f(1) > p_{HT1} + (1 - p_{HT1})(b_1/d_1).$$

(Again, I discount second-season effects. It is not hard to see that including them may make it easier for some Competitors to be maintained in the population.) Emlen computes b_1/d_1 (or, to be exact, something that is probably a bit larger than b_1/d_1) to be about 0.25. Hence, if p_{HT1} is about 0.2, our condition simply requires that the probability of finding a territory for a rare Competitor be greater than 0.4.

I introduce this model not because I take it to be particularly realistic but because it underscores the point that contemporary evolutionary theory has the resources to offer precise analyses of fitness that display clearly what parameters may be important and that reveal unsuspected forms of dependence. It should be apparent that qualitatively similar outcomes can be achieved by assigning very different sets of values to the parameters, and that attention to detail can enable us to tease apart biological situations that would otherwise have been conflated.

breeders with the gain in genetic representation resulting from assistance at the parental nest. The estimate is only accurate if the second-season reproduction of a first-season helper (who is, of course, a new breeder in the second season) is the same as the second-season reproduction of a first-season breeder (who will be an experienced breeder in the second season). Emlen supposes that the helper loses one breeding season, and takes it that the appropriate season is the first. But if inexperienced breeders always do worse than those who have had a year's experience, then the helper will always be playing catch-up.

Fourth, and most important, consideration of the details suggests the possibility of a shift of perspective on the entire situation. Emlen's work reveals that the expected number of descendants for a breeding bird is substantially greater than the gain in inclusive fitness that a bird achieves by helping at the parental nest. (More exactly, it tells us that the expected first-season contribution to genetic representation for a breeder is greater than the *mean* inclusive-fitness effect of helping. There is reason to think that the effect of first-season helping is less than the mean, in that unseasoned helpers are likely to contribute less.) Woolfenden's observations show that new territories are relatively scarce, so that the probability for either a helper or a competitor of acquiring a territory in the first season will be low. Together, these results yield an interesting conclusion.

Think of competitors as active house hunters, looking for rare bargains. They are likely to be more successful at finding and acquiring any bargains that arise than are their stay-at-home cousins, who do a day's work helping the family and only venture out for a survey of the field on a Sunday afternoon. Given very little competition from other active bargain hunters, the few bargains that are available may be enough to satisfy the needs of the group of competitors, *if this group is small.* Hence, if the frequency of helpers in the population is relatively high, a rare competitor's chances of winning a new territory will probably be greatly increased.

So what? Answer: Because of the fitness advantages of breeding directly, competitors will be favored when they are rare. Similarly, we can argue that helpers are favored when competitors predominate. Under the latter conditions both helpers and competitors will have the same (small) chance of acquiring a territory, and helpers stand to achieve greater lifetime reproductive success as a result of their efforts in the first season. (They may also have a higher probability of obtaining a territory later through inheritance.)

We could easily complicate matters further. If inheritance is an important factor, and if most territories are occupied by birds with helpers, then a competitor's chances of grabbing a vacancy may be diminished. If there are some territories with many helpers and some territories with no helpers, then rare competitors can be expected to win enough of the latter. Some of the resultant models would leave us with the conclusion of the preceding paragraph. Selection would be frequency-dependent, and we would expect to find a polymorphism (not necessarily a stable polymorphism) in nature. Different models would return us to our initial description of the situation.

Qualitative analyses can be very delicate. When they are precisely articulated in one way, they may yield one expectation; elaborated differently, they may suggest something quite distinct. (See technical discussion D for details.) Plainly, it is not enough to stop when we have achieved something more or less like what seems to go on in nature. The details need to be worked out, or we may run the risk of missing what is really going on. Kepler's ellipses elude those who are satisfied too soon.

Should we go back to the Florida scrub and expect to find a polymorphic population of jays? Perhaps. But perhaps not. Consider some of the ways in which our expectations could be defeated. Maybe fighting for a new territory would have so damaging an effect on a bird that its reproductive life would be truncated. Maybe the set of strategies envisaged as available to the jays is much too restricted. Scrub jays might have a continuum of possible strategies, depending on the division of their time between territory hunting and helping. Maybe the idea that competitors can establish *and maintain* territories when there is a preponderance of helpers is a mistake. New territories might be vulnerable to invasion by gangs of helpers at adjacent nests.

Although I shall not delve deeply into the complications, it is worth looking at the last possibility. In one sense, it reestablishes the original conclusion. Helpers again come out ahead. Yet as we begin to ponder the envisaged scenario, helping behavior per se begins to lose its significance. We might even discover that what seems like filial cooperation results from an evolutionary accident.

Consider the following possibility. Birds who fail to establish their own nests are forced to live somewhere. Suppose that scrub jays have a propensity to return to familiar surroundings. Suppose that they also have a disposition to imitate the behavior of others. Then an initial accidental tendency for siblings to assist in rearing the young would tend to be preserved in a nest and to spread in the population. For if the behavior promotes the welfare of offspring and if jays often acquire territories by expansion and fissioning, then nests that produce a larger number of young will send forth larger gangs of would-be imperialists. As a result, the habitat would tend to be divided into territories, each occupied by a small number of related jays, consisting of one breeding pair and the remnants of the male's army. The latter birds, with their tendency to emulate the behavior of their older siblings and with their history of rearing in a nest with sibling assistance, would follow the pattern. We should see helping behavior everywhere—even though there are no "genes for helping."

I do not suggest that any of the possibilities that I have alluded to tells the true story about the scrub jays. However, when we see that even a slight amount of attention to the formal models devised by mathematical sociobiologists uncovers complications galore, we may take a less rosy view of qualitative analyses that appear to have some connection with observed animal behavior. The Florida scrub jay is a bird whose behavior has been paintakingly studied. In the last decade there has been an exciting combination of mathematical analysis and careful observation on "helping behavior" in a variety of animal groups (see Emlen 1984). The sociobiologists who have worked on the problem have shown great skill in formulating alternative evolutionary possibilities and in gathering the observations needed to discriminate among them. Even in this much-studied case, we do not know all the things that we acknowledge ourselves as needing to know. The complexities of animal behavior refuse to satisfy those who crave fast, definitive answers. Nonetheless, the marriage of precise analysis and detailed field research offers further exciting prospects.

Emlen concludes his review with remarks that sum up the perspective of many sociobiologists. Alluding to his contribution to an earlier edition of the same volume, he writes,

> In the short time since the publication of that edition, considerable advances have been made both in the development of the-

> ory pertaining to cooperative behavior and in the collection of empirical field data. We now have before us a preliminary set of models and testable hypotheses. The decade ahead should be an exciting one as we begin to see vigorous testing of these various hypotheses. (1984, 338–339)

Such comments are appropriate not only for the best achievements of narrow sociobiology but also for the most interesting proposals in contemporary evolutionary theory. They deserve to be remembered when we consider pop sociobiology's haste in trumpeting grand conclusions about human evolution and human nature.

Malaise in the Menagerie

Not all of the research in nonhuman sociobiology is as precise and as careful as the studies that we have reviewed so far. Attempts to understand the evolution of forms of animal behavior can easily be infected by various errors. I shall now take a brief look at some of the ways in which things can go wrong. This examination advances our project in two distinct ways. First, it reveals the kinds of mistakes we shall discover in the bold claims about how people maximize their fitness. Second, it undermines the confident tone that pop sociobiologists use to report the results of animal studies. Many of those who are interested in charting the evolution of behavior in particular animal groups are well aware of the kinds of difficulties that I shall identify. Their main goal is to overcome the difficulties, not to announce final truths about animals that can then serve as a springboard for conclusions about ourselves.

We can start at the top, with the "king of beasts." George Schaller reported a puzzling fact: sometimes male lions are more inclined to share food from kills with cubs than are females (1972). In an interesting article, Brian Bertram (1976) tries to explain the puzzle. On the basis of his estimates of the relatedness of males and females to cubs within the pride—estimates that have been refined in subsequent studies (for example, Packer and Pusey 1982)—Bertram concludes that an arbitrary cub in the pride is likely to be more closely related to an adult male of the pride than to an adult female. He proposes to apply this finding to explain why females are less inclined to share.

There seems to be an obvious problem with the application. Females, unlike males, can be expected to recognize their own offspring. We would thus expect that females would be more likely to share food with some cubs and not to share at all with others. Convinced that inclusive-fitness effects will resolve the puzzle, Bertram responds by invoking an ad hoc hypothesis. "It is likely that a female generally can recognize her own offspring, although it may be difficult in practice for her to discriminate among cubs during competition among hungry animals at a kill" (1976, 171). The picture painted is of a horde of starving lions gorging themselves in a melee around the carcass of a prey, and of the poor female failing to exercise her normal powers of discrimination in all the hubbub.

Whether or not females can be so confused—and the hypothesis does pose problems for the daring experimentalist—the picture is irrelevant. Schaller's lengthy observations of lion kills, which pose the problem that Bertram intends to address, portray individual lions grabbing parts of the carcass for themselves and frequently retiring to places where they can eat undisturbed. Yet even when the feasting takes place on the spot, the behavior of male lions is more complex. Males tend to allow cubs to feed only after first discouraging them "with a slap" (Schaller 1972, 152). Females typically stuff themselves before going to fetch their own cubs, so that quite often the cubs arrive too late to find anything (151). Schaller also reports some curious incidents in which females seem to take meat back to their cubs, only to refuse to share it once the issue actually arises (152). There is no doubt that the food-sharing propensities of male and female lions are puzzling. But to suppose that the behavior is to be explained through the citation of inclusive-fitness effects that are greater in males, and to back this up with a hypothesis about temporary confusion of lionesses, is surely to replace the obscure with the implausible.

In this example we find an attempt to resolve an apparent failure of fit—resulting from the assumption that female lions can be expected to recognize their offspring and thus to share food with them—that directly conflicts with the data that generate the anomaly. Usually the troubles are not so glaring. There are many instances in which an ingenious defense can be found, the rot temporarily stopped, but in which there is a more subtle methodological error. The saving hypothesis introduces a new factor into the account, and the effects of this new factor on prior analyses are left unexplored. If the first requirement on explaining apparently anomalous data is that the job really be done—the details of the discrepancy considered and the discrepancy eliminated—a second demand is that the solution should not perturb the examples previously hailed as successes.

Even some suggestive sociobiological studies have an air of choosing to mention important selective factors where convenient and ignoring them elsewhere. In a pioneering article on mating systems in birds and mammals, Gordon Orians (1969) develops a simple model designed to specify when it will be to the advantage of a female to mate with a male who has already mated. He considers the possibility that the benefits of settling in a high-quality habitat might suffice to outweigh the loss of help in raising young, so that a propensity to polygyny could evolve in a species. Orians is sensitive to the possibility of countervailing pressures. Perhaps overcrowding could attract predators. Indeed, a full analysis of the male and female strategies would be quite complicated. To mention one feature among many, there would be an apparent tendency for selection to produce competition among females, which, if sufficiently strong, could lead to the establishment of monogamy.

Orians indicates how his model yields qualitative predictions, and he acknowledges that the predictions are not always satisfied. His reaction to one of the failures is instructive. According to the model, "polygyny should evolve more readily among precocial birds than among altricial species" (Orians 1969, 125). Precocial species are those whose young are relatively self-sufficient at birth. The young of altricial species depend on some kind of care to survive. Orians's prediction is based on the simple idea that when parental care is required, females should be more inclined to mate with males who are unmated and can therefore be expected to do more of the work with the offspring. If little care is needed, then it is a matter of indifference to the female whether or not the male with whom she mates also mates with others. Factors such as habitat and, perhaps, "good gene combinations" become more important.

The prediction seems to encounter numerous exceptions, and Orians draws attention to its failure in the case of swans, geese, and ducks: "in most species of ducks the female alone cares for the young, and yet monogamous pair bonding seems to be the rule" (125). Why? Orians notes that tropical species of ducks are, as expected, polygynous, and he proposes that "perhaps the prevalence of monogamy among high-latitude species may be the result of the advantage of pair formation on the winter ground and rapid initiation of breeding, which give a stronger advantage to monogamy for the male than would otherwise be the case" (125).

Possibly. But once the deus ex machina is onstage, it is necessary to think about how it will affect the fates of the entire cast. If there is a premium on quick breeding in high-latitude species, then this is a factor that ought to operate in other groups of high-latitude birds. Maybe we should stop thinking of mating systems as responses to females' need to secure male care for their young, and view males and females as working to get their brood well developed before the cold weather returns. Of course, birds are subject to both kinds of selective pressures. An argument is needed to motivate the idea that the short northern breeding season interferes with the behavior of only some species. Lurking behind Orians's model is a much more complex set of selective factors, and the complications only become prominent when the phenomena fail to fit.

There are many similar examples. In their research on the "castes" of social insects, Oster and Wilson consider the possibility that worker insects may retain their ovaries and lay unfertilized eggs that develop into males. Apparently, it is to the advantage of individual workers to produce sons. From the queen's perspective this is a bad bargain, and she can be expected to oppose the production of worker eggs. Since it seems to be easiest for workers to escape the queen's surveillance if the colony is large, Oster and Wilson reach the hypothesis that large colonies should favor the retention of ovaries by workers and should thus have monomorphic workers (workers of only one "caste").

> Unfortunately for the hypothesis, the reverse is true. However, there is a competing explanation for this relation: species with large colonies are also those with the most complex adaptations, requiring an array of castes, for example, leafcutters and army ants. They may also be those with the greatest tempo and turnover of workers, requiring a finer division of labor. (Oster and Wilson 1978, 103)

Such remarks foster the idea of an enterprise whose central maxim is "Heads I win, tails you lose."

Oster and Wilson are correct in supposing that there are competing selection pressures—that the advantages to an individual worker are not simply to be read off the coefficients of relationship. The real problem is that certain explanations, in which some but not all of the relevant evolutionary factors are considered, are heralded as complete, and the rest of the factors are invoked only when things go wrong. Theorists consider the problem of sexual strategies in one place, the problem of division of labor in another, the problem of foraging elsewhere. It is then pointed out that difficulties with favorite analyses can be overcome by considering some neglected factor. The excuses are convincing only if we have a right to confidence that the combined analysis in which everything is given its due would sustain the conclusions drawn from the simplified analyses.

The problem arises again and again in sociobiological work. One attempts to account for the size of a troop of primates by considering the distribution of items of food, and only later takes into account the difficulties of avoiding predators. (See Kummer 1971, 43–45, 52, for an explicit recognition of the problem.) Or one supposes that spacing distance between animals is fixed and that troop size has to adjust to the animals' tolerance of one another. Or one suggests that the animals have a particular sexual strategy and that this determines patterns of mutual tolerance and of foraging. When we are dealing with complicated vertebrates—especially birds, carnivores, ungulates, and primates—it is especially hard to juggle all the possible factors simultaneously. So there is a temptation to fix on one particular form of selection, applying it ruthlessly to the problem at hand, until the exceptions turn up. At that stage the theorists remind themselves and their readers that there are, of course, other factors that might be taken into consideration. But unless it can be shown clearly that those factors were rightly neglected in the earlier cases, then there is no reason to think that the selective accounts produced are correct.

Not only are there different selective pressures that act on animals, but in any complex social group it will be possible to consider the situation from the perspective of many different animals or of groups of animals. Consider the phenomenon that primate watchers used to call "agonistic buffering." Observers have witnessed a common pattern of behavior in a variety of species. In macaques, vervets, olive baboons, and langurs, males have sometimes picked up infants in the presence of "dominant" males of the same troop or have "used" infants as "appeasement devices" in gaining access to a new troop. (See Wilson 1975a for a brief review.) The old explanation supposed that the behavior was adaptive for the male because the infant served as a shield against attack by a possibly dangerous rival. By chance, however, a group of zoologists discovered that infants are at considerable risk from some males in neighboring troops, or even in the same troop, if they lack adult protection. Hence a new explanation. The males who carry infants are guarding them against potential murder, and they do so because those they pick up are probably relatives.

Sarah Hrdy offers an eloquent statement of a moral that she draws from this example. "The history of our knowledge about primate infanticide is in many ways a parable for the biases and fallibility that plague observational sciences: we discount the unimaginable and fail to see what we do not expect" (Hrdy 1981, 89). Hrdy's moral captures only part of the import of the example. There are many participants in the behavioral situations in which males carry infants—the threatening male, the carrying male, the infant, and onlookers, including sometimes the infant's mother. To understand the situations, it is not sufficient to note that a particular form of behavior would bring advantage to one of the participants. Ultimately, the effects on each of the various possible strategies must be considered, and it must be explained why there are no conflicts, or how any conflicts are resolved. Sociobiological analyses will always seem less than fully rigorous as long as they appear to pick their spots, to concentrate on whatever evolutionary factors occur to the investigator in the case at hand, to choose to optimize the behavior of whatever participants strike the fancy. Interactions among animals are just that—interactions. If we are indeed to think of evolution as a process in which animals have arrived at optimal forms of behavior, we must appreciate that on any given occasion different animals will be adjusting their behavior to the actions of others, that there is always potential for conflict, and that no response can be dismissed as a passive part of the evolutionary landscape.

Let us take stock. We have uncovered several types of difficulty that beset nonhuman sociobiology. Accounts of animal behavior may not be articulated with precision. Attempts to remedy the deficiency may reveal that a plausible scenario blots out a host of interesting possibilities. There may be discrepancies with the behavior actually observed. Efforts at explaining them may not account for the actual data. Anomalies may be resolved by invoking new factors whose role in other cases is not investigated. There may be a taking refuge in the recognition of conflicting optima, and a hopeful suggestion that observed behavior results from a resolution of the conflict. There are many ways in which the proposed Darwinian history may fail to achieve that confirmation from the evidence that we admire in the best evolutionary theorizing.

These problems are hardly fatal. Sociobiology, like other young sciences, should not be chided for failing to deliver flawless answers to the questions it addresses. There are sociobiological accounts and sociobiological accounts. Some show commendable precision; others are inexact. Some are suggestive; others appear to multiply excuses out of necessity. By considering a range of examples, I hope to have revealed the kinds of difficulties that may underlie superficially plausible scenarios.

Those sociobiologists who are not dedicated to the promulgation of pop sociobiology are often clear on what has already been accomplished, and on what needs to be done. In a review of mating systems that discusses the numerous refinements that have been made since the pioneering work of Orians and others, Sandra Vehrencamp and Jack Bradbury emphasize the variety of models now available, the complexity of the behavior under study, and the need for extensive tests of hypotheses that have been suggested in cases where the models fail to fit. (See Vehrencamp and Bradbury 1984, especially

260–264, 277–278.) For sociobiologists of this sort, the methodological points I have been making are old news. They are actively involved in trying to overcome the difficulties.

Pop sociobiology, however, hopes for quick returns. It is also after bigger, if not better, game.

Violent Defenders

Local studies that acknowledge the particular constraints on a particular species are most likely to be pursued to rigorous conclusions. However, a justified account of copulation time in dung flies is hardly what is needed for the grand program of human sociobiology. To ascend Wilson's ladder, we need generalizations about behavior in groups to which members of *Homo sapiens* belong. What groups? Two kinds of animals have been popular. Humans are evolutionarily related to primates, so it seems natural to take the primates, particularly the great apes, as the models for ourselves. But we are ecologically unusual primates, and on the basis of the fossil record there is good reason to think that we have been ecologically unusual for a long time. Perhaps we should focus on a different group, the animals that hunt. Maybe the right model for "man the hunter" is the social carnivores. The two approaches involve different styles of explanation of human behavior. Those who believe that we share behavioral tendencies with the primates can maintain that these tendencies evolved in some common ancestor and were passed on to the descendants. When our propensities are compared to those of the social carnivores, the claim must be that similar evolutionary forces have shaped the behavior independently in different lineages.

Students of animal behavior regularly, and understandably, fall in love with the animals studied and come to see their behavior as reflecting our own. Even the hyena can come to seem attractive and can be hailed as a source of information about ourselves (see Kruuk 1972, 3). From the perspective of pop sociobiology, squabbles among the partisans of different animals are wrongheaded. Each animal species is what it is, and not another thing (to paraphrase Bishop Butler), but there can be crisscrossing patterns of similarity among various clusters of species. Pop sociobiology hopes to generalize, assembling a composite picture of human nature by formulating and justifying generalizations that ascribe certain rules of behavior to all the species in a particular group (or, possibly, subsets of those species—the juvenile females, for example). The inclusive group may vary from case to case. All that is required is that it be *relevant*—that is, that it overlap *Homo sapiens* (or, possibly, one of our hominid ancestors)—and that the overall assignment be *consistent*. Trouble will threaten when we draw generalizations about the members of different groups with common members and find that these generalizations attribute incompatible forms of behavior.

Consistency does not come easily. Barash offers soothing words about the unaggressiveness of our evolutionary relatives, chimpanzees and gorillas, "reflecting an adaptive strategy for these species in which contest competitions are relatively unimportant" (1977, 236). Because they are our closest living evolutionary relatives, Barash consoles us with the thought that we may be unaggressive too. (Alas! the picture is subverted by recent discoveries of gang warfare in chimpanzees.) Others who wish to draw conclusions about people from studies of animal behavior have a different model in mind: "In many respects the hamadryas baboon's society of closed but coordinated family units is a better model of human social structure than that of the chimps" (Kummer 1971, 152). But the baboons are less close to us than the chimps. No matter. Baboons are adapted to the environment in which the primitive hominids evolved. When we descended from the trees, we became subject to the same selective pressures that have shaped hamadryas society. Small wonder, then, that our social system resembles that of the hamadryas baboon.

The generalizations compete. One tells us that species with characteristic *A* (including chimps, gorillas, and humans) have a certain behavioral disposition. The other tells us that species with characteristic *B* (including hamadryas baboons and humans) have an incompatible behavioral disposition. One of the generalizations must be wrong. One of the predictions about ourselves must be mistaken. But which? If we retreat to the suggestion that for a species that is *A*, there is a high probability of having the first behavioral disposition, and for a species that is *B*, there is a high probability of having the second disposition, we may achieve true premises. But what conclusion should we draw? The generalizations are useless in deciding what to do in the crucial cases—species like us that are both *A* and *B*. If you know that almost all the people in a small town in Vermont are Democrats and that almost all the farmers in Vermont are Republicans, you have no basis for deciding the political affiliation of the town's only farmer.

If the findings of generalizations about groups of species do not prove consistent, then the strategy of assembling the generalizations will be inconclusive. We shall be left asking for the behavior that is to be expected, given a unique constellation of factors that pull in different directions. At that point responsible human sociobiologists will have to conclude that the beasts speak with many tongues and that the only way to understand our behavior is to take our own species as the object of serious study.

Pop sociobiology, however, strives for large generalizations that can be instantiated in conclusions about the behavior of human beings. One recurrent suggestion is that our ancestors were grouped together in small bands, that the males of each band defended their group, and that leading (or "dominant") males were prepared to react with violence to the approach of their neighbors. (See, for example, Wilson 1975a, 553, 564–565; 1978, 107ff.; and, for an especially forthright version, Barash 1979, 188–189.) The picture is supported by generalizations about the defensive behavior of animals. Postponing worries about how such generalizations are to be applied to human beings, let us look at an example of the kinds of broad claims about animals that figure in pop sociobiology.

Consider the following discussion of defense against predators.

> Although vertebrates are seldom suicidal in the manner of the social insects, many place themselves in harm's way to defend relatives. The dominant males of chacma baboon troops (*Papio ursinus*) position themselves in exposed locations in order to scan the environment while the other troop members forage. If predators or rival troops approach, the dominant males warn the others by barking and may move toward the intruders in a threatening manner, perhaps accompanied by other males. As the troop retreats, the dominant males cover the rear (Hall, 1960). Essentially the same behavior has been observed in the yellow baboon (*P. cynocephalus*) by the Altmanns (1970). When troops of hamadryas baboons, rhesus macaques, or vervets meet and fight, the adult males lead the combat (Struhsaker, 1967a,b; Kummer, 1968). The adults of many ungulates living in family groups, such as musk oxen, moose, zebras, and kudus, interpose themselves between predators and the young. When males are in charge of harems, they usually assume the role; otherwise the females are the defenders. This behavior can rather easily be explained by kin selection. Dominant males are likely to be the fathers or at least close relatives of the weaker individuals they defend. Something of a control experiment exists in the large migratory herds of ungulates such as wildebeest and bachelor herds of gelada monkeys. In these loose societies the males will threaten sexual rivals but will not defend other members of their species against predators. A few cases do exist, however, that might be open to another explanation. Adult members of one African wild dog pack were observed to attack a cheetah and a

> hyena, at considerable risk to their own lives, in order to save a pup that could not have been a closer relation than a cousin or a nephew. Unattached Adelie penguins help defend nests and creches of chicks belonging to other birds against the attacks of skuas. The breeding colonies of penguins are so strikingly large and the defending behavior sufficiently broadcast to make it unlikely that the defenders are discriminating closely in favor of relatives. However, the possibility has not to my knowledge been wholly excluded. (Wilson 1975a, 121–122)

Sociobiology is rightly seen as an encyclopedic work. (Wilson is not given his due when Lewontin, Rose, and Kamin describe him as an "expert on ants"—1984, 227.) The sheer scope of the examples to which Wilson alludes in this passage is impressive, and can easily foster belief that the underlying explanation must be equally strong. A closer look reveals that the appeal to "kin selection" is founded on a quagmire of difficulties.

Wilson is obviously not concerned with the general explanation of aggressive behavior. His focus is on the evolution of "altruistic" mechanisms that release aggressive tendencies in situations where the aggressor seems to run an increased risk of damage and in which other individuals seem to be blessed with reduced risks as a result of the aggressive behavior. We are bound to wonder about the different forms that the altruistic mechanisms will take across so heterogeneous a group of animals, and to suspect that the underlying "altruistic" genes will interact with the rest of the genome in very different ways in different cases. Let us suspend our doubts, conceding for the moment the possibility of the hypothetical mechanisms and the hypothetical genes. Let us ask why such genes might have been selected.

Wilson's response is that individuals with the altruistic genes have greater inclusive fitness than those who lack them. But is there any reason to believe this? Further, is there any reason for thinking that individuals who do not exhibit the behavior would not also have greater inclusive fitness if they were to behave similarly? A look at some of the examples shows that these questions are appropriate. The hypothesis that there is a simple inclusive-fitness effect for the altruistic genes seems plausible only when we refrain from giving the details of the costs and benefits.

Three kinds of encounters figure in Wilson's list of examples: there are situations in which a group of animals is approached by a predator, situations in which a group is approached by a rival male of the same species, and situations in which two groups of conspecifics meet and fight. Moreover, there are two kinds of questions that Wilson seems to be addressing with respect to these situations: why some individuals in the group put themselves at risk by behaving aggressively (the chacma baboons who place themselves in exposed locations, the males who lead the charge in intergroup battles) and why, under particular circumstances, the role is fulfilled by particular individuals (the "dominant males"). Simply distinguishing the issues should begin to make it clear that there is no single behavioral phenomenon for which we might seek a single evolutionary explanation. The potential losses from an encounter with a group of conspecifics or the approach of a rival male are likely to detract from the individual fitness of a male in the group. But it is also clear that the evolutionary accounting is highly complex. To apply Hamilton's inequality, $B/C > 1/r$, to the *general* case of a group member "defending the group" across so large a set of contests presupposes that we can identify some expected values of B and C so that, for each species considered, the values of B and C, together with the expected value of r for groups in that species (that is, the coefficient of relatedness of the defending individual to the member of the troop to whom the defensive benefit accrues), satisfy Hamilton's inequality. To point to the defenders' relatedness to those they defend is only to make the most casual attempt to show that the explanation is even *consistent* with the phenomena.

But there are further complications. Consider the question why it is the "dominant male" who assumes the defensive role. Wilson assumes that, among the males, it will be the "dominant male" who has the most to lose if one of the members of the troop is attacked by an interloper. Let us suppose, for the sake of concreteness, that the threat comes from a predator and that adults in the troop are relatively safe, while young are in danger. Wilson's sketch of an explanation supposes (1) that young are likely to be related more closely to the "dominant male" than to any other adult male, (2) that the "altruistic genes" are accompanied by a mechanism that enables the animals to determine when their own reproductive interests are threatened, and (3) that in the presence of "dominant males" any female defensive behavior is inhibited. However, as we shall see in a later chapter, there are reasons to doubt that leadership of a troop necessarily coincides with greater production of offspring. Hence, (1) is by no means invariably true. Wilson's own qualification to the effect that "dominant males" are likely to be close relatives of the individuals they protect shows sensitivity to the point; but it raises the question why, under these circumstances, closer relatives do not assume the defensive role. With respect to (2), there is no evidence that all of the animal groups Wilson considers have any such mechanism. That the mechanism exists is simply an assumption, required by the explanation. Finally, there is the mystery of the suppression of the behavior in females. Presumably they too bear the "altruistic" genes, so it appears that there must be some inhibitory mechanism against the defensive behavior when appropriate males are there to defend their young.

To see the force of these concerns, let us press the analysis in two different cases. First, let us consider a multi-male troop of animals in which all the young are sired by one male, Dom. We might explain Dom's behavior of positioning himself at the edge of the troop on the following grounds. For any other male in the troop, the probability of losing one of his offspring through attack from a predator is zero. (These other males have no offspring to lose.) For a female in the troop, the probability of losing one of her young in a successful attack by a predator is the number of her offspring in the troop divided by the total number of young in the troop. (Here I assume that a successful predator would remove exactly one of the young and that all the young are equally likely to be victims.) Then the cost to a female of a successful attack might be identified with this probability. However, the male is, by assumption, father of all he surveys, so the expected cost to him of a successful attack by a predator is always one offspring. Hence there is a greater inclusive-fitness effect to his behaving in a way that promises to ward off predators than there is to others' behaving similarly.

This approach to the issue of inclusive fitness is too simple, however. One of the great themes of sociobiology is that the loss of young may affect males and females differently. Thus we ought to ask what difference a successful attack by a predator will make to Dom's lifetime reproductive output. If, as the story suggests, Dom is in reproductive control of the troop, then lost offspring may be replaceable at small cost. For a particular female in the troop, the loss of a single offspring may make a much more dramatic difference to her lifetime reproductive output. Hence it is far from obvious that the simple calculation of the last paragraph is accurate. Furthermore, if the females in the troop are related, then there will be indirect losses to females if any of the young are killed. Finally, if we try to take refuge in the idea that females are simply incapable of defense, then it will be necessary both to show that this is plausible and to explain why it is true. As we probe the example, we begin to see that the assumption of a larger fitness effect for the "dominant" male needs careful investigation and defense.

Matters become worse when we consider a single-male troop of animals containing several females and their young. If the troop is attacked by a group of predators, then defense by the male alone may well be ineffective. While we might argue that sentry duty requires only one animal to expose itself to danger, we can expect that defense of a troop will be more successful the more adult members participate. Hence there should be no reason for differentiating between the roles of males and females in defense: all can increase their inclusive fitness by vigorous resistance. Ironically, one of the examples cited by Wilson raises exactly this concern. Zebra herds fare very poorly against attacks from packs of hyenas, precisely because the females do not join the male in the defense (Kruuk 1972, 183–185, 208). (Of course, there may be some number less than the size of the troop, such that adding further defenders would not improve the defense. If this is the case, then we can envisage an interesting game-theoretic analysis: for each animal in the troop, fitness is maximized if exactly the critical number of animals rally to the defense—and if it is not among them.)

Finally, in single-male and multi-male troops alike, it would be wrong to suggest that an evolutionary analysis can begin from whatever sexual dimorphism and troop composition is now observed in the species under study, taking this as the starting point for the explanation. We have no basis for assuming that troop size and troop composition were selected first, that males and females evolved to their present relative sizes, and that natural selection *then* operated, subject to these constraints, to select troop members for the propensity to defend their fellows. Are we to assume that the groups in question managed without any defensive arrangements while they were sorting out their optimal troop size, and that they faced the problem of predation only when this question was settled? Surely not.

I conclude that there is no simple appeal to inclusive fitness that will underwrite Wilson's attempt to explain defensive behavior across so large a variety of groups and in so large a variety of situations. It *may* be possible to develop some of the examples in detail—but that remains to be shown.

Aficionados might protest that I have missed the obvious. Defense of others by a troop member is found when some member of the troop is related to many of the potential victims, and it is absent when there is no such individual who might figure in such a defensive role. Surely this correlation must amount to something? Doesn't it suggest that, despite all the complications that occur in particular cases, inclusive fitness has been an important factor in the evolution of defensive behavior?

No. Wilson's alleged control experiment yields very murky results. Recall that Wilson emphasizes the differences in defensive behavior among various groups and suggests that in "loose societies," such as wildebeest herds, adult males will not defend their young. In the first place, adult wildebeest, male and female, are very poor at defending themselves—let alone their offspring—against attacks from hyenas. (Kruuk reports that hyenas are major predators of wildebeest, at least in the Ngorongoro crater, killing 75 percent of the wildebeest calves born annually; see Kruuk 1972, 166.) Hyenas are often successful in hunting adults. Second, females do defend their own offspring, both by attempting to interpose themselves and by charging hyenas who have caught one of their calves. True, female wildebeest do not defend the young of other females—but then the hamadryas baboons and zebras to whom Wilson compares them are, by hypothesis, defending their own young. Third, male wildebeest *will* sometimes help out in defense. Kruuk reports, "Sometimes hyenas are attacked while chasing a calf not only by the mother, but also by bull wildebeest in their territories, or sometimes even by zebra stallions if the calf has taken refuge near a zebra family" (1972, 171). Although Kruuk goes on to note that such intervention is unusual, its occurrence at all suggests that defense is not simply a matter of counting kin. (Perhaps such cases can be explained away as animal errors. The misguided defenders operate according to an imprecise program that occasionally leads to bizarre results.)

So the example of the wildebeest admits of two explanations that are quite different from that given by Wilson. Perhaps cooperative defense in wildebeest is rare because wildebeest are rather bad at defending themselves against their predators. Or perhaps cooperative defense is present in wildebeest to the extent that it is present in several of the cases that Wilson takes as examples of kin selection at work—the zebras and the baboons, for example. Nor does the instance of the gelada monkeys do much better. Here we are dealing with a very different type of "society," a loose organization of young adult males. Gelada monkeys seem to have a social structure based on interactions among females. A single male forms a social unit with several females, but the male is "a social supernumerary whose role in the group is mainly that of breeder" (Dunbar 1983, 301). Gelada males have little to do with one another once they have managed to achieve a small band of their own, and the social interchanges between the expectant bachelors are not extensive. Young male geladas do not really form themselves into troops that are comparable to the other units mentioned by Wilson. Comparing their behavior with that of a group of chacma baboons or of a hamadryas band is somewhat like comparing the activities of strangers who find themselves waiting for the same bus with the behavior of a family on an outing.

Then, of course, there are the exceptions, the African hunting dogs and the Adelie penguins. (We might also add the occasional bull wildebeest and even the confused zebras.) What are we to say in these cases? Wilson's answer is unclear. Perhaps he intends to see a different selective force at work in such examples. If so, then it would be interesting to identify that force and to see if it might not also be operative in many of the allegedly unproblematic instances. Or perhaps Wilson hopes to take refuge in that bare possibility, the hope that helper and helped will turn out to be related after all.

The passage from *Sociobiology* we have been scrutinizing exhibits some of the troubles of the studies of nonhuman animals that are used as if they were unproblematic in much pop sociobiological work. There is a suggestion that certain forms of behavior have been selected because of a particular benefit they bring to those who exhibit them, but neither the behavior nor the benefit is precisely specified. What is the behavioral rule that chacma baboons are supposed to follow? What is the benefit that escalated fighting among hamadryas males is supposed to bring? Cases are lumped together as if there were some single system of evolutionary accounting that could apply to them all. Constraints are produced out of thin air. Exceptions are accompanied by the expression of hope that they will not prove to be exceptional after all. A consistent account is suggested—but only because the picture is very softly focused.

The attempted generalization falls victim to all the troubles that we noted in the previous section. Yet, unlike those sociobiologists who recognize the problems and set about working out the details, pop sociobiologists propose to use their premature generalizations about animals as the bases for grand accounts of human nature. By doing so, they not only found controversial claims about people on speculative generalizations about animals but also distort the accomplishments of their colleagues who attempt to do justice to the full complexity of animal behavior. (A detailed treatment of the evolution of cooperative defense in even one of the species mentioned by Wilson would be an impressive achievement.)

One pimple does not make a case of the pox. To forestall the objection that I overestimate the gravity of the disease, more examples are needed. I shall begin with a case that offers an even more provocative conclusion and that involves even more dubious arguments.

Generalizations about Gender

Sex, while not universal, is very common. One of the most celebrated—or notorious—proposals of early Wilsonian sociobiology is the claim that we can understand gender roles in humans in terms of the fundamental sexual asymmetry. From the biological point of view, males are the producers of large numbers of small, mobile gametes; females form a much smaller number of large gametes. On this basic difference major aspects of the behavior of men and women are supposed to turn.

For the main outlines of the argument there is no better source than Wilson.

> The anatomical difference between the two kinds of sex cell is often extreme. In particular, the human egg is eighty-five thousand times larger than the human sperm. The consequences of this genetic dimorphism ramify throughout the biology and psychology of human sex. The most important result is that the female places a greater investment in each of her sex cells. A woman can expect to produce only about four hundred eggs in her lifetime. Of these a maximum of about twenty can be converted into healthy infants. The costs of bringing an infant to term and caring for it afterward are relatively enormous. In contrast, a man releases one hundred million sperm with each ejaculation. Once he has achieved fertilization his purely physical commitment has ended. His genes will benefit equally with those of the female, but his investment will be far less than hers unless she can induce him to contribute to the care of the offspring. If a man were given total freedom to act, he could theoretically inseminate thousands of women in his lifetime.
>
> The resulting conflict of interest between the sexes is a property of not only human beings but also the majority of animal species. Males are characteristically aggressive, especially toward one another and most intensely during the breeding season. In most species, assertiveness is the most profitable male strategy. During the full period of time it takes to bring a fetus to term, from the fertilization of the egg to the birth of the infant, one male can fertilize many females but a female can be fertilized by only one male. Thus if males are able to court one female after another, some will be big winners and others will be absolute losers, while virtually all healthy females will succeed in being fertilized. It pays males to be aggressive, hasty, fickle, and undiscriminating. In theory it is more profitable for females to be coy, to hold back until they can identify males with the best genes. In

> species that rear young, it is also important for the females to select males who are more likely to stay with them after insemination.
>
> Human beings obey this biological principle faithfully. (1978, 124–125)

Wilson continues with a catalog that includes many of the stereotypes of the corner bar: men initiate "most of the changes in sexual partnership"; sons "sow wild oats"; daughters "risk being ruined"; men are usually the political leaders. What we always knew about men and the girls (nudge, nudge) turns out to be good science.

Wilson makes the scenario sound very plausible. For a very large collection of species we are apparently able to show that certain kinds of behavior will maximize the fitnesses of males, certain other kinds of behavior will maximize the fitnesses of females. When we compare the theoretical optima with the actual behavior of males and females, it seems that there is a beautiful match. So we can begin to climb Wilson's ladder toward a conclusion that will be anathema to feminists but may console many of the folk—specifically the menfolk—in the corner bar.

Let us look more closely. What are the theoretical optima? Are Wilson's claims about actual behavior in "the majority of animal species" true? The answer to the latter question is "Not really." At a superficial level the host of exceptions to Wilson's generalizations seem only to reinforce his basic point. There are some groups of animals in which males care for the young and in which females compete for males. Prime examples are pipefish and some species of birds (see Trivers 1972, 59). However, there are also numerous cases—particularly among fish and amphibians—in which males care for the young, even though females do not compete for males. Such instances are sometimes explained by the suggestion that males are able to defend the young without forgoing opportunities to compete for mates (see Gross and Shine 1981, especially 781).

These examples reveal that Wilson's awestruck emphasis on the sheer numbers of sperm that a male produces is not the main point. If the sociobiological analysis succeeds, it is because of the dynamics of copulation and fertilization. In species with internal fertilization, it is hard for females to escape substantial investment in the young. When fertilization is external, the male can sometimes be left holding the brood while the female goes off to spawn elsewhere. Wilson's claims about sex rest on a generalization that is not explicit in his presentation. When one sex invests more in the offspring, then members of that sex will become a "limiting resource" for individuals of the other sex, who will compete for access to this limiting resource. In humans as in most mammals, females are the limiting resource for males. So goes the story on which Wilson really relies.

The source of the story is a provocative paper by Trivers. Central to Trivers's account is an extended economic metaphor. Trivers begins by defining *parental investment* as "any investment by the parent in an individual offspring that increases the offspring's chance of surviving (and hence reproductive success) at the cost of the parent's ability to invest in other offspring" (1972, 55). Suppose now that the sexes of a species invest differently in offspring. On the average, members of one sex invest more in each offspring than do members of the other sex. The total number of offspring produced is equal for both sexes. So the sex that invests less is getting the same evolutionary "return" at less evolutionary "cost." Assuming that both sexes have equal amounts of "capital" to invest, the sex that typically invests more becomes a limiting resource for the sex that invests less. So we should expect that members of the sex that invests less should compete to breed with members of the limiting resource. However, because of the different sizes of male and female gametes, there is always a greater *initial* female investment, even if this is sometimes outweighed by the investment activities of the male.

Even sympathetic readers quickly point out that the economic terminology has its difficulties. For a start, we may note that Trivers's notion of investment seems to be defined in terms of itself and that the intention seems to be to understand how various strategies of behavior with respect to the other sex would yield different expected numbers of offspring. (The latter point was made in a lucid paper by Maynard Smith, 1976b, who urged a "prospective analysis" of the notion of parental investment.) But now we encounter a banal arithmetical truth. The expected number of offspring for a male is the total number of offspring produced in one generation divided by the number of males. Assuming that females and males are about equinumerous, then the expected number of offspring for a male will be the same as the expected number of offspring for a female.

Suppose that a population starts from a situation in which all males play strategy M_1 and all females play strategy F_1. Assume that mutant males, playing M_2, arise and are favored in competition against males playing M_1; similarly, assume that mutant females, playing F_2, arise and are favored against the females who play F_1. Finally, assume that there is some environmental limit to the number of offspring that can be produced in a generation, and that the population always manages to produce this limit. Even though M_2 may invade the initial population and become the prevalent male strategy, and even though F_2

may do likewise, there is one obvious sense in which these strategies are no better than those they replace. Given that the population maintains the same size and produces the same number of offspring throughout the replacement process, the expected payoff to males at the end is exactly the same as the expected payoff to males at the beginning.

These observations suggest that the Trivers argument is best approached from the perspective of game theory. We need to inquire whether assertiveness is a male strategy that might have been favored *against certain rival male strategies in the context of certain female strategies* and might have been evolutionarily stable. Perhaps assertive males are now no better off than their evolutionary forebears. But the assertive mutants who arose in the original uncompetitive population were favored. Their seed inherited the earth.

We can begin to make progress by descending from Trivers's grand perspective. Many animals, especially birds and mammals, engage in numerous kinds of intersexual encounter, and it may be a ludicrous oversimplification to assume that there is a single behavioral strategy of sexual interaction that each sex is to choose. So let us pose a limited problem. Abstracting from other aspects of the relations between the sexes, let us investigate the possibilities of desertion by either sex. In making this abstraction, we should be aware that different kinds of fidelity should not be confused with one another. An animal that stays with its mate is not necessarily averse to some copulation on the side. Indeed, the relationship of a detailed model of desertion to the sexual activities of complicated animals—especially carnivores, primates, and humans—is tenuous. So many patterns of behavior are possible that, in Maynard Smith's phrase, the models "have an obvious air of unreality." Yet Maynard Smith's justification of model building is apt. It forces us to make our assumptions clear. (For details of the analysis see technical discussion E.)

Suppose that two animals have mated. Each of them now has a choice: to stay and rear the offspring ("guard") or to desert and find a new mate ("desert"). It is assumed that the parent who deserts stands a chance of mating again and that doing so will yield a greater number of offspring than would have been obtained by guarding. We also suppose that offspring are most likely to survive to maturity if they are cared for by two parents, less likely to survive if they are cared for by one, and even worse off if they are forced to struggle on their own. Given these assumptions, should we expect to find deserting males and guarding females?

No. Game-theoretic analysis exposes some of the assumptions on which the qualitative argument, originally given by Trivers and

Technical Discussion E: When To Be Fickle

Two animals have mated. Each of them has two options—Desert and Guard. P_0, P_1, and P_2 are the survival probabilities of the young if cared for by 0, 1, or 2 parents respectively. We assume that

$$P_0 < P_1 < P_2.$$

Let p and p' be the probabilities for a male of mating with a second female if he deserts or guards; let V and v be the number of offspring produced by a female if she deserts or guards. We suppose that

$$p' < p, \qquad v < V.$$

The payoff matrix can be written as follows:

		Female	
		Guard	Desert
Male	Guard	$vP_2(1 + p')$ \ vP_2	$VP_1(1 + p')$ \ VP_1
	Desert	$vP_1(1 + p)$ \ vP_1	$VP_0(1 + p)$ \ VP_0

(From Maynard Smith 1982a, 127. The payoffs to females are in the upper right corner of each square. Payoffs to males are at the lower left.)

Depending on the parameter values, *any* of the four combinations of male and female strategies can be an ESS. Suppose, for example, that P_2 is much bigger than P_1 and that p and p' have comparable size. (Two parents are much better than one, and the probabilities of remating are not substantially increased by desertion.) Then it will be in the evolutionary interests of both parents to guard. Similarly, if P_0 is not much less than P_1, then a guard mutant could not invade a population of deserters. Single parenting will be an ESS if one parent does much better than none and about as well as two. (See Maynard Smith 1982a, 127–128, for further analysis.)

Let us now consider a *Wilsonian population*, a group of animals in which males desert and females guard, and in which the sex ratio is approximately 1:1. Assume that the probability of a deserting male's finding a new mate is $f(r)$, where r is the frequency of deserters; that $f(1)$ is extremely small, $f(0)$ is 1, and f decreases as r increases. Suppose also that males stay with their second mates, if they desert and mate again. Then, in a population of females who guard, with a frequency r of male desertion, the expected payoff to a male deserter is

$$P_1v + f(r)P_2v.$$

The expected payoff to a male who guards is

$$P_2v + p'P_2v.$$

In a Wilsonian population, r is 1. Since $f(r)$ is extremely small, it is at best negligibly greater than p' (the probability of remating for males who guard). Hence, even given small differences between P_2 and P_1, a newly arising mutant male who guards can be expected to invade. At the other extreme, if all the males guard and if p' is negligible, then the payoff to a newly arising mutant male who deserts will be greater than that of the guarding majority if

$$P_1v + P_2v > P_2v.$$

Since this inequality always holds, we may expect to find a mixture of guarding males and deserting males.

simplified by Wilson, depends. (See technical discussion E for details.) Crucial to the idea that it pays males to be fickle is the supposition that a deserting male will have a relatively high probability not only of copulating with a second female but also of fertilizing some of her eggs. Another important assumption is that the benefit of extra offspring will outweigh the decreased viability of the offspring already produced. It is also taken for granted that the survival probabilities of the young of the second female are comparable to those of the young of the first female. That could easily be wrong. If there is female-female competition that results in a situation in which females who mate on the first round obtain resources that are important to the successful rearing of young, or if the second round of matings occurs relatively late in the breeding season, then the probabilities of survival for offspring of second matings will be substantially reduced.

It is also easy to overlook the possibility of delayed fitness effects. If our animals confront a situation in which the competition to find mates is very severe (for either sex or for both sexes), then the number of surviving offspring is no longer an appropriate measure of fitness. Increasing the number of copulations, of fertilizations, or even of young who survive to sexual maturity is of no avail. It's the quality that counts. We can do justice to such considerations by reinterpreting the payoffs. Instead of thinking just of the probabilities of survival, we can consider the expected payoffs for the young in whatever competition they face. So, if the struggle is for mating, then the appropriate measure is the probability for an offspring raised by no parents (or by one, or by two) that it will survive to maturity with the resources needed to succeed in this struggle. Intuitively, we shall expect that the introduction of this type of delayed fitness effect will increase the value of care by both parents and thus lead to a situation in which the ESS for both males and females is to guard.

A further complication necessitates an extensive elaboration of the game. At least one of the probabilities that figure in our account is likely not to be a constant. If fickle males are to stand any chance of doing better than faithful ones, then they must be able to monopolize the reproductive output of their first-round mate(s) and still find second-round mates later. The chance of remating depends on the extent of the competition from other males. As the frequency of deserters goes up, there is a reduction in the probability that any individual deserter will find a second mate. So, in a population dominated by deserters, the chances of remating may be too small to offset the loss in viability of first-round offspring. Hence, a rare mutant male who elects to guard will have a greater payoff than the deserting majority, and we shall expect this mutant to be able to invade. Thus, given a very plausible assumption, a population of deserters would not be evolutionarily stable. To a first approximation, the constellation of behaviors envisaged by the barroom stereotype would tend to be replaced by a mixture of male strategies, with some males guarding and others deserting.

While the analysis is recognizably only a first approximation—we have not even considered the slightly more complicated game in which there are both male and female mutants—it does have the virtue of breaking the grip of the idea that there must be some single male strategy or some single female strategy that evolution will establish and maintain. Moreover, once we begin to consider the behavior of complex vertebrates who can assess their own positions, their own advantages, the strategies of those around them, and so forth, it becomes clear that the qualitative account sketched by Wilson is ludicrous. Males and females can be expected to play highly complicated conditional strategies whenever evolution has equipped them with the cognitive capacities required by those strategies. Just as pure Hawks are likely to be replaced by combatants who assess the strength of each rival, so pure Deserters (and pure Guarders) can be expected to be replaced by animals who adjust their actions to the behavior of the males and females around them. Whether or not a full analysis of the dynamics of desertion, which did justice to all the possibilities for male and female animals equipped with the ability to assess their competitive chances, would lead us to the conclusion that there is a single optimal male strategy and a single optimal female strategy is not obvious. (As we have seen, the first steps at analysis lead us to the suggestion of polymorphism.) Yet even if were to conclude that there are unique optima for males and females, there is absolutely no reason to think that they are captured in Wilson's slick scenario.

In this case we have discovered that the ladder cracks at the first rung. Wilson's allegedly optimal strategies are not optimal. The analysis that needs to be given for "the battle of the sexes" turns out to be extremely complex—indeed, there is room to doubt whether there is a single analysis that will represent all the possible alternative strategies available in all the groups of animals about which Wilson wants to generalize. The results depend crucially on the assumptions about parameter values. Moreover, this evaluation applies not only to Wilson's presentation but also to the more careful pop sociobiological accounts of Trivers and Dawkins (Trivers 1972; Dawkins 1976, 162–165). However, there are several careful discussions of models of parental desertion in which sociobiologists recognize and explore the dependence on particular assumptions. (See, for example, Grafen and Sibly 1978; Schuster and Sigmund 1981.) Once again, sociobiology cannot be indicted for the oversights and simplifications of pop sociobiology. The difference in attitude is nicely captured in the tongue-in-cheek conclusions of Schuster and Sigmund, who remark that "the behaviour of lovers is oscillating like the moon, and unpredictable as the weather," and admit that "people didn't need differential equations to note this before."

Failure at the first step is matched by failure at the second. The barroom stereotypes can only be applied to animal behavior if the actual doings of the animals are kept in soft focus. Monogamy is very popular with birds, although it is not just for the birds. Among the higher primates some of our closer relatives, the six species of gibbons, are monogamous. Moreover, female chimpanzees are sexually assertive. Chimps in estrus actively solicit males. Female-female competition, as well as complex patterns of male-female cooperation, occur in baboons (Seyfarth 1978) and in chimps. Female hyenas are both larger and more aggressive than males.

The catalog of instances in which animals fail to conform to the paradigm of the hasty-fickle-aggressive-promiscuous male and the demure-passive-uncompetitive female could be extended far beyond these familiar examples. Those in the grip of the stereotypes will have ways of excusing these lapses. Ecological conditions sometimes impose monogamy, wrenching the basic biology of sexual differences into a new pattern. But for the distribution of their food resources (fruit trees), female gibbons would not have to space themselves as they do; but for the spacing, the old primate pattern of male-male competition for access to many females would reassert itself (Hrdy 1981, 55–56). The legendary promiscuity of female chimps may have evolutionary advantages. In any group of animals in which males have the tendency to kill young whom they do not "perceive" as their own offspring, females gain by confusing as many males as possible into thinking they have a chance of paternity. Again we have a perturbation of the underlying sexual pattern.

We arrive at a curious situation. Our articulation of the theoretical analysis of optimal male and female strategies showed that there is no reason to think that the stereotypes describe the optima—or even that there are optima to be described. We now discover that the mythical optima are used as standards against which "deviations" in behavior are judged. Wilson's grand generalization, the claim that "in the large majority of animal species" males are assertive and fickle, females passive and coy, turns out to fragment. In the absence of a motley collection of conditions—compiled from a review of the presently known exceptions—the generalization is expected to hold. But we should have no faith in so gerrymandered a generalization. There is no theoretical basis for seeing the promiscuity of the female chimpanzee, the dominance of the female hyena, the cooperation of Seyfarth's baboons, as modifications of The Fundamental Male and Female Behavior in the light of special conditions. Moreover, we should have no faith that the haphazard list of "deviations" at which we have so far arrived is anything like complete.

We should not quit the topic of sexual strategies without a brief look at one obvious source of embarrassment. Gaudy males and drab females are often cited as signs of intense male competition for access to the "limiting resource." Yet, in humans, females are often more ornamented than males. Why? Does this betoken a "role reversal"? Should we replace the stereotype of the competitive male with that of the competitive female, attributing to her the traits that go with the role of chooser, to him the characteristics of the object of choice?

Darwin seems to have flirted with the idea (1871, 369–371), and it has recently been revived (Low 1979, 463). Low suggests that there are species (including our own) in which "male quality varies greatly" and in which we should expect female-female competition aimed at securing something (sperm? parental care for the young?) from the most desirable males. The shrinking violet gives way to the femme fatale.

Low's suggestion exposes the poverty of the underlying theoretical analysis. The suggestion that there are delayed fitness effects, second-generation benefits, to females who mate with "high-quality" males is not something that can just be tacked onto a simple qualitative analysis of the kind favored by Wilson or Trivers. We cannot argue for general claims about male assertiveness, male fickleness, and male competition by choosing one measure of fitness and then add, as an afterthought, that selection of a different measure of fitness can give us some female competition (specifically, the struggle to allure) just where we would like to have it. Low's suggestion breaches the dam. It shows us that even the relatively careful attempts to understand the evolutionary stability of sexual strategies are going to have to come to terms with delayed "returns on investment" (for example, by recognizing that it is not simply the *survival* probabilities of the young but their *expected competitive abilities* that count). Before we pile one casual optimization analysis on top of another, it is well to look at the foundation on which we build.

Pop sociobiologists often charge their critics with dewy-eyed sentimentalism. Recall the daring with which Barash and van den Berghe defend unpopular stereotypes (of course, they are not members of the class that has the most to lose from those stereotypes). I freely admit that I would prefer to inhabit a world in which males and females have opportunities to interact on equal terms, in which the roles are not those prescribed by the sages of the barroom. But my opposition to the defense of stereotypes is not a childish refusal to accommodate the facts. It is based on a diagnosis of the allegedly scientific arguments.

Critics of pop sociobiology frequently charge that the devotees are all too willing to overlook rival hypotheses (they fail to appreciate the impact of a history of male dominance, for example) and that there are dubious assumptions about the relation between optimization and selection and about genes and phenotypes. These are important criticisms, and their turn will come. But critics should begin at the very bottom of Wilson's ladder. They should ask if the stories about optimal behavior really do identify forms of behavior that maximize inclusive fitness, and they should ask if those stories are even consistent with the evidence we have. I have argued in previous chapters that individual sociobiological accounts should be judged on their merits, according to the canons that govern evolutionary theorizing generally. Having weighed one influential account in the balance, I find it wanting.

The contrast should be heightened by our consideration of earlier examples. Nonhuman sociobiology is not perfect. But *it is often highly promising.* We have seen the merits of the evolutionary analyses of aspects of the behavior of dung flies and scrub jays, analyses that seek to combine painstaking observation with precise models. In the case of larger groups of organisms, groups that are to include animals as complex as ourselves, it is more difficult to achieve either the observational care or a precise model that will do justice to the complexity. However, that should not be taken to imply, by some curious logic, that we can content ourselves with anecdotal information and casual optimizations. If it belongs anywhere, barroom gossip belongs in the barroom.

Opium for the People

Homer is entitled to an occasional nod. Even the inaugurator of a "new synthesis" can be forgiven for a lapse, a sloppy piece of argumentation in a popular book. However, those who lull the conscience of the people by supplying palatable palliatives are less easily pardoned. In popular presentations of human sociobiology, misreporting of observations and speculations about optimal behavior are all too common.

Wilson's discussion of male and female sexual strategies has the merit of a foundation in Trivers's outline of a model. Other discussions have the air of spur-of-the-moment thoughts about advantages. After stressing the adaptive value of play in the "higher mammals," Wilson continues,

> At its most potent, in human beings and in a select group of other higher primates that includes the Japanese macaques and chimpanzees, playful behavior has led to invention and cultural transmission of novel methods of exploiting the environment. It is a fact worrisome to moralists that Americans and other culturally advanced peoples continue to devote large amounts of their time to coarse forms of entertainment. They delight in mounting giant inedible fish on their living room walls, idolize boxing

> champions, and sometimes attain ecstasy at football games. Such behavior is probably not decadent. It could be as psychologically needed and genetically adaptive as work and sexual reproduction, and may even stem from the same emotional processes that impel our highest impulses toward scientific, literary, and artistic creation. (1975a, 167)

Maybe. Then again, maybe not. There is little doubt that primates sometimes learn things, and transmit information, through the playful manipulation of objects in their environment. There is an impressive, growing literature on animal play, in which possible adaptive advantages are carefully explored (see, for example, Fagen 1981; Smith 1982). However, even if we grant that the tendency to play is adaptive, there is only a loose connection between the activities of the higher primates and the forms of behavior to which Wilson alludes. The idea that American adults are prompted to catch and display fish, to flock to the ringside and the football stadium, by the same propensities that spurred them to their profitable childhood explorations is a thesis to be argued. Conclusions about adaptation are not to be established by free association. Here, as before, the words of those who are concerned with the details stand in marked contrast to Wilson's speculations. Fagen admits that "little is known about effects (and even less about functions) in any species, including humans" (1982).

In discussing some experiments designed to show animals' predilections for the familiar, Wilson offers a guess about the adaptiveness of the preference.

> The more something stays around without causing harm, the more likely it is to be part of the favorable environment. In the primitive lexicon of the emotive centers, strange means dangerous. It is perhaps adaptive to become homesick in foreign places, or even to suffer culture shock. And, for animals, it would seem prudent to treat a familiar and relatively harmless enemy as dear. (1975a, 274)

Again, there is no serious attempt to explore the way in which fitness is supposed to be maximized. One might equally argue that, for many animals, there will be a need to cope with new surroundings—perhaps because juveniles are routinely expelled from the parental nest, troop, or range—and that there will thus be selection for an attraction to the novel, with perhaps a concomitant propensity to be bored by the familiar. More sensibly, we might suggest that animals are sometimes attracted by new stimuli and new surroundings and sometimes not. (As pop sociobiologists never tire of reminding us, higher primates do not choose to mate with those with whom they have been associated most closely.) The real problem in understanding adaptation is to develop precise analyses of the kinds of novelties that can be expected to prove attractive. Anecdotes about homesickness are no substitute.

Wilson is not alone. Although the troubles I have remarked are thickly strewn throughout *Sociobiology* and *On Human Nature*, they are even more noticeable in the writings of some of Wilson's followers. Here is Barash on an important, and unsolved, problem about animal forbearance in combat.

> A more realistic explanation for restraint in fighting [more realistic than the idea that restraint is for the good of the species: PK] is that this too is selfish. If the winner can prove his superiority without pressing home his advantage, he is probably better off refraining, since he otherwise risks possible injury to himself. His opponent might also be a relative or an individual who benefits the victor's group so much that it is selfishly advantageous for the victor to be "magnanimous." In addition, a magnanimous winner might even profit directly in the future through reciprocity. (1979, 183)

This passage is the prelude for some further speculation on why humans do not have the kinds of inhibitions that might be needed to check our self-destruction. Although the lines of the further argument are tangled enough, it is worth emphasizing the looseness of the discussion of nonhuman restraint. While Barash is probably right to reject the old ethological account—that animals exhibit magnanimity to promote the welfare of their species—we simply do not know how to provide detailed accounts in terms of individual benefits. Appealing to the idea that victors might press home their advantages only at cost of injury simply fails to account for the observed phenomena: animal restraint is most striking in those examples in which the defeated animal exposes vulnerable parts of its anatomy, as in the famous case of the defeated wolf that presents its throat (described by Lorenz and cited by Barash). In many instances of male-male combat we have reason to believe that the males are not related, and in some of these cases the defeated male does not belong to any social unit that includes the victor. Ritual combat is common in relatively asocial animals, such as lizards and ungulates. Finally, the suggestion of possible reciprocity breaks down once it is seriously examined. If the "return" is to take the form of "magnanimity" should the defeated animal prove victorious on some future occasion, then the present victor would surely maximize his fitness by preventing the possibility of such future occasions. If the return takes some other form, then we shall have to suppose a network of social relationships that is conspicuously absent in many of the examples in which ritual combat occurs.

The truth of the matter is that Barash has no serious model of restraint. He scatters some suggestions about possible benefits, suggestions about possible benefits, suggestions that face severe obstacles if they are to be coherently developed into a cogent analysis of the observed behavior. As he eventually admits (1979, 184), nonhumans are not as forbearing as ethologists once believed. There is a serious problem in charting the exact conditions under which restraint is to be expected. There is no clear line of solution. There is no firm basis for fantasies about human aggression, human lack of inhibitions, and possible dire consequences in a world in which someone has a "finger on the Button" (185).

Barash is also not entirely punctilious in reporting the phenomena. In a section designed to support his thesis that humans are naturally polygynous, he begins by explaining why monogamy is a rare condition among mammals: males are not good sources of food for the young. Once again, a serious and important scientific controversy lurks in the background, unnoted. Because there are many mammalian groups in which it would seem that inclusive fitness could be maximized by using the male as a source of food, the question why male mammals have not evolved the ability to lactate has been seriously posed (Maynard Smith 1976b, 110). But Barash is not content to ignore complications. He goes on to remark that "monogamy . . . is almost unheard of in primates" (1979, 65). Of course, this "fact" helps to reinforce the idea that as primates go so goes *Homo* and thus to buttress Barash's stereotype of male behavior. However, the actual observations of lemurs, tamarins, gibbons, and siamangs undermine the confident dismissal. Sarah Hrdy serves as a reminder that one can write an accessible book without misleading the reader.

> Monogamous breeding systems are four times more common among primates than they are in mammals generally (on the basis of current information). Out of the 200-odd species of primates, as many as 37 (possibly more), roughly 18 percent, live in breeding pairs in which male investment in offspring is substantial and focused exclusively on the offspring of a single female. (Hrdy 1981, 36; see also Kleiman 1977)

Whether or not humans should be included in the club is, of course, debatable. However, it is not quite kosher to cut off the debate by pretending that a condition found in almost a fifth of a group is "almost unheard of" there.

For a final example of the way in which pop sociobiology relates to actual scientific practice, we can do no better than to look at some remarks of Barash's colleague van den Berghe.

> Male primates who exert dominance tend to mate and reproduce much more. Politics (in the broadest sense of the struggle for dominance) is thus primarily a male game, the end of which is ultimately reproduction (Tiger and Fox 1971). Sexual politics, indeed, though not quite as Kate Millett (1970) would like!
>
> Ideological passions unfortunately contaminate our way of looking at data and interpreting them. I am not suggesting that male dominance is good, but merely that it *is*. Nor do I deny that individual women *can* be dominant over individual men. On the average, however, males are dominant and the more dominant have, throughout our past evolutionary history, been the more reproductively successful. Dominance displays in men have been selected as sexual "turn ons" for most women and, conversely, female dominance is a sexual "turn off" for most men. (1979, 197)

We have come a long way from Parker's dung flies and Woolfenden's scrub jays. We are even in a different realm from Wilson's speculations about defensive behavior, flawed though they may be.

What is supposed to be adaptive? Struggles for dominance (whatever that may turn out to be) among men? Male dominance over women? Women's receptivity to dominance displays? What do these things have to do with one another? Why do they preclude a struggle for dominance among females or an adaptation for receptivity to female dominance displays among males? What are the costs of male strugglings? How do they relate to male inclusive fitness? How does dominance (in van den Berghe's broad sense) relate to reproductive success? (As we shall discover, the dominant male does not by any means always achieve the maximal number of copulations.) What does all this have to do with politics (narrowly understood)? What does van den Berghe make of the numerous species of primates (including chimpanzees and macaques) in which there is pronounced competition among females? How does he account for the abilities of male lemurs to be "turned on" by their dominant females? There is surely ideological passion in this area, and some of it is to be found in the passage I have quoted. It seems to me legitimate to ask whose eye contains the mote and whose the beam.

Pop sociobiology is unsatisfactory when its practitioners rely on inadequate models and when they fail to relate their analyses to the observed forms of behavior. It is worse when they have no models at all, when they substitute spur-of-the-moment thoughts about possible advantages for rigorous analysis, when they omit uncomfortable facts and misreport the findings of students of animal behavior. This chapter has run the gamut of sociobiological theorizing, good, bad, and downright ugly. My suggestion is that we prize the good, improve the not-quite-so-good, abandon the bad, and expose the irresponsible.

It is worth closing with a reminder of the caution that the best workers in the evolution of animal behavior have revealed on occasion after occasion. Clutton-Brock and Harvey conclude their extensive and much-respected study of social organization in primates by emphasizing the difficulties of constructing explanations that will account for the full range of variation that they document (Clutton-Brock and Harvey 1979c, 368–370). Peter Jarman explores in detail the problems that confront attempts at making interspecific comparisons in sociobiology (1982). Parker follows an exceptionally refined analysis of animal assessment and fighting behavior with an explicit disavowal of the idea that there is any direct conclusion to be drawn about humans (Parker 1974, 292). Like the insalubrious examples at the other end of the continuum, cases of healthy caution could be multiplied. The spirit of Kepler is not dead. It can be found in the writings of many sociobiologists. It is only missing where we ought most to expect it.

Chapter 6
Misadventures in Method

The Sins of the Sociobiologists?

Most methodologists have their own catalogs of favorite sins. Previous critics of sociobiology are no exception, and the literature on sociobiology is heavy with accusations of methodological turpitude. Sociobiologists, it is suggested, fall prey to vulgar anthropomorphism. They misleadingly generate conclusions about groups of animals that include human beings by describing human and nonhuman behavior in the same, sometimes colorful, language. Vulgar anthropomorphism is joined by equally vulgar forms of reductionism. In sociobiological constructions of evolutionary scenarios important rival possibilities are routinely ignored, so that outsiders are gulled into believing that the proposed story provides the only possible evolutionary explanation. Finally, in arguing for their doctrines about human nature, sociobiologists neglect important cultural factors that might all too easily shape the form of the behavior that we observe. So there abide a number of methodological sins, and those whose methodology is pure will abide none of them.

As criticisms of sociobiology these charges are overblown. Many of those who study the evolution of animal behavior are sensitive to all the issues that the critics raise. They are careful to purge forms of language that might facilitate unwarranted transitions between claims about animals and claims about people. They explore alternatives to their own favorite evolutionary scenarios. Finally, since theirs is not the project of fathoming human nature, they are not tempted by quick arguments for the fixity of human behavior and the inevitability of our social institutions.

Nevertheless, the points made by the critics are relevant both to pop sociobiology and to *some* work in the sociobiology of nonhuman animals. Particular attempts to climb Wilson's ladder presuppose descriptions of animal and human behavior couched in similar terms. Moreover, such descriptions are sometimes found in discussions of the behavior of nonhumans. Are these linguistic practices legitimate?

As we have already seen, the confirmation of ambitious Darwinian histories depends on careful exploration of alternatives, and special attention is needed in cases where the history of a behavioral characteristic may involve processes of cultural transmission. Do pop sociobiologists consistently overlook potential explanations that appeal to human culture and its history? Lastly, to chart the limits of human nature and human social institutions, those who climb Wilson's ladder must negotiate the final step. Can they do so? Are the appeals to the efficiency of adaptation and to cognitive impenetrability successful?

I shall argue that, properly formulated, these questions tell decisively against the early Wilson program. (Some of them raise important objections to other versions of pop sociobiology as well.) We shall also discover that there are discussions of the evolution of animal behavior that lapse into methodological error. However, it is worth repeating at the outset that sensitive sociobiologists are attentive to the possibilities for going astray. The situation is similar to that which we discerned in the last chapter. There are internal disputes in sociobiology that raise points analogous to those I shall make. Such debates prepare the way for an improved understanding of the evolution of animal behavior. They spell doom for pop sociobiology.

Original Sin

Before the Fall, so the story goes, Adam called the beasts together and named them. Students of animal behavior are true descendants of Adam. In recording the activities of a species, they must give names to what they see, and herein lies the possibility of methodological error. For the classification of behavior, like the classification of anything, involves assumptions. Inevitably, scientists must suppose that those things they call by the same name will be alike in the crucial respects—that this sample of oxygen will have the same chemical properties as the ones that have previously been prepared, and so forth. Yet there seems to be a particular danger in the discussion of animal behavior. Because we have so rich a vocabulary for describing the activity of fellow human beings, it is tempting to use similar expressions to discuss animal behavior that seems very like pieces of human behavior. So there may grow up a largely unexamined collection of unsubstantiated hypotheses, latent in our linguistic usage, that allow us to pass freely from conclusions about nonhuman animals to conclusions about ourselves. And of this there may come much mischief.

Some critics have viewed anthropomorphism as the original sin of pop sociobiology. The sin lies in neglecting to investigate the kinship of forms of behavior that are superficially similar. We find in an animal species some pattern of premating behavior that reminds us of something that people do. We call both pieces of behavior by the same name. We then acquire evidence for the existence of a genetic predisposition to the behavior in our animal subjects. We announce a general result: there is a genetic basis for this type of behavior. Critics who worry about whether there is a genetic basis for the behavior in humans are silenced with the riposte that they are soft-headed victims of the old delusion that humans are not animals. There is a result about behavior that applies across the board, and the truly tough-minded will not falter at the application to ourselves. But the critics should protest. If the evidence for the grouping of human and nonhuman behavior together is just superficial similarity, then the thesis that the human and nonhuman behaviors are alike in having a genetic basis has never been seriously tested. The claim about humans is a piece of magic—a rabbit pulled out of a well-prepared hat.

Let us begin with a blatant example, the use of "rape" to cover behavior in scorpion flies, mallards, and humans (see Barash 1977, 67–69; Barash 1979, 54–55; Krebs and Davies 1981, 131, 153, 256). The celebrated instance is the case of mallards. Barash defends his use of the term "rape" in a footnote:

> Some people may bridle at the notion of rape in animals, but the term seems entirely appropriate when we examine what happens. Among ducks, for example, pairs typically form early in the breeding season, and the two mates engage in elaborate and predictable exchanges of behavior. When this rite finally culminates in mounting, both male and female are clearly in agreement. But sometimes strange males surprise a mated female and attempt to force an immediate copulation, without engaging in any of the normal courtship ritual and despite her obvious and vigorous protest. If that's not rape, it is certainly very much like it. (Barash 1979, 54)

The very people who might bridle at the notion of rape in animals might also wonder about how clear the agreement is and how vigorous the protest. Barash could easily defend himself on this score—females who are being "raped" often try to escape from the male or males who are forcing the copulation, and they sometimes suffer considerable injury in the process. But we ought to ask why this term is being used and what inferences will later be based on the usage.

More is at stake here than Humpty Dumpty's privilege of using words as he chooses or than a praiseworthy attempt to brighten dull scientific prose with colorful language. In his more scholarly account Barash is at some pains to avoid one obvious potential implication. Showing the occurrence of "rape" in nature and explaining how "rape" might maximize a male's fitness should not be viewed as endorsements of the behavior (Barash 1977, 68–69). The use of "rape" in connection with mallards and men is intended to suggest affinities between human and mallard behavior.

> Rape in humans is by no means as simple, influenced as it is by an extremely complex overlay of cultural attitudes. Nevertheless, mallard rape and bluebird adultery may have a degree of relevance to human behavior. Perhaps human rapists, in their own criminally misguided way, are doing the best they can to maximize their fitness. If so, they are not that different from the sexually excluded bachelor mallards. Another point: Whether they like to admit it or not, many human males are stimulated by the idea of rape. This does not make them rapists, but it does give them something else in common with mallards. And another point: During the India-Pakistan war over Bangladesh, many thousands of Hindu women were raped by Pakistani soldiers. A major problem that these women faced was rejection by husband and family. A cultural pattern, of course, but one coinciding clearly with biology. (Barash 1979, 55)

Barash's speculations about what men find stimulating are not fit for serious discussion. The crucial scientific issue is whether he has any basis for his conclusions. The central idea—though coyly expressed—seems to be that male mallards are selected to maximize their fitness by having some disposition to engage in "rape," a disposition that can be exercised if they lack mates or are able to surprise the mate of another duck. We are led to think that something similar goes on in human males. Allegedly we too have a fitness-maximizing genetic predisposition to engage in rape. In criminals this breaks out into actual behavior.

The entire story depends on Barash's evidence for thinking that a superficial similarity between the behavior of mallards and human rape warrants the attribution of a genetic predisposition in humans. There are two possible lines of reasoning. On the first, the forms of behavior would be regarded as produced by similar causal processes. Superficial similarity would be rooted in similarity of mechanism. On the second, the behaviors would be seen as beneficial both to the mallards and to the humans that exhibit them. Behind the surface kinship we would perceive a common adaptive advantage.

The first argument is a nonstarter. It is not mere pride that makes us think that the mechanisms of sexual behavior are very different in humans and in ducks (not to mention scorpion flies). The second is equally implausible. Even if we were to accept Barash's casual story about the ways in which mallard males maximize their fitness—a story about as well grounded as Wilson's anecdotes about the advantages of being hasty and fickle (or coy and passive)—it would obviously be premature to say "And the same goes for humans." Let us give Barash two controversial assumptions. Suppose that male mallards would maximize their fitness by "raping" females when opportunity presented. Suppose that we are entitled to conclude that there is a gene, or combination of genes, that gives a predisposition to engage in the behavior. We still have no basis for the crucial idea that human males—or, more pertinently, male hominids—would maximize *their* fitness, in the environment in which our tendencies to behavior evolved, by behaving in similar ways.

If we take that idea seriously, in its own terms, it is easy to see that we are woefully ignorant about all kinds of potentially important factors. At first sight it might appear that rapists have more babies than nonrapists. But a moment's thought reveals complications that might interfere with the spread of some hypothetical "gene for rape." If rapists are frequently attacked or punished, then the propensity to rape may operate against long-term reproductive success. The payoffs to rapists may be negligible if the raped women rarely conceive or if the offspring of the rape are usually killed, abandoned, or maltreated. We have little idea about whether these potential complications were operative in the primitive hominid environment. Barash's terminology is founded on the quite unjustified assumption that they were not.

We do know some things about rape in humans. It frequently takes place on juveniles, on women past the age of menopause, and on members of the same sex. Sometimes the victims die as a result of the rape. Actions of all these kinds contribute nothing to the spread of the rapist's genes. Of course, pop sociobiologists could contend that such behavior is a by-product of mechanisms that were selected under different conditions. The propensity to rape maximized fitness under hypothetical ancestral conditions. Now, under changed conditions, it prompts some people to actions that do not enhance fitness (and may radically detract from it). Such special pleading takes cover in our ignorance of the hominid social environment, in an effort to accommodate troublesome facts that are all too obvious.

We also know some things about so-called "rape" in waterfowl (many of those who study it seriously prefer to call it "forced copulation"; see McKinney, Derrickson, and Mineau 1983). Forced copulation is usually performed by males who are already paired with females (McKinney, Derrickson, and Mineau 1983, 283; this contradicts the expectations and assertions of Barash, as in 1977, 68). It usually occurs during the season at which eggs are being fertilized and is usually directed at females who are in reproductive condition. However, there is no firm evidence on the extent to which forced copulation issues in reproduction. Even with a wealth of careful observations, sociobiologists interested in patterns of copulation in waterfowl emphasize that they are not yet able to offer firm conclusions about fitness maximization and adaptive significance (McKinney, Derrickson, and Mineau 1983; for other arguments against the use of "rape" by fellow sociobiologists, see Estep and Bruce 1981 and Gowaty 1982).

Once we take a careful look at the behaviors Barash lumps together, superficial similarities are seen to mask important differences. Selective pressures on hominid behavior and mallard behavior are obviously very different: we need only recall the significant point that waterfowl have breeding seasons. There is simply no argument for thinking that enforcing a copulation would maximize fitness in waterfowl and in humans, but not in the host of other cases in which "rape" goes unrecorded. Insofar as Barash has any reasons for thinking that "rape" enhances fitness, they are reasons that would apply to almost any animal group.

The entire squalid story rests on mistakes about mallards and wild assumptions about humans and their past environments. But the use of anthropomorphic language serves not only to conceal logical lacunae. It also does harm by underwriting a pervasive stereotype about rape in humans, the idea that rape is a piece of *sexual* behavior. In this respect the speculations of other sociobiologists about "rape" in nonhumans and about the origins of human rape and rape laws are equally unfortunate. Krebs and Davies, who are usually sensitive to the methodological issues surrounding sociobiology, make an atypical lapse in this regard (1981, 256; they do, however, manage to resist the "True Confessions" style). Alexander and Noonan also share the idea of rape as a sexual strategy (Alexander and Noonan 1979, 449–450); and Alexander adds the idea that rape violates the sexual rights of a male with a proprietary interest (Alexander 1979, 242).

In recent years it has been suggested that rape is first and foremost a crime of violence. What is essential to the act is the infliction of pain and humiliation. The fact that copulation takes place is peripheral. (Of course, people can be raped without there being any contact with their genitals.) For some social theorists, the source of the crime is to be traced to societal attitudes toward women, to prevalent conceptions of the role of women and their value, to equally prevalent conceptions about male status and the ways in which this status is determined in relation to women. Such theorists hold two theses: (1) the behavioral mechanism that leads to rape is a disposition to violence, and (2) individuals acquire this mechanism because of features of our social environment. It is worth noting that even if (2) were false and (1) were true, even if the propensity that issues in rape were the product of evolution, this would pose a very different evolutionary question from that considered by Barash. The puzzle would be to explain why evolution had favored the disposition to violence.

Whether or not those who assimilate rape to crimes of violence are correct, their views cannot be dismissed a priori by linguistic legerdemain. Ironically, Barash's own discussion alludes to a type of case in which the conception of rape as an aggressive act is most plausible. The proverbial conquering soldier—whether he be subjugating Bangladesh or Belgium—is a threat to the conquered women (women of all ages). The rapes that occur in the wake of conquest are naturally seen as violent assertions of the subordinate status of those defeated, further ways of inflicting pain on the subjugated. The explanation—even the evolutionary explanation—of such violence is hardly to be undertaken by noting a minute increase in a soldier's number of achieved copulations.

What's in a name? Sometimes nothing. There are many occasions on which we can smile indulgently at the whims of the Humpty Dumpties. But in situations like this, where the entire "scientific" case is borne by question-begging terminology, the methodological rebuke is richly deserved.

There are numerous other instances. Barash (following Larry Wolf 1975) is prepared to talk of "prostitution" in tropical hummingbirds (1979, 78; 1977, 159–60); Dawkins and Wilson suggest that "coyness" is likely to be a trait of "the courted sex" (usually females) (Wilson 1975a, 320; Dawkins 1976, 161); and there are a host of similar anthropomorphisms, each with its associated ability to smooth a transition from claims about nonhumans to conclusions about human sociality. Yet these blatant examples do not exhaust the difficulties of the language of sociobiology. There are more subtle cases, in which the choice of a word allows the reader to draw unwarranted conclusions.

In Lumsden and Wilson's most recent book, pages of exposition are occasionally interrupted by self-conscious story telling. The authors hope to provide insight into "the social worlds" of primitive hominids by constructing tales about common incidents in the lives of our ancestors. These vignettes are not supposed to be pure fiction: in each case Lumsden and Wilson supply references to scientific studies from which they have derived their descriptions of the imagined behavior. Here is an instance of the technique.

> In several days the band will move on toward a winter rendezvous point. There they will join a friendly group, composed of familiar faces. Some of the adults will be recognized generically as kin. In the ancient hominoid manner there will be an exchange of young females. (Lumsden and Wilson 1983a, 100)

Our primitive ancestors are here envisioned as exchanging women like objects of barter. Readers are left to draw their own conclusions about possible genetic bases of certain contemporary attitudes toward women. But where does the idea that "exchange of females" is a primitive hominoid form of behavior come from? Lumsden and Wilson cite an article by Anne Pusey (1979) that describes patterns of "intercommunity transfer of chimpanzees." Pusey's main point is to document the fact that, unlike many other primates, chimps have a social system in which the juveniles who migrate from the home troop are the females, rather than the males. (In many primate species young males are driven out of their natal troops or simply migrate to another troop; see Hinde 1983, especially 309–311.) There is no suggestion of a system of *exchange*. Young female chimpanzees just leave their home troops, going off to seek mates elsewhere. So the use of the word "exchange," with its attendant implications for the evolutionary basis of current male-female relationships, is entirely unwarranted. If Lumsden and Wilson were seriously concerned to use primate studies to tell illuminating stories about our ancestors, they would talk about young females who occasionally wander off and are usually encountered later as mothers in another band. The implications of that narrative would be somewhat different.

Verbal tricks abound in pop sociobiology, serving as substitutes for argument or vehicles for misleading suggestion. Yet even if they were all expunged, there would still be an issue about anthropomorphism. When we look closely at some central concepts that are used in studying animal behavior, we discover important difficulties.

How Not to Talk about Sex and Power

Among the major theoretical notions of sociobiology and ethology there are at least two families that deserve especial scrutiny: those that relate to animal mating systems and those that attempt to encapsulate relations of power. The categories in the common taxonomy of mating systems receive their names from human systems of marriage. Numerous studies divide up species as "monogamous," "polygynous," "polyandrous," or "promiscuous," paving the way for subsequent speculation about whether humans are "monogamous," "promiscuous," "mildly polygynous," or whatever. What theoretical assumptions lie behind the usage?

Human beings share with other animals certain obvious forms of behavior: people eat, fight, and engage in sexual activity. In the context of human affairs these patterns of behavior are transformed. They occur in different situations, they fall into different relationships with one another and with the rest of our behavior. We do not have to be very imaginative to recognize the variety of ways in which copulation, sexual activity, reproduction, marriage, and parental care can be combined or separated. Contemporary technology and the contemporary social scene provide ample illustrations of all kinds of possibilities. Hence, when we look at animal mating systems and attempt to draw conclusions about what is "natural" for *Homo sapiens,* we should understand which aspects of human sexual behavior are our concern. Blind application of blanket concepts can easily lead to confusion. In the movie *Tom Jones* there is a memorable scene in which eating becomes a vehicle of sexual gratification. That scene serves as a vivid reminder that human eating behavior (or sexual behavior, or fighting behavior) may defy any attempts to fathom its significance by looking at what the animals do. (I am grateful to Richard Lewontin for suggesting the example.)

When terms like "monogamy" are applied to humans, they are usually employed to designate the social or legal arrangements found in a group. A society is said to be monogamous when there is some available arrangement (legal, economic, or ceremonial) that pairs males with females; at any given time the pairings have to be one-to-one. Legal monogamy can obviously coexist with sexual promiscuity. There have been many societies in which humans had a system of official pairings that was one-to-one (even permanently so, so that individuals once paired were paired for life) and in which the hefty majority of copulations took place between unpaired people (there are societies, such as the Dobu and the Abron, in which husband and wife live apart). In any case, despite the intricacy of some courtship ceremonies in the animal kingdom, animals do not exchange rings or vows, or warn one another not to split asunder those who have been joined together. So when we talk about animal monogamy (or polygyny, or whatever), we must be discussing sex and reproduction.

But we must tread carefully. When we are concerned with legal monogamy, we are focusing on the characteristics of a society, the existence within that society of certain rituals or enforced rules of conduct. When we begin to consider sexual relations, we have moved to consideration of the behavior of individual animals. Our taxonomy is intended to identify patterns of copulation among the individuals of a species. We want to know the typical behavior of members of the species, the way in which most of them join together to copulate. Or do we? Are we simply interested in patterns of reproduction, or in patterns of reproduction and parental care? What should we say about a species in which two individuals join to reproduce and cooperate in rearing young, but in which there is a large amount of copulation outside the pairing? What is the appropriate description of a species in which males mate with many females but only assist with the offspring of one? How—to take one of the complex cases that occur in nature—shall we describe a species (of baboon) in which a "dominant" male has been observed copulating with many females and doing little in the way of parental care, while his "subordinate" has been found to copulate primarily with one female and to assist with the offspring of several (Seyfarth 1978)?

The definitions usually offered do not resolve such queries. Krebs and Davies provide one of the better explanations of the terminology.

> *Monogamy.* A male and a female form a pair bond, either short or long term (part of or a whole breeding season or even a lifetime). Often both parents care for the eggs and young.
> *Polygyny.* A male mates with several females, while females each mate only with one male. A male may associate with several females at once (simultaneous polygyny) or in succession (successive polygyny). With polygyny it is usually the female that provides the parental care.
> *Polyandry.* This is exactly the reverse of polygyny. A female either associates with several males at once (simultaneous polyandry) or in succession (successive polyandry). In this case, it is the male who does most parental care.
> *Promiscuity.* Both male and female mate several times with different individuals, so there is a mixture of polygyny and polyandry. Either sex may care for the eggs or young. *Polygamy* is often used as a general term for when an individual of either sex has more than one mate. (1981, 135)

These are the definitions. Now let us perform an exercise to test our understanding of them. We plan to fathom human sexuality by identifying the sexual system that typifies our animal relatives, and we choose chimpanzees as the model for humans in their natural state. The question, then, is this: are chimps promiscuous, polygynous, polyandrous, or monogamous?

The problems in answering our question are not purely empirical. The concepts are so loose that almost any answer could be justified by an appeal to our knowledge of chimpanzees. Given the frequency of changes in sexual partners, it would strain credulity to count chimps as monogamous—the parts of the breeding season for which pairings are stable are very short indeed! But we could make a case for applying each of the other concepts to the chimpanzees.

Early investigators labeled chimps as promiscuous (see, for example, Wilson's summary, 1975a, 546). They did so on the basis of observing that a chimpanzee female in estrus often copulates with many males. Males endeavor to copulate with each female who comes into estrus and are only rarely unsuccessful, so that chimpanzee society seems to be a union of all with all (to invert Hobbes's famous phrase). To justify the application of the definition, we need only interpret "mates with" as "copulates with." But we can also view the system as one in which males engage in successive polyandry, while females practice simultaneous polyandry. From the male's perspective, he is successively a member of polyandrous associations that are monopolized by a female. The male belongs to a sizable number of such associations during his reproductive life, with the frequency determined by the rate at which females in his home range come into estrus. (Occasionally he may belong to two polyandrous associations at once, if two females come into estrus at the same time, but these occasions will be viewed as atypical.)

Quite a different approach is possible. Chimpanzees, we may claim, are really polygynous. Even though all males in a troop may copulate with an estrous female, there is some reason to believe that the timing of the copulations is not so democratic. Some males—*perhaps* those who are considered "dominant," perhaps those who are most attractive to females—may succeed in copulating with estrous females close to the time of actual ovulation, and these males may sire a disproportionate number of the offspring. So if we interpret "mates with" as "reproduces with," there may be grounds for saying that chimps are polygynous. There is dalliance aplenty, but the copulations that really count involve a small subset of the males in the troop; and we may choose to assign these individuals a number of female "mates" and their less fortunate fellows none at all.

There are two sources of difficulty in the definitions. One is the fact that decisions to talk of polygamy or serial monogamy or promiscuity can be justified by choosing a particular time scale. Anatomy demands that for almost any given species we can find a period, however short, during which the sexual pairings are one-to-one. For many species, by extending the time period we can find pairings that are one-many or many-many. Partitioning the lives of the animals under study in carefully selected ways, we can fit whatever terms we choose to the phenomena. Sometimes the partitions will seem strained and unnatural. But that is by no means always the case. The second problem derives from the ambiguity of "mates with." When reproductive pairings do not reflect the distribution of copulations, the interpretations of "mates" as "copulates" and as "reproduces" give different results.

There are further problems that are not exposed by our simple exercise. Mating systems may also be regarded as systems for rearing young; and in species that provide substantial amounts of parental care, it is often suggested that certain parental roles go hand in hand with certain reproductive roles. But there are many cases in which sexual polygamy may accompany parental monogamy. There seem to be species of birds in which paired males (or females) sometimes copulate with or reproduce with outsiders. Snow geese are a probable case in point (McKinney, personal communication).

At the beginning of his masterly account of social insects, Wilson is at considerable pains to distinguish various grades of sociality and to identify various combinations of features that may accompany one another (1971, 4–6). Ironically, despite the fact that sexual and parental systems, conceived generally for all species of animals, appear more complicated, some sociobiologists and ethologists have remained content to apply labels that are both vague and ambiguous. It would be possible to resolve some of the problems we have uncovered by proceeding as Wilson does in the case of insect sociality —by distinguishing patterns of copulation, reproduction, and cooperative parental care. (One might also want to allow for cooperation in mutual defense or in foraging.) But even a refined version of the notions of monogamy, polygamy, and so forth would still leave us with one last problem.

When we categorize the mating pattern that typifies a species, should we consider the actual mating behavior of members of the species under their natural conditions? The answer seems obvious: Yes. But if our interest lies in understanding the dispositions of individual humans for various forms of sexual behavior, it is not clear that the mating patterns of related species are relevant. Descriptive anthropology provides us with an account of the ways in which social and ecological conditions affect human sexual relations. If we are interested in understanding the extent to which variant sexual arrangements are available to us—or, perhaps, available to us without "significant costs"—then what we should apparently be concerned with are the sexual predilections of individual animals. However, what the study of mating systems reveals is the outcome of a process in which ecological constraints act on the sexual predispositions of members of a species to produce a typical mating behavior. Hence, one might suggest, what we really ought to focus on are the mating patterns that would be exhibited under ideal conditions, conditions in which the ecological constraints that interfere with the free exercise of dispositions to sexual behavior are removed.

A story about William James (probably apocryphal) makes the point. James is supposed to have dreamed, on a number of successive nights, that the secret of life had been revealed to him. The opportunity was too good to miss, and James resolved to keep a notebook and a pencil by his pillow. The dream recurred; he awoke, scribbled down the message he had been vouchsafed, and fell back asleep. In the morning he was slightly disappointed to find the couplet

> Higamus hogamus, woman is monogamous;
> Hogamus higamus, man is polygamous.

Some people seem to believe that the secret of human sexuality, if not the secret of life, resides in this couplet. (See, for example, Wilson 1975a, 554; Wilson 1978, 125; Barash 1979, 64.) Yet it would be fallacious for such people to defend their claims by appealing to the existence of a polygynous mating system in some group of primate species. The transfer of a group characteristic (hamadryas baboons are polygynous) to individuals (hamadryas baboons are polygynous) passes almost unnoticed. But it should be scrutinized. The behavior of a group of animals will not necessarily reflect the predispositions of any of the members of the group.

Let us consider the gibbons. The gibbon species are usually classified as monogamous—and with good reason. A male and a female gibbon typically mate for life. They cooperate in rearing offspring and in defending the territory that they occupy. However, neither males nor females are averse to copulations with passing strangers. As a group, gibbons are monogamous. But what about individual gibbons? It seems possible that male and female gibbons do not differ individually, in their dispositions to copulate, from other great apes. They simply lack the opportunities. One plausible account of their mating system is that it results from the distribution of food, which forces females to disperse and thus prevents males from trying to monopolize a cluster of females. Scarcity is the mother of fidelity.

What we learn when we look at primate mating systems is what compromises our relatives make when ecological conditions interfere with their sexual proclivities. To draw conclusions about ourselves would be unjustified. For if we take a mating system to be a property of a group of individuals, our premise is a statement about the ecologically constrained behavior of typical individuals; since humans are under rather different ecological constraints (to say the least), the significance of this premise is unclear. If we take terms like "monogamous" to denote predispositions of individual animals, then we seem to come closer to a justification for a useful conclusion; if we knew the dispositions of our near relatives, that might help us to draw conclusions about our own dispositions and thus to understand what systems of sexual relations are possible for us. But we cannot reason in any obvious way from the mating systems of groups to the dispositions of individuals. Claims about the predilections of individual hamadryas females do not follow from premises about hamadryas social organization. So the idea that we can read the monogamous tendencies of human females from the mating systems of favored primate species (always excluding those promiscuous chimpanzees, of course) is simply a gross mistake. We might as well rely on the deliverances of dreams.

What, then, does the theoretical concept of monogamy do for us? It covers the various ways in which animal species have arrived at a certain type of mating arrangement, as a result of the interaction between the sexual predilections of individuals and the ecological constraints on the species (the latter including the history of social arrangements). Seen in this way, monogamous species are a motley collection, who have achieved a common end state in response to very diverse ecological conditions. The individual sexual predilections of members of monogamous species may have no more in common with one another than they have with those of animals belonging to species with radically different mating systems.

The conclusion I have reached is an articulation of a point made by S. L. Washburn. After noting that even in humans monogamy is not a unitary behavior that might have a genetic basis, Washburn continues, "The use of such words by sociobiologists shows a total misunderstanding of social science. Even ape behavior is far too complicated to be analyzed by labels and guessing" (1980, 261). Washburn's remark is best read as a charge against pop sociobiology. Wilson and his followers speculate about whether polygyny is our "natural condition." Others note the difficulties of the terms for mating systems and attempt to caution their fellows. (There are recent studies that draw attention to the variety of ways in which a species may arrive at "monogamy" [Wittenberger and Tilson 1980] and that stress the possibility of intraspecific variation in mating systems [Lott 1984]).

Analogous points apply to another favorite concept of sociobiology and ethology, the concept of dominance. Dominant animals are those who are able to displace other animals in their groups from valuable resources. More exactly, the primitive concept of dominance is relative. For any two animals Dom and Sub, Dom is dominant with respect to Sub if Dom can displace Sub from valuable resources. But this is vague. Should we require that Dom can displace Sub from all resources? On every occasion? Attempts to remedy the vagueness disclose problems with the concept of dominance.

Kummer introduces the notion of dominance in the standard fashion: "The term 'dominance' is widely used to describe a particular type of order in organized groups. Its most general criterion is the fact that an animal consistently and without resistance abandons his place when approached by a more dominant group member, a sequence called 'supplanting' " (1971, 58). The crucial ideas behind the concept of dominance are those of consistent displacement and the avoidance of violence. Dominance was initially regarded as an important device for promoting group well-being: to avoid damaging and time-wasting conflicts, animals in a group quickly arrange themselves in a dominance hierarchy (or "pecking order"), and the relations of dominance are used to settle who gets the good things that are going. But once we abandon the idea of a group benefit and of associated group selection, it should be clear that "supplanting" behavior is likely to be complicated. Assume, for the sake of argument, that such behavior has indeed evolved so that the present behavioral strategies are evolutionarily stable. Then the notion that an animal will always behave in the same way toward another in any interaction with respect to a contested resource is highly implausible. We might expect that Sub's disposition to go quietly would vary from context to context, depending on the value of the resource, the estimated determination of the opponent, the estimated probabilities of winning an escalated contest, and so forth. Sub might be expected to give way to Dom when the resource is easily replaced and to put up a fight when the resource is not easily replaced. And to recognize this minimal context dependence is only the beginning of the complications.

There are thus theoretical reasons for believing that the notion of dominance may be a poor explanatory concept, that the simple appeal to dominance relations may fail to account for why animal contests go the way they do. Once we recognize the real possibility that dominance relations will sometimes prove effective and sometimes not, we shall not be satisfied to be told that an animal displaced another from a sleeping position because the former was dominant. We shall also want to understand why this was one of those cases in which so-called dominance relations were effective.

The theoretical reasons are backed up by empirical findings that indicate variability in the outcomes of contests. Within a page of giving the definition of dominance, Kummer remarks,

> . . . baboons sometimes supplant low-ranking group members from grass plants that have already been dug up by a subtle approach-avoidance sequence without any threatening gestures. This advantage is usually of little importance, since primate vegetable food occurs mostly in small bits scattered over an area that can accommodate all members of the foraging party. With large items of food, however, dominance becomes decisive. The young antelopes that baboons sometimes kill are almost exclusively eaten by the adult males, and fighting over such prey is frequent. (1971, 59)

Dominance is effective when the resource is easily replaced. A baboon is well advised to look for roots elsewhere if it is approached by another who would probably beat it in any fight that might ensue. But the outbreaks of fighting at antelope kills reveal that dominance is *not* decisive in these cases. With respect to a valuable resource, it is sometimes worth a subordinate's while to put up a fight. *If* the subordinate loses, then the loss cannot be ascribed to the effect of dominance. The victor triumphed, not in virtue of its being dominant, but because of the qualities that probably made it dominant in the first place. But there is no reason to think that subordinates do always lose—unless, of course, one makes the mistake of thinking that victors are, ipso facto, dominant. To conceive of dominance in this way is to define dominance so that, for any given time, *A* is dominant over *B* only if *A* either wins a contest with *B* that occurs at that time or has won the most recent contest with *B*. Adopting this definition, we can always be sure that the dominant animals win. However, we can hardly explain their winning by citing the fact of their dominance—for such an "explanation" would suggest that the victorious animal won because . . . it was victorious.

The problems with the concept of dominance multiply. We should not only raise the question whether all contests in all contexts will exhibit the same patterns of victors and vanquished, but also ask if dominant animals will prove more reproductively successful, if they will lead their groups, and so forth. The connotations of the ordinary usage of "dominant" in connection with humans are hard to avoid. However, the connections among leadership, victory in contests, and reproductive success often break down. Kummer's observations of hamadryas baboons provide a beautiful example. He reports the attempt of a young male ("Circum") to determine the direction in which a band consisting of himself, an older male ("Pater"), and "their" females would move to seek food. Pater's plans for the trip ran counter to those of Circum, and Pater ultimately decided the direction in which the band went. A victory for age, experience, and dominance? Perhaps. But if we think of hamadryas male dominance as measured by the number of females that a male can monopolize, then the situation changes. Pater "owned" a single female, while Circum "owned" several. Ability to lead does not coincide with ability to monopolize females.

Hamadryas baboons, sage grouse, red deer, and elephant seals all encourage the picture of females as a valued resource, over which males fight. In such cases we would expect dominance to correlate with reproductive success. Although many sociobiologists know better, much pop sociobiological writing proceeds on the assumption that the correlation holds generally—dominant males are supposed to be those that win the battles and reproduce most prolifically (and, perhaps, act as leaders of their groups). Consider the following quick explanation by Trivers: "A second reason why choosing to mate with more dominant males may be adaptive is that the female allies her genes with those of a male who, by his ability to dominate other males, has demonstrated his reproductive capacity" (1972, 88; see also Dawkins 1976, 170). If the idea is that the female's male offspring will be likely to inherit the ability to dominate other males and *therefore* to prove reproductively successful, then Trivers is simply assuming that dominance (measured by victory over other males) correlates with reproductive success. The assumption seems to hold for cervids—deer, sheep, and so forth—to which Trivers immediately refers. (See Geist 1971; Clutton-Brock, Guinness, and Albon 1982, 152–156.) But it will only be reasonable for species in which certain kinds of female sexual strategies are excluded. Of course, one of the leading themes in pop sociobiology has been the notion of the passive female, bereft of any serious sexual strategy of her own. Some recent studies, for example that of Clutton-Brock and his associates, are sensitive to the point; despite the fact that they are concerned with groups in which females seem relatively passive, they explore possible female strategies.

There is a mounting body of evidence that male dominance does not always correlate with reproductive success. Wilson notes the breakdown (or *apparent* breakdown) of the correlation in chimpanzees: "The great majority of hostile acts involve adult males. Yet curiously in view of this fact, the dominance system appears to have no influence on access to females" (1975a, 546; see also 531 on lemurs). Kummer also notes some deviations from the dominance system in the formation of consortships in olive baboons (1971, 93—but see also Packer 1977 for a more detailed analysis). Barash takes a stab at explaining these departures from the natural order of things. "Depending on the species and the circumstances, this may simply reflect the fact that dominant animals cannot be totally dominant all the time" (1977, 241). Of course, this comes perilously close to the trivialization of the concept of dominance that we envisaged above: the temporarily dominant are to be the temporarily victorious. Perhaps Barash senses that he is headed for vacuity, for he follows up on this explanation with a brief suggestion about noblesse oblige—subordinates "garner crumbs" from the dominant's table. (Whether there are genuine benefits to the subordinates or genuine costs to the dominants is never specified.)

Many students of animal behavior are aware of problems with the concept of dominance. Hrdy writes, "Typically, dominance is difficult to assess and highly dependent on context; furthermore, dominance is not necessarily related among different spheres of activity" (1981, 3). And, in the most sensitive defense of the (restricted) use of the concept of dominance that I know, Clutton-Brock and Harvey suggest that game-theoretic considerations are useful in accounting for the problems involved in a generalized notion of dominance (1979b, 301–303; see also Clutton-Brock, Guinness, and Albon 1982, 216, for a careful demonstration that dominance rank among red-deer hinds does not correlate with reproductive success). The moral of empirical studies and of game-theoretic analyses of competition, simple though they may be, is that the greater the variety of conflicts among individuals and the greater the variation in power, in evaluation of the resource, and in ability to assess the ability of others, the more complicated will be the outcome of animal interactions. Sensitive researchers respond to the problem by suggesting that "dominant" be given a precise, contextually localized, operational specification. We are to evaluate dominance relations solely in the context of the behavior of stags in the rut or of grooming relations in a primate species. It is not to be assumed that there is some single characteristic that stands behind the various uses.

Conceptual progress in sociobiology thus undermines the casual references of pop sociobiology. Wilson's blanket claims about "aggressive dominance systems" in humans (and in primates generally) presuppose notions that have to be refined beyond recognition once we take a serious look at the interactions of complicated animals. (See Wilson 1975a, 551, 567, for representative remarks.) The predicament is familiar. Many scientists are now convinced that there is no single measure of intellectual ability—no unitary intelligence. Their suspicion of the concept of intelligence is based on the view that various intellectual capacities are not well correlated. Psychology now faces the serious task of mapping the various cognitive capacities and their interrelations. Because the search for intelligence has been predicated on the idea of a unitary ability and because the search has been deemed important in the construction of social policy, it is useful to continue to expose the myth of "general intelligence" (see, for example, Block and Dworkin 1976b, Gould 1981). Similarly, while careful sociobiologists begin to map the ways in which animal conflicts are resolved, it is well to explode the myth of "general dominance." Some leading themes of pop sociobiology vanish in the process.

For Hard Heads Only

Pop sociobiologists pride themselves on their intellectual daring. Unlike those who try to shackle the scientific study of human behavior, they are not prepared to admit that such study is impossible (or, perhaps, even difficult) or to withdraw from the unromantic conclusions to which their inquiries might lead them. They defend their anthropomorphic language by accusing their critics of attempting to set human beings apart from nature. Similarly, as virtuous reductionists, they delight in portraying their opponents as wallowing in a slough of mushy incomprehensibility. To deny that biology is the key to understanding human behavior and human society is allegedly to take refuge in a nonphysical "mind" or "will" or to appeal to a fictitious entity, "culture," with peculiar and ill-defined characteristics. (See, for example, Lumsden and Wilson 1983a, 172–173.) Hardheaded scientists—genuine scientists, that is—will have none of these consoling myths. They see that nothing stands in the way of extrapolating from animals to humans. Nothing, that is, except clouds.

There are varieties of reductionism and varieties of antireductionism. In neither instance should all versions of a doctrine be charged with the sins of a particular sect. We should start by understanding where pop sociobiologists and their opponents agree, and where they differ.

Physicalism is the thesis that all things, processes, states, and events are ultimately physical things, processes, states, and events. When a cell divides, there is a very intricate rearrangement of molecules, not the action of some "vital force" or the intrusion of some "vital substance." When somebody writes a letter to a friend, falls in love, or thinks of a new melody, there are processes involving the firings of vast numbers of neurons. There is no mysterious nonphysical "mind" that hovers behind the goings on in the brain, that really thinks the thoughts, feels the emotions, and serves as the center of creation. When England expects that every man this day will do his duty, there is a complex aggregate of attitudes of English people toward those engaged in battle. The ghost of Britannia does not hover over the scene, secretly surveying the actions of every British sailor.

Physicalism is true. No antireductionist should deny it. Sensible antireductionists have no truck with the discredited entities postulated by discredited sciences. They repudiate the *élan vital*, the ghost in the machine, and mysterious entities called "Cultures." Nevertheless, they may accuse pop sociobiologists of reductionist errors (Lewontin, Rose, and Kamin 1984, 9–10). Is this simply confusion?

Reductionist proposals take the form of claims that the features of certain sorts of things, *X*'s, can be explained by reference to the properties of certain other sorts of things, *Y*'s. The merits of the proposals vary with the identifications of the *X*'s and the *Y*'s. Reductionism may be very tempting when there is universal agreement that the *X*'s are made up of nothing but *Y*'s. A moment's thought dispels the temptation. Zygotes are nothing but complicated combinations of molecules. We cannot, however, explain the development of a zygote solely by reference to the properties and arrangement of the molecules of which it is composed. How the zygote develops depends on the surrounding environment. Similarly, human societies are nothing but collections of people; yet it would be folly to infer that the characteristics of the society can be understood simply by focusing on the attitudes of the people who compose it. The past may be crucially important to the presence of various social institutions.

Critics assert that pop sociobiologists slide from the innocuous truth that human societies are composed of animals belonging to a species with an evolutionary history to the controversial thesis that human social behavior and human social institutions are to be explained by appeal to evolutionary biology alone. Two issues arise here. The first concerns a variant on mistakes that we have seen earlier. One vulgar form of reductionism is the assumption that social characteristics can be identified in terms of the psychological traits of the majority. To claim that a society has a propensity for male dominance is just to say that most individuals within it have the propensity. To view a nation as aggressive is merely to see it as a collection of bellicose individuals. Vulgar reductionists shuttle back and forth across levels, using the convenient fact that we sometimes use the same terms for attitudes of groups and attitudes of individuals.

The second form of the antireductionist charge is that pop sociobiology overlooks certain kinds of explanations. Suppose we are satisfied that a certain form of human behavior maximizes the inclusive fitness of those who exhibit it. Suppose further that there is no doubt that this form of behavior is prevalent in *Homo sapiens*. Are we entitled to conclude that there has been a history of selection in which genes promoting the behavior have replaced hypothetical ancestral genes? Or is it possible to claim, without invoking some mysterious cultural force, that the behavior has become prevalent because of the history of human culture? One sort of antireductionism hails such possibilities as genuine and chides pop sociobiology for neglecting them. I shall consider the two antireductionist complaints in order.

There are ample signs that pop sociobiologists like the convenient strategy of the vulgar. Consider Wilson's discussion of the mating propensities of primates. "There is a tendency for males to be polygynous and aggressive toward one another, although pair bonding and pacific associations are permissible minority strategies" (1975a, 515). The conclusion is based on what we know about *group* mating systems in primates. Similarly, Wilson uses an analysis of the behavior of tribes and nations to conclude that "humans are strongly predisposed to respond with unreasoning hatred to external threats and to escalate their hostility sufficiently to overwhelm the source of the threat by a wide margin of safety" (1978, 119). Whether or not Wilson's pessimistic reading of the ethnographic record is correct, the argument involves a totally unsubstantiated assumption. We can only view intertribal warfare as evidence for innate human aggression by supposing that the behavior of the group reflects directly the propensities of the participants. If warfare is seen in the context of an intricate arrangement of social institutions, traditions, and present needs, then there is no reason to think of intertribal battles as signaling our native thirst for violence against strangers. Perhaps it takes coercion and intimidation by rulers and authorities to send people off to war.

Sophisticated reductionists as well as antireductionists will scorn the notion that a nation is aggressive because it is composed of individually aggressive people, just as they will scorn the idea that an animal is intelligent because its brain is made of intelligent cells. This issue need occupy us no further. The remaining area of debate is more subtle.

Imagine that we have arrived at the second stage of Wilson's ladder. We have discovered a case in which people exhibit a form of behavior that maximizes their inclusive fitness. Is it possible to understand the situation by supposing that there is no genetic difference between those who exhibit the fitness-maximizing behavior and hypothetical ancestors who did not? Our discussion of the behavior of the Florida scrub jay indicates the possibility. We envisaged a possible scenario in which helping behavior spreads through imitation. In effect, the idea was that birds recreate the nest environment of their own early experiences; and, because nests with helpers send out more young into the scrub, nests with helpers come to prevail through the available territory.

We can easily generalize the idea. Suppose that common combinations of alleles in a species are expressed in one way in one family of environments and are expressed in a different way in another family of environments. If the members of the species who inhabit the second family of environments leave more descendants than those who inhabit environments in the first family, and if there is a tendency for offspring to rear their young in environments similar to those in which they were reared, then, without any genetic change, the environments of the second family may come to predominate and the behavior associated with these environments may become prevalent. What is crucial to this type of explanation is that there should be variation in behavior dependent on environment and that there should be some form of cultural transmission that correlates the environment in which parents rear their young with the environment in which the young rear their own offspring.

Let us add a further wrinkle. Suppose initially that the species is divided into populations, some of whom inhabit one type of environment, others of whom inhabit a different type of environment. It is entirely possible that a population should introduce a social arrangement that is disastrous from the point of view of members of that population and that creates strong pressures on the other populations to modify their own social structures. The result may tip the balance in favor of one of the types of environments originally present, so that the behavior associated with it becomes prevalent in the species.

More concretely, we can envisage the possibility that the prevalence of intertribal hostilities is to be explained not through supposing that there has been selection of "genes for aggression against strangers" but through seeing the environments in which our genes are expressed as shaped by a process in which occasional hostile groups force other populations to develop social institutions that promote violence toward outsiders. Plainly, it is not a trivial task to work out the details of any such scenario. My point here is to draw attention to the possibility, and to note that it appears to be overlooked in the enthusiastic rush up Wilson's ladder. As critics of the early Wilson program have noted (Sahlins 1976, and especially Bock 1980), it is possible to appreciate the importance of historical events and of cultural transmission without supposing that Clio herself needs to be added to the biologists' inventory of evolutionary forces.

To underscore the point, let me explore a possible alternative to one of the pop sociobiologists' favorite stories. Human incest avoidance is supposed to be genetically based. Ignoring some complications, the idea seems to be that selection has favored genes that dispose people not to copulate with those with whom they have been reared. The offspring of close relatives who mate and reproduce have a much higher incidence of physical and behavioral abnormalities, so that we may expect that selection will operate against those who have a disposition to copulate with relatives and in favor of those who have an opposite propensity.

There is a different story. Suppose that human sexual behavior is extremely flexible, that its adult expression is highly dependent on features of the environment in which the child develops. Assume further that children tend to rear their own children in environments similar to those in which they were reared. It is now possible to argue that incest avoidance will spread in human populations even if there are no genetic differences among individuals with respect to incest avoidance, provided that initially there are some environments in which incest is tolerated and some in which incest avoidance is promoted. Children who are reared in the latter will, by hypothesis, leave more offspring who attain maturity, so that the environment promoting incest avoidance will become more prevalent; derivatively, incest avoidance itself will become more widespread. The case is exactly analogous to that of the scrubjays. (Elliott Sober has independently offered a similar scenario to make a similar point about sociobiological explanation; see Sober 1985.)

The moral to be drawn from these speculations is that we must be very careful in supposing that the prevalence of a behavioral trait together with a convincing argument that the trait maximizes inclusive fitness entitle us to conclude that the trait has a particular kind of evolutionary history. Appeals to the importance of culture or history cannot just be dismissed as muddle-headed mush. We have seen, in outline, how it is possible to make sense of them.

Canons of Convenience

Let us scramble higher, to the final stages of Wilson's ladder. Imagine that an analysis of some piece of human social behavior begins with unprecedented success. The devotees manage to show that there is a tendency for the behavior, found across a group of animals including human beings; that the tendency can be expected to maximize inclusive fitness (given the appropriate environments); and that there are good reasons for supposing the existence of genetic differences between those people who manifest the tendency and their ancestors who did not. Does it follow that the human tendency to exhibit the behavior cannot be modified by altering the environment in which humans develop and in which they interact with one another?

Surely not. Finding out how human genes combine with one range of environments to issue in a behavioral phenotype does not yield a conclusion about the fixity of the behavior in all environments, unless we suppose that the collection of possible human environments is fully represented in the range already studied. The gap must be bridged in one of the ways canvassed in chapter 4. Pop sociobiologists who want to reach conclusions about human nature must emphasize the efficiency of evolution or the idea that the behavioral tendency is an adaptation that is unlikely to be affected by our cognitive state and thus unlikely to respond to changes in the social environment. Or they can buttress their case with an appeal to the diversity of the human societies in which the behavior has developed. As we scrutinize these arguments, we shall find that we return to some of the issues canvassed in the last section.

Consider first the appeal to the efficiency of evolution. We start from the premise that the tendency we have identified maximizes inclusive fitness in a range of animal environments. If these environments are exactly those encountered by members of the group of animals under study, then, it may be suggested, the simplest way for selection to operate is to favor those animals who have the tendency to exhibit the behavior, irrespective of the environment in which they develop. For what could be simpler than to associate with a gene a behavioral rule prescribing that the animal always have a tendency to exhibit the behavior? If baring teeth at strangers is fitness maximizing for primates in environments in which the animals forage in small groups, then would it not be most efficient for evolution to select primate genes that cause me to bare my teeth at strangers, whatever my developmental environment?

Not necessarily. Natural selection may scrutinize the behavior of organisms in a collection of different environments, but if those are the environments actually encountered, then environments outside the collection are of no concern. What is required is a mechanism for succeeding in the actual environments; and animals who have tendencies that would break down under radically different circumstances will not be exposed as inferior until those circumstances actually arise. Moreover, there is no saying a priori, without consideration of the developmental biology of the animals in question, which solution to an evolutionary "problem" is the simplest. To appeal to the idea that it is always "simpler" for an animal to behave in the same way, independently of its developmental environment, is to invent a canon of convenience.

Next let us examine the second tactic for negotiating the top step of Wilson's ladder. Some behavioral tendencies are old adaptations, fixed in animals with rudimentary cognitive abilities and inherited by us. Why should we view these tendencies as modifiable by alterations of the environment that act through our burgeoning mentality? Change the ideals of a society and its educational system as you like, you will still fail to affect the primeval centers in which the roots of male aggression, female submissiveness, distrust of strangers, and love of familiar places are found.

The obvious answer is that this is a completely unsubstantiated story. There is no reason to think that as humans evolved the ability to make finer discriminations of the objects and situations around them, our older tendencies to behavior remained unaffected. On the contrary, the evolution of enhanced cognitive capacities would seem to call for an ability of the more perceptive mind to interfere with mechanisms of behavior already favored by selection and thus to use the increased cognitive abilities in helping us to cope better with our environments. This is more speculation than sound argument. We simply do not know enough about the springs of our behavior to be able to define the extent to which social and cultural factors might affect these hypothetical primeval tendencies. If we did, the very issues that pop sociobiologists congratulate themselves on addressing would already be resolved.

A homely analogy may prove helpful. Beguiled by advertisements, a family buys a home computer in the hope that it will help the children in their schoolwork. Once in the household, the computer is put to work in unanticipated ways. It takes over the functions of the typewriter and the family filing system. Index cards yellow and molder. They are not replaced. The typewriter falls apart and is consigned to the garbage. The marketer's dream family ends up doing all the old jobs in new ways. Similarly, in the course of the evolution of the human brain, tasks once performed by other mechanisms may be taken over by the cognitive system, leaving the older mechanisms to wither and disappear. If this is so, then our ways of acting so as to maximize our fitness (avoiding incest, reacting to strangers) may be very different from those that are present in our nonhuman relatives. Our behavior may be cognitively penetrable, even though it seems to achieve the same ends as behavior in animals with only rudimentary powers of cognition. To assume that selection never fashions new means to old adaptive ends would simply be to construct another canon of convenience.

The third tactic, the device of appealing to the constancy of human behavior under diverse cultural conditions, leads us back to the issue of reductionism. The natural response to the recitation of the ethnographic record is to claim that the societies actually developed do not come close to exhausting the range of possibilities. If we find universal male domination, universal aggressiveness, universal institution of private property—and I shall leave to the anthropologists the task of deciding whether this is indeed what we find—then that only reveals that the societies we have achieved may share some common condition, which we could change and thereby liberate ourselves from the social behavior alleged to be our lot. Pop sociobiologists may try to cast a shadow on this optimistic picture. Followers of Wilson will argue that the common condition itself reflects the demands of our biology, so that we have no genuine option of removing it from the social sphere. Because our social arrangements express our biological heritage, we can rid ourselves of institutions that many people now regard as unjust only by incurring "costs that no one can yet measure."

The form of the fundamental dispute is obvious. Those who resist the pop sociobiological conclusions argue that the alleged features of human nature would no longer be manifested if people developed and lived in different social environments. The reply is that achieving the alternative social environments would require us to eliminate characteristics of our social situation *whose current presence is to be explained in terms of our biological properties.* Ultimately, we confront a series of reductionist proposals.

Two examples will illustrate both the character of the debate and the pop sociobiological eagerness to foreclose crucial issues. On the way to his conclusions about sex-based differences in behavior, Wilson considers differences in smiling between young boys and girls. Assuming that it is legitimate to omit any investigation of variation in the social environment, he remarks, "Several independent studies have shown that newborn females respond more frequently than males with eyes-closed, reflexive smiling. The habit is soon replaced by deliberate communicative smiling that persists into the second year of life" (1978, 129). So we are launched on a story about behavioral differences between the sexes that can be eradicated only by fighting our nature.

Is there an alternative explanation for the variation in smiling behavior? Of course. Even very young boys and girls may behave differently because of differences in the ways people care for them. (See, for example, Money and Ehrhardt 1972, 12, 119.) Wilson's neglect of the obvious point is revealing. It may result from any of three reductionist views.

First version: The caring behavior of parents makes no difference to the young infant; all that is required is that the child be fed, kept warm, and so forth. According to this version, the smiling reflex is simply the result of the interaction between the child's genotype and physical factors in the environment. Boys and girls are exposed to common physical factors. Hence the difference in timing of smiling reflects a genetic difference in propensities for behavior.

Second version: The caring behavior of parents might make a difference to the behavior of the young infant, but the fact that parents respond differently to male and female children should itself be seen as indicating sex differences. We naturally treat girls more gently, and we do so because this behavior has proved adaptive in our evolutionary past. So, although male and female infants are responding to slightly different environments, the differences in the environments signal a psychological difference that is ultimately to be traced to our genes.

Third version: The caring behavior of parents might make a difference to the behavior of young infants, and this behavior might itself be the expression of the socialization of the parents. Perhaps the parents now treat male and female infants differently because they have grown up in a society that encourages differentiation of boys and girls. However, the fact that we live in such a society cannot be treated as an accident. It is the result of our basic biology. The evolution of *Homo* fixed in us genes that predispose us to develop social institutions of particular kinds, and specifically to give different treatment to boys and girls. To attempt to neutralize these differences would require us to create a social environment within which many of our adaptations would no longer fit.

We have no evidence for the details of any of the three. In each case the attempt to explain the status quo as an expression of our biological nature jumps to conclusions about issues that deserve careful empirical investigation. Nor can pop sociobiologists propose that some explanation of this general type *must* be correct. As we saw in the last section, there are possibilities for giving cultural or historical explanations of our current practices and institutions without personifying either History or Culture.

There is no reason to think that, by multiplying versions, we shall eventually reach a stage at which we confront some pure expression of our biological nature. Antireductionists should temper pop sociobiological enthusiasm by offering some straightforward claims. We do treat boys and girls differently. Our treatment of them makes a difference to their development. Our propensity to make gender distinctions reflects the social environments in which we ourselves were reared. Those social environments are responses to prior social arrangements, developed ultimately from the culture begun by our hominid ancestors. Our social institutions appear to perpetuate themselves. That does not imply that we are incapable of changing them, of inaugurating a new tradition in which we achieve equal treatment of the sexes without finding ourselves at odds with our adaptations. We need only remind ourselves of the possibilities for cultural explanations reviewed in the last section.

A more blatant reductionism is found in the emphasis that some pop sociobiologists place on the failure of the "kibbutz experiment." Here are the facts of the case. After an initial period in which women in the kibbutz abandoned "traditional roles" in favor of other forms of work, there has been a reversion to significant maternal care. Second-generation women of the kibbutzim were apparently unwilling to relinquish entirely the care of their children during the day, and they demanded a midday period with their children—the "hour of love." What accounts for this return to "traditional behavior"?

Wilson considers a possible explanation. "It has been argued that this reversion merely represents the influence of the strong patriarchal tradition that persists in the remainder of Israeli society, even though the role division is now greater inside the kibbutzim than without" (1978, 134). Wilson, too cautious to maintain that the explanation stands refuted, contents himself with a tone of skepticism. Van den Berghe, however, is disinclined to allow any alternative to a genetic explanation. Having elaborated the story of changing patterns of child care in the kibbutz, he declares,

> Kin selection triumphs after half a century of ineffective suppression. Lest the change be interpreted as a sinister male chauvinist plot to put women in their place, let it be emphasized that the reversion to the standard family groups was overwhelmingly a response to the mounting dissatisfaction of *women*. (1979, 74)

The kibbutz "experiment" assumes pride of place in pop sociobiological writings because it seems to be a rare analog of the kinds of tests that plant geneticists routinely perform. The alleged environmental variables are varied and one inspects the variation in phenotype. Analogs are obviously rare—we have to await occasions on which people choose to manipulate their environments in the appropriate ways. (Manipulating the environments of others seems to be the prerogative of kings—and their modern successors. Few are as disinterested as James I, who hoped to determine if children brought up in isolation would learn to speak Latin. Unfortunately for the king, the children refused to cooperate. They died.)

It is not only the paucity of cases that leads to emphasis on those that are available. Vulgar reductionists can have a field day with the kibbutz story. It is tempting to project the attitudes of the society from those of its members, to suppose that the social practices that emerge depend only on the biological propensities of the second-generation women and on the attitudes of the first-generation people by whom they were reared. The daughters who reverted were brought up inside the kibbutz by ideologically committed care givers. They were thus immune from the impact of preferences in the surrounding society. Those preferences belong to individuals remote from the scene, whose attitudes cannot be expected to outweigh the contrary views of the care givers. Hence the "experiment" is interpreted as showing that the allegedly crucial environmental factor has been varied and that the variation has failed to produce a change in behavior. Van den Berghe concludes that the emancipation of women from roles as primary care givers is a genetically precluded state.

The argument depends on a reductionist premise. The attitudes of the daughters could only be explained in terms of the attitudes of the ideologically committed parents or in terms of the daughters' genetic propensities. Since the parental attitudes favored the social practice that was abandoned by the daughters, we are forced to conclude that the filial revolt testifies to the power of the genes. Antireductionists should reply that we are forced to no such conclusion. There are any number of alternative cultural explanations for the daughters' behavior, many of which will appeal to the influence of the surrounding patriarchal society. We know far too little about the ways in which human goals and aspirations are influenced by complicated social institutions and by the attitudes of other people (including people with whom there may be no direct contact) to pronounce confidently on the causes of the return to "traditional roles." (The very fact that we describe what occurred in those terms itself suggests ways in which a society that is dedicated to preserving its traditions might influence the attitudes of its young citizens.) The failure of the "kibbutz experiment" reveals that a particular, very direct, approach to the modification of gender roles did not achieve its intended goal. Without a tacit—and highly controversial—reductionist premise, that is all it shows.

Many of Wilson's early critics campaigned vigorously for recognition of the role of society in human development. Marshall Sahlins reminds the reader of an "anthropological commonplace": "the reasons people fight are not the reasons wars take place" (1976, 8). The criticism has not escaped Wilson's notice. The shift from the early Wilson program to the study of gene-culture coevolution, carried out by Wilson and Lumsden, is explicitly designed to respond. We shall inquire in a later chapter if the accusation has really been met.

For the moment, we can add to the list of misadventures in method that abound in Wilson's pop sociobiology. Conclusions about human nature cannot be drawn by invoking canons of convenience. Nor can they be reached by introducing unargued reductionist premises. History matters.

Chapter 7
Dr. Pangloss's Last Hurrah

Making the Best of Everything

Dr. Pangloss looked at the world and saw that it was good—so good, in fact, as to testify everywhere to its divine origins. Contemporary adaptationism, in its extreme form, is heir to Pangloss's rhapsodies about optimal design. Where Pangloss discerned traces of an omnipotent designing hand, some evolutionary biologists hunt out perfections of organismic design, previously hidden from the eyes of the base vulgar, and so claim to fathom the workings of natural selection. As with Pangloss, there are dangers of excessive enthusiasm.

Optimization arguments promise a way of cutting through some of the difficulties in confirming adaptationist histories. Surely, if we could successfully "play God," as Oster and Wilson put it (1978, 294), redesigning the biological system under study according to what we conceive as the relevant parameters, and if we found that our optimal design closely matched what nature has produced, we would be too skeptical if we dismissed the result as mere coincidence.

Here we approach a controversial and technical set of issues that affect not only pop sociobiology but sociobiology and evolutionary theory generally. I shall try to show that the questions are delicate and that it is all too easy for evolutionary biologists to overlook important possibilities. However, it is important to make clear at the start that many of those who advance hypotheses about the course of evolution, including many who contribute to nonhuman sociobiology, are acutely aware of the difficulties in this area. Pop sociobiology is another matter. The idea of a simple connection between optimal design and a history of selection is integral to step 2 of Wilson's ladder; it figures prominently in other versions of pop sociobiology as well.

When, having discovered that the behavior of a group of animals coincides with the deliverances of an optimization argument, we infer that natural selection has fixed the behavior because it is optimal, there is a tacit presupposition. We are assuming that evolution, or perhaps evolution under selection, fixes the optimal traits. It is worth asking what this means. Oster and Wilson offer a prudent review of the use of optimization arguments (1978, chapter 8). But there is one assumption that they do not subject to careful scrutiny. They suppose that, almost as a matter of definition, selection will be an optimizing process, and they express concern that this thesis will degenerate into "tautological nonsense." In this section I shall consider two questions: In what sense (if any) does natural selection inevitably optimize? In what sense (if any) does evolution inevitably optimize?

The answer to the first question seems obvious. Selection spreads the fitter genes and, if other evolutionary agents are absent, will ultimately fix the fittest genotype. There may be constraints on the process. Some genes of superior quality may not be available in the population. Some alleles producing beneficial effects may give rise to damaging side effects. However, subject to constraints like these, selection seems to be an optimizing process. Spencer's tag "survival of the fittest" may have been slightly misleading, but we are inclined to think that it was on the track of an important truth.

Matters are not so simple. *Naive* versions of the tag face immediate difficulties. Selection will not inevitably fix the fittest gene at a locus—for there may be no such thing to fix. Cases of balanced polymorphism, such as sickle-cell anemia (chapter 2), pose immediate complications. Nor can one appeal to selection as fixing the entire fittest genome: sexual reproduction would break it up as soon as it appeared. A more promising approach is to run to the middle ground of elementary population genetics, thinking of selection as fixing the fittest allelic pair at a locus. But the problem recurs. If we consider sexually reproducing organisms and a locus at which a heterozygote has greatest fitness, then, even given temporary achievement of a population of heterozygotes, mating would automatically produce some homozygous offspring.

Elementary considerations expose a difficulty. If we think of the claim that natural selection optimizes as a thesis that selection leads to the fixation of some property of individual organisms (as opposed to a property of the population, then what is the property? If we do not, then how is pop sociobiology to turn its accounts of fitness successfully maximized into histories of selection?

There are a number of possible responses. First, we might point out that if the optimal pair of alleles at a locus were heterozygous, then we could not construe the idea of selection as an optimizing agent by saying that selection fixes the fittest pair of alleles at that locus. But all that this shows is that when we triumphantly produce our optimal design and find it to correspond with the structures and behavior present in nature, we simply have to assume that the prevalent trait is directed by a homozygote. Given this assumption, it makes perfect sense to suppose that selection has optimized by fixing the fittest allelic pair. But the new proposal rests on a presupposition: if there are a number of rival alleles at a locus, if each of the allelic pairs formed from them is present in the population, and if one of the homozygotes has superior fitness, then selection will work to fix the pertinent allele in the population. Is this assumption true?

Unfortunately it is not. Selection can even lead to the demise of the fittest. Population geneticists have long recognized that when there are three or more alleles at a locus, the course of evolution cannot simply be read off the qualitative features of the ordering of fitnesses (see Roughgarden 1979, chapter 7, especially 107). Here is a beautiful example, described by Alan Templeton (1982, 16–22). Suppose that we have a population in which three alleles, *A*, *S*, and *C*, are all initially present. *AA* is found in virtually every member of the population, and the following conditions hold:

AS is fitter than *AA*;
SS is lethal;
C is recessive to *A* (that is, *AC* and *AA* have the same phenotype and, in consequence, the same fitness);
CS is inferior in fitness to *AA*;
CC is the fittest allelic pair.

What will happen to the population? Answer: *C* is eliminated; thus the fittest combination, *CC*, although initially present in the population, not only fails to become fixed but indeed is driven out. As Templeton remarks, "So much for the phrase 'survival of the fittest' " (1982, 20). Because of the initial preponderance of *A* alleles, *S* alleles occur most frequently in *AS* combinations, and *C* alleles turn up in *AC* combinations, which, because *C* is recessive, display the *AA* phenotype. The population thus moves toward a balanced polymorphism between *A* and *S*, with a few *C* alleles still present. Once the polymorphism is reached, selection then works to drive out the rare *C* alleles. This is because the average effect of incorporating a *C* allele into a zygote is negative: *C* alleles do no good when they occur in the *AC* combination, and they are inferior when they turn up in the company of *S*. So, despite the fact that *CC* is the best available genotype, natural selection works to displace *C* from the population.

What is the significance of this result? First, it is not the idle fancy of a model builder. The evolutionary story just sketched describes the course of evolution at the hemoglobin locus in a Bantu population. Second, the result indicates a general difficulty for attempts to view the optimizing power of natural selection as issuing in the fixation of the best available genotype. One is inclined to think that the power of selection will be constrained by availability. If the beneficial mutant never arises, then selection cannot carry it to fixation. Further, once the possibility of an advantageous heterozygote is recognized, we see that selection cannot necessarily fix the optimal genotype. Templeton's case shows that these are not the only problems of selection. The mere presence of an allele that would be favored in homozygous condition is not enough. Even if the allele is available in the population, it may be debarred from following the trajectory that we would have anticipated for it.

The story has a further moral (noted by Templeton). Alter the conditions of the case so as to allow a slight amount of inbreeding. Under the new specifications—which, of course, do nothing to change either the initial composition of the population or the effects of the genotypes on the individuals who have them—the *C* allele is fixed by natural selection. Thus, even if we were to allow for the effects of the initial distribution of alleles and for determinate consequences of the possession of particular genotypes, the evolution of a population under selection might yield qualitatively different outcomes depending on small differences in breeding structure—differences that might themselves depend on the alleles present at different loci, on features of the external environment, or on a combination of the two.

When we turn our attention to genetic systems that involve more than one locus, matters only become worse. Interaction effects among loci can defeat the enterprise of fixing the fittest allelic pair at any locus, thus making a mockery of Barash's claim that organisms are like all-star teams—living things come to be composed of individual genes each of which tend to have consistently high batting averages [sic] in the game of reproduction" (Barash 1979, 21). Genes do not compile impressive statistics by their own unaided efforts.

Let us say that there is a *trajectory problem* for a population if the population initially contains a collection of alleles at a number of loci and if the fittest combination of those alleles cannot be attained under selection. (Intuitively, the idea is that you can't get there from here.) A classic example of a trajectory problem when different loci are involved is the investigation of chromosome polymorphisms in the Australian grasshopper *Moraba scurra.* (See Lewontin and White 1960 and, for an exposition, Lewontin 1974, chapter 6.) In this case two pairs of homologous chromosomes are involved. For each pair there are alternative arrangements of the genes—*inversion systems*—present in natural populations. The inversion systems interact, in the sense that the phenotypes depend on the combination of orderings on the two different pairs. Lewontin and White used measurements of frequencies in successive generations to estimate the fitnesses of various combinations.

Four arrangements were considered. There is a *Standard* arrangement on the chromosome pair CD, a Standard arrangement on the pair EF, a rival arrangement to Standard on CD (*Blundell*), and a rival arrangement to Standard on EF (*Tidbinbilla*). The names for the deviant arrangements derive from the populations in which those arrangements were first found. Lewontin and White discovered that the overall fittest combination was Blundell/Blundell on CD and Tidbinbilla/Tidbinbilla on EF. Hence, if selection were always to operate so as to maximize the fitness of individual organisms, we would expect the Standard arrangement to be eliminated on both chromosome pairs and the deviant arrangements from Blundell and Tidbinbilla to become fixed. However, the trajectory of a population under selection depends crucially on the initial composition. Assuming that the population initially contains the Standard arrangement on both chromosome pairs, the Tidbinbilla arrangement, and the Blundell arrangement, so that all four arrangements are available, the course of evolution under selection is determined by the initial frequencies. There are two stable final states. One is the state in which Standard and Blundell reach a stable polymorphism on CD and Standard becomes fixed on EF, and the other is the optimal state, in which both Blundell and Tidbinbilla become fixed. The suboptimal state is reached from initial conditions in which substantial frequencies of Blundell and Tidbinbilla are present. (For example, an initial value of 75 percent for Blundell and 10 percent for Tidbinbilla will drive the population to the polymorphic equilibrium at which Tidbinbilla is eliminated and the frequency of Blundell is 55 percent.)

Everything we know about the molecular basis of heredity suggests that interactions among genes are inevitable, so that the fitness of an allele will always be a function of the company it is likely to keep. Thus, we should expect that trajectory problems may be prevalent in nature—that there may be plenty of situations in which all the alleles that together make the fittest combination of the available alternatives are present originally in the right combinations but are eliminated or reduced in frequency under selection. To defend the thesis that selection is an optimizing agent in always producing the fittest available genotypes—Barash's "all-star team"—it will be necessary to plead that the optimal genotypes are not really available in such cases. But what is the concept of availability that is being used here? The alleles are all present, and they are all present in the right combinations. Superanimal is there from the start. Hence, champions of the optimization thesis must declare that a genotype is available to selection only if selection is able to fix it in the population under study.

Now we reach that abyss of vacuity from which Oster and Wilson hoped to draw back. Selection, as it is being considered in our discussion, is a deterministic process: given the initial state of a population, the end state under selection is fixed. This mean that if a population *can* become fixed for a particular genotype under selection, then it *must* become fixed for that genotype. The optimization thesis has been reduced to the claim that selection fixes the best available genotype, that is, the best genotype that can be fixed in the population, that is, the best genotype in a collection with exactly one member, the single genotype that selection inevitably fixes in that population. (Perhaps I should note explicitly that this spirited defense of a result of crushing triviality is predicted on the assumption that there is a unique fittest homozygous genotype.)

Not only is this result tantamount to the claim that selection does what it does, but it is also entirely useless for the purposes for which pop sociobiologists hope to use the notion of optimization. Recall the predicament of ambitious Darwinians. Recognizing that adaptationist histories are hard to confirm, we might be tempted to short-cut our problems by "playing God," looking to see if our optimal design is to be found in nature, and claiming dramatic support if our expectations are fulfilled. The procedure rests on a connection between the notion of optimal available design that we can use on the basis of our prior knowledge of the possibilities accessible to ancestral organisms and the constraints imposed upon them, and the notion of optimal available design that results from selection. That connection is destroyed when we gerrymander the thesis of optimization in the way I have described.

There is a different tack to try. Instead of focusing on the maximization of individual fitness, we can consider the mean fitness of a population. A well-known theorem in mathematical population genetics (derived by Sewall Wright) tells us that under certain simple conditions selection acts to maximize the mean fitness of a population. The theorem holds when we consider only a single locus, for which there are determinate constant fitnesses of the allelic pairs at that locus. It can be extended to *some* situations in which loci interact (see Crow and Kimura 1970, chapter 5, especially 232–236). Ignoring these complications, let us note a different kind of limitation.

In cases of frequency-dependent selection the mean fitness of a population may actually decrease under selection. Suppose that the allele A is dominant with respect to a, that the fitnesses of the genotypes vary directly with the frequency p of A alleles in the population, and that at each value of p the fitness of the *aa* genotype is greater than that of the *AA* (or of the *Aa*, given the assumption that *A* is dominant over *a*). Then, under selection, the population will go to fixation of *a* from an initial mix of *A* and *a*. However, the mean fitness of the population will continually decrease. (See technical discussion F for a concrete example.)

A relatively realistic interpretation of this selection regime can easily be given. Suppose that *A* is a gene that promotes cooperative behavior among animals and that *aa* individuals are willing to be beneficiaries but are never inclined to be helpers. Then the assignments of fitness appear reasonable: all individuals in the population benefit to the same extent from the presence of altruists, in direct proportion to the frequency of the helpful animals; but selfish types have greater fitness because they do not incur any of the costs of giving. The population would be better off as a whole if there were altruists around and would achieve maximal fitness if everyone were to cooperate. But, sad to say, any entering animal with a genotype promoting selfishness would be at a competitive advantage, and the genes of this animal would be spread through selection to the detriment of the group.

Wright's theorem suggests a simple interpretation of the view that natural selection is an optimizing agent. Selection maximizes the mean fitness of a population in certain elementary situations; yet there are circumstances under which this result breaks down. As my hypothetical example reveals, certain kinds of frequency-dependent selection would *minimize* the mean fitness of the population. Hence, in their discussion of populations subject to frequency-dependent selection, proponents of optimization analyses who show that a certain design would maximize mean fitness may not automatically assume that selection can produce this design.

Let us take stock. There are two kinds of cases in which an optimality model might be invoked to justify an adaptationist history. In the first type of case the properties of individual organisms are the focus of study. The adaptationist proudly lays before us a set of constraints on the population under study and shows us that, given these constraints, a particular trait would maximize something that is taken to be a measure of the fitness of the organisms. We go to nature, and lo! the trait (or, more realistically, something like it) is found in the actual population. Should we conclude that the adaptationist has identified the conditions under which selection operated to produce the trait? The answer is far from clear. Lacking any knowledge of the genetics of the process, we have no firm basis for judgment as to whether this is one of those cases in which selection might have been expected to

Technical Discussion F

Let the unnormalized fitness of AA (and Aa) be $bp + c$ and that of aa be $bp + d$, with $0 < c < d$ and $b > 2(d - c)$. The mean fitness of the population, $\bar{w}$, is given by:

$$\begin{aligned}\bar{w} &= p^2(bp + c) + 2p(1 - p)(bp + c) + (1 - p)^2(bp + d)\\ &= (bp + c)[p^2 + 2p(1 - p) + (1 - p)^2] + (d - c)(1 - p)^2.\end{aligned}$$

It follows that

$$d\bar{w}/dp = b - 2(d - c)(1 - p).$$

Under the conditions given above, this is always positive. Hence $\bar{w}$ increases as p increases. So, as a becomes prevalent—that is, as p goes to 0—the mean fitness of the population decreases.

fix the fittest combination of alleles originally present or whether this is a situation in which selection has achieved the fixation of a suboptimal available genotype. In the latter event the concurrence of the model design with the trait actually found would be evidence against the correctness of the model, and it would show that our prior judgments about the constraints (or about the measure of fitness) were mistaken.

Similar considerations apply to the second type of case, in which we focus on some property of a population and reveal that the characteristic actually found in the population corresponds to a design that would maximize mean fitness. Maximization of mean fitness is only to be expected under certain kinds of genetic conditions. When frequency-dependent effects are to be anticipated (or when we have reason to believe that there are severe complications due to linkage), then there is no easy transition from the match between optimal design and actual findings to conclusions about the operation of selection.

A homely analogy makes the point. Parents are commonly inclined to believe in the perfection of their children. In more realistic moments, however, each of us appreciates that there are some faults present in even the most angelic-seeming offspring. Once we have this point consciously in mind, behavior that is too perfect can make us suspicious. The same ought to be true of our understanding of evolution. Even when we restrict ourselves to the consideration of selection, we have to face the fact that selection does not necessarily produce the best available organism. Hence, a string of optimality models in which nature accords perfectly with our views of good biological design ought to provoke suspicion. Optimality models are likely to be too good to be true.

So far, we have considered the relation between selection and optimization. But selection does not have complete authority in evolution. As every biologist knows—and as some proclaim before going on to concentrate on selection—chance effects can play a big role in the evolution of small populations. Imagine a small population with three alleles initially present at a locus. The fitness ordering of the genotypes is $AA < AB < BB < AC < BC < CC$. Under these circumstances, given an initial prevalence of *A* alleles and small frequencies of the *B* and *C* alleles, selection will work to fix the *C* allele. Suppose, however, that a flood destroys all the organisms that possess the *C* allele before they mate. Under these circumstances the subsequent course of selection in the population will fix the intermediate *B* allele. Assume further that the molecular constitution of the alleles makes direct mutation from *B* to *C* highly improbable. (This could occur if, for example, both *B* and *C* are obtainable from *A* by single base substitutions, so that they differ from one another by two base substitutions. The most direct mutational path from *B* to *C* would then be by way of the inferior *A* allele.) Because of an extraneous event (the flood) a population that would have become fixed for the best available allelic pair reaches a different final state. The conditions of the original optimization are changed. Chance diverts a population from the optimum it would have reached.

The same effect can be obtained by eliminating the extraneous event in favor of a particular history of mutations. Suppose that mutations from *A* to *B* and from *A* to *C* are both improbable, but that it is slightly more probable that the population should include mutants who have the *C* allele than mutants who have the *B* allele. Nevertheless, despite the fact that, in an obvious sense, the population is evolutionarily closer to the optimum allelic combination *CC*, it may move to fixation of the intermediate allele *B*. This is simply because the less probable sometimes occurs, and a history in which the *B* individuals appeared and were successful, so that the *B* allele swept to fixation before any appearnce of the *C* allele, is perfectly possible. (Notice that a slight amendment of the fitness ordering, making *BC* inferior to *BB*, would then provide a substantial block to the careers of any mutant *C* alleles that happened to arise later.) A seldom noticed corollary of the familiar point that rare mutant alleles are liable to be lost through drift is that the elimination of rare alleles need not coincide with their relative fitnesses. Another understressed consequence is that, when the genetic system under study is one like that which Lewontin and White found in *Moraba scurra,* where the initial frequencies of particular combinations of alleles (or gene arrangements) make a critical difference to the evolutionary trajectory, populations that are close to critical points can have their evolutionary histories dramatically altered by small chance fluctuations in gene frequency. At certain initial mixes between Blundell, Standard, and Tidbinbilla, a minute perturbation can change the outcome of the evolutionary process (see Lewontin 1974, 280).

It is natural to protest that, while the workings of selection can be affected by the slings and arrows of outrageous fortune, this is not likely to be a significant factor in the history of life. There are two versions of this response. The first claims that the occasions on which chance enters the evolutionary picture are too rare to be worth considering seriously. The second contends that the perturbations of chance have only temporary effects, that in the long run selection prevails. In both instances I think that the response is too confident about the power of selection.

How likely is it that chance will change the conditions and thereby affect the outcome of selection? The general question is almost impossible to answer. What we can do is to address a more specific issue. First, let us restrict our attention to small populations. Second, let us confine our attention to populations for which there is a non-negligible probability of reduction in numbers through some cataclysmic event. I doubt that populations meeting these two conditions are rare or that their role in the history of life is insignificant. (Many evolutionary biologists have emphasized the idea that the evolutionary action takes place in small populations in marginal habitats.) We can now pose a more specific question: If the population contains n individuals, some small number k of whom carry the B allele and some small number m of whom carry the C allele, what is the probability that, in a flood that destroys r of them, all of the C alleles will be lost and at least one of the B alleles will be retained? (See technical discussion G for a preliminary analysis of the problem.)

Consider a particular case. Imagine that we have a population of 10 individuals, one of which bears the B allele and one of which bears the C allele (that is, $n = 10, k = m = 1$). The probability of preserving B and eliminating C, given that a flood destroys some subset of the population, turns out to be about 0.18. This is hardly a negligible probability. Hence, *if* we are entitled to conclude that catastrophes sometimes eliminate subsets of small populations, then it would be wrong to neglect the possibility that such events only rarely change

Technical Discussion G

Let us disregard the complications due to sex, assuming that the population can reproduce asexually, so that it can survive if even one member survives the flood. This assumption does not seriously alter estimates, but it does make the computation easier. We shall also assume that if a flood occurs, the probability of losing any number of organisms in the flood is the same as the probability of losing any other number. More exactly, for any numbers r and r', both of which are greater than 0 and less than n, the conditional probability that exactly r members perish, given a flood, is exactly the same as the conditional probability that r' perish, given a flood. The probability that B is preserved and C is eliminated, given a flood that kills exactly r members, is

$$\frac{k \cdot {}^{n-m-1}C_{n-r-1}}{{}^{n}C_{r}} = \frac{k \cdot (n - m - 1)! \cdot (n - r) \cdot r!}{n! \cdot (r - m)!}.$$

(Recall that there are initially k B alleles and m C alleles.) The probability that the flood kills exactly r members of the population, given that there is a flood, is $1/(n + 1)$. Hence, the probability that B is preserved and C eliminated, given that there is a flood, is

$$\sum_{r=m}^{n-1} \frac{k \cdot (n - m - 1)! \cdot (n - r) \cdot r!}{n! \cdot (r - m)! \cdot (n + 1)}.$$

the conditions of selection. (This conclusion is reinforced if there are a large number of possible ways in which extrinsic factors might affect a population, as in our mutation scenario.)

I conclude that it is rash to dismiss the possibility that chance effects play a part in shaping the course of evolution. However, it might still appear that such disruptions of the smooth action of selection will be only temporary. Examples in which chance eliminates the best available allele and some less fit allele becomes fixed in a population inspire an obvious response: sooner or later the unlucky allele will reappear by mutation, and when it does, selection will have a new opportunity to move the population from a suboptimal condition to the genuine optimum. But this optimistic assessment of selection's power in the long run may prove quite unrealistic. Typically the fixation of an allele at one locus will affect the fitnesses of mutants that arise at other loci. If these mutations arise and become prevalent before the unlucky allele receives its second chance, then the new opportunity may come too late. In the altered genetic environment the previously optimal allele may no longer be superior. The originally less fit allele that displaced it may have armed itself with alleles at other loci, so that the invasion of the unlucky allele is now resisted by selection.

Sewall Wright introduced a helpful device for thinking about evolution. Imagine a surface with peaks and valleys. Each peak is a *local maximum* of mean fitness—the state achieved by a population that begins from a genetic composition in the neighborhood of the state. (The notion of a *local* maximum is explicitly designed to represent the end state of the evolutionary trajectory available for a population and thus to sidestep some of the difficulties we have noted.)

Wright and, more recently, Templeton have argued that, properly understood, stochastic factors work in harmony with natural selection. Templeton's formulation is lucid and elegant. He invites us to invert the adaptive landscape, "transforming peaks into pits and valleys into ridges" (1982, 25). Imagine populations as balls rolling around the landscape. They tend to fall into the nearest pits (achieve the local maxima of mean inclusive fitness). A population may not land in the deepest pit. So without stochastic factors, accidents of mating and death, populations become stuck at local but not global optima. The situation changes when we allow for such factors. We can think of them as lateral shakings of the landscape, shakings that cause some balls to roll up the sides of pits, over ridges, and into new pits. Because balls are more likely to roll out of shallow pits than out of deep ones, prolonged shaking will tend to produce a situation in which balls are located in the deepest pits.

Templeton's analogy seems to give the friends of optimizing selection just what they sought. In the short run, the forces of chaos may disrupt the action of selection. In the long run, they enable selection to work more efficiently than it would otherwise. Panglossians should relish the solution. They have always preferred to think *sub specie aeternitatis*.

The Wright-Templeton analogy involves an assumption, however. If the assumption is violated, the analogy proves misleading. It is crucial that the selective pressures be relatively constant, that they change only slowly compared with the population movements caused by drift. If the adaptive topography changes sufficiently quickly, then the local maxima at which populations arrive may leave their mark on subsequent evolution. Finding itself in a particular pit, a ball may now be adjacent to another, newly formed pit, which it would not have been able to reach from other positions on the prior landscape.

Even more straightforward is the reliance of the Wright-Templeton argument on the absence of frequency-dependent selection. If the selection pressures depend on the ways in which populations are distributed on the adaptive topography, then the effects of chance will be felt in altered fitnesses of particular allelic combinations, so that the subsequent course of evolution under selection may be substantially modified. It is as if the falling of balls into pits itself affected the contours of the adaptive landscape, deepening some of the pits and making others shallower, or perhaps eliminating them. Under these circumstances it would be incorrect to propose that the effects of chance are only temporary. For want of a competitor allele, eliminated as the result of a freak accident, an unusual allele may become fixed in a local population. Because it does so, the relative fitnesses of alleles at other loci may be altered. As they are altered, the population may evolve in unsuspected ways, and its evolution may impose new pressures on other populations of the same species and even on populations of different species.

For any sufficiently large ensemble of populations, at whatever time we choose, it is likely that some proportion of those populations will not be in the state that selection would have marked out for them. Those who play down the role of stochastic factors in evolution are inclined to think that such populations will eventually reach that very distribution of phenotypes to which selection, pure and untrammeled, would have led them, and from which they are temporarily estranged. But since the previous optima, once missed, may never be available again, it is not only true that there are always some populations that are diverted from the optima to which they would have gone under different conditions, but it is also false to think that any population ultimately achieves the optimum that it would have achieved without the disturbances of chance. It is wrong to suggest that evolution necessarily optimizes.

Friends of optimization have a last response. They can counter the theoretical possibilities to which I have drawn attention with the claim that the kinds of traits in which they (and, perhaps, the pop sociobiologists) are most interested are influenced by many different loci, which act independently of one another and which behave in more or less similar fashions. There are thus an enormous number of possible ways for an evolving population to achieve the optimal distribution of phenotypes. Even though historical contingencies may disturb the course of evolution at some loci, there is sufficient genetic variance elsewhere to ensure that the optimum will ultimately be achieved. Here the dialogue ends in conjecture. One large task of contemporary evolutionary theory is to explore the amount of genetic variance that is available to an evolving population. We know that the theoretical possibilities that I have discussed do occur. We do not know how widespread they are, and consequently we do not know the extent to which the evolution of populations is affected by stochastic factors. Rampant Panglossians, who assume that sufficient genetic variance is there whenever they need it, beg a major empirical issue.

The organisms produced by the evolutionary process can thus be expected sometimes to violate our views about optimal design: how often remains an unsolved question. When we attempt to reconstruct their evolutionary histories by finding a set of constraints and alternative characteristics within which nature selected the optimal traits, we ought to expect our analyses sometimes to be wrong. A complete adaptive story for everything would certainly be mistaken. The Panglossian vision of a world in which all organisms are optimally designed in all respects is not an accurate vision of our world. Nor should we be beguiled by the modern disciples of Pangloss who insist that, at each stage of the historical process and subject to the constraints then present, optimal design always prevailed. Evolution is not the best of all possible architects.

Where Every Prospect Pleases

In considering the "strict and philosophical" approach to optimization, we have surveyed some relatively subtle concerns about the strategy of reading evolutionary history from optimality analyses of prevalent traits. It is now time to look at the "loose and popular" employment of optimization, at the foibles of those proposals that ignore the methodological strictures elaborated in reflective discussions (Oster and Wilson 1978, chapter 8).

Critics of pop sociobiology sometimes charge that the enterprise results in a collection of "Just-So-Stories" (Gould and Lewontin 1979): pop sociobiologists (and some misguided evolutionary theorists) focus on properties of animals that catch their fancy, and struggle to show that, viewed from the right perspective, the properties in question display the optimizing hand of evolution. (Sociobiologists sometimes express similar misgivings; see Rubenstein 1982, 87; Thornhill and Alcock 1983, chapter 1.) However, Gould and Lewontin charge that, despite an explicit commitment to recognizing the ways in which the power of selection may be limited (the influences of linkage, pleiotropy, allometry, and genetic drift), pop sociobiological practice is prey to the lure of a good story.

Gould and Lewontin substantiate their charge by considering an account proposed by Barash. Intrigued by the thought of "testing the proposition that animals act so as to maximize their fitness," Barash conducted an investigation into the behavior of mountain bluebirds:

> . . . while the male was away getting food, having left his mate at their newly constructed nest, I attached a model of another male bluebird to the tree, close to the nest and to the female. I was curious as to how the male would respond when he returned and discovered the "adulterous" couple. In particular, I wanted to compare the male's behavior when he caught his female *in flagrante* during the breeding season with his response later in the year, once the eggs were already laid. Early in the season, when breeding was taking place, all hell broke loose when the husband returned; as expected, he attacked the dummy male quite aggressively. But—and this I found especially interesting—he also attacked his own mate, in one case even driving the suspected adulteress away. She was eventually replaced by another female, with whom he successfully reared a brood. What happened when males were presented with the identical situation *after* the eggs had already been laid? There were still attacks on the intruding male, but of much lower intensity, and no further aggression was directed toward the female. (Barash 1979, 52; see also Barash 1976)

In this behavior Barash sees the designing hand of selection: the male was optimally designed in that he was aggressive when it was useful to be aggressive—that is, when the paternity of the female's offspring was at stake—and less aggressive when the presence of the intruder no longer mattered. If we were to design an optimal male, then, it is suggested, we would design a male hot to defend his honor when there are serious evolutionary consequences of being cuckolded (to wit, having the bearers of alien genes thrive in his nest), but more tolerant when there is no danger of such dire results. And look! That is (more or less) the male we find in nature. More power to selection.

The first point to make is that there is no genuine thought of testing the proposition that animals act so as to maximize their fitness. Our belief in this proposition is grounded in our understanding of the genetics of the evolutionary process; as the previous section suggested, a more refined understanding would lead to a more refined proposition. What Barash has put to the test is a particular idea about the way male mountain bluebirds maximize their fitness. (Would

Barash have abandoned the proposition that animals act so as to maximize their fitness if he had found no difference in behavior during and after the breeding season?) Second, the study seems to have been undertaken with no information about the kinds of behavioral alternatives that might have been available to ancestral bluebirds or the kinds of constraints that might have operated in the evolution of this behavior. We can easily imagine alternative ways for a bluebird to put an end to hanky-panky in the nest, but Barash makes no serious attempt to explore which of these alternative strategies might have been evolutionarily available. Third, there is no obvious connection between the behavior exhibited and the maximization of the number of expected descendants. We are tempted to forget that an attack on another male might prove very costly to a returning bluebird, that the risk of injury might outweigh whatever benefits are achieved by driving the interloper away from the nest.

What is most noteworthy about Barash's scenario, however, is its blissful neglect of an obvious rival explanation. As Gould and Lewontin point out, there is a simple way to account for the diminution in hostility after the breeding season. Mountain bluebirds, like people, can be fooled for a time, but they eventually catch on. After a while a dummy on the doorstep no longer poses a serious threat and no longer incites the returning male to defend his wounded honor. The two accounts could easily be distinguished by conducting further experiments. By varying the time at which the dummy first appears, one could determine whether hostility wanes with familiarity or with the end of the breeding season. Apparently this was not a trial that Barash thought of making, and therein Gould and Lewontin see an important moral.

If the moral is simply that it is easy to do optimization analyses carelessly and that adaptationists, in their enthusiasm for a particular scenario, may fail to explore obvious rival hypotheses, then many adaptationists and sociobiologists will surely agree. Barash should have given a clearer specification of the alternatives that he took to be available in the evolution of bluebird behavior, he should have formulated the constraints that operated during the evolution of this behavior, he should have given a more precise account of how the behavior hailed as optimal really promotes fitness, and he should have tested rival accounts against one another. These are familiar methodological points, and they are given their due in the review of optimization models conducted by Oster and Wilson.

According to Oster and Wilson, "optimization models consist formally of 4 components: (1) a state space; (2) a set of strategies; (3) one or more optimization criteria, or fitness functions; and (4) a set of constraints" (1978, 297). The choice of state space is supposed to reflect the variables that will be considered; in many biological cases only a few aspects of the animals under study will be taken into account. The set of strategies is the collection of alternatives taken to be available to selection. So, for example, we might envisage various kinds of behavior as open to the animals we are studying. The optimization criteria are intended to pick out some quantity that is correlated with fitness, so that maximization of this quantity will correspond to maximization of fitness. In the example of the mountain bluebirds, Barash tacitly assumes that the number of zygotes produced during one breeding season is an appropriate fitness criterion—probably a reasonable assumption in the case that he investigates (though not in all cases). Finally, the constraints are imposed by all kinds of features of the organisms in question—their genetics, their physiological needs, their environments, and so forth.

It is not difficult to describe a careful optimization analysis by reference to the components that Oster and Wilson identify. Choice of state space will be justified by background biological knowledge of the animals one is investigating. Similarly, the selection of strategy set should reflect understanding of the animals: we are justified in not considering the possibility that a male mountain bluebird might devise some avian chastity belt as a solution to his problem with interloping males. Our choice of fitness criteria must reflect an understanding of the long-term consequences of present behavior. To cite the number of zygotes that an organism produces as an appropriate fitness criterion is to suppose that the number produced now will have little impact on the breeding success achieved by the organism in future seasons and that the strategies included in the strategy set do not vary in their effects on the probabilities of survival of the zygotes. For one who wondered whether male aggression toward the female might interfere with the successful rearing of offspring, the number of zygotes produced would not be a proper fitness criterion. Finally, one must acknowledge the constraints that are set by the animal's physiology, ecological requirements, and so forth. Using our knowledge of daily food requirements, energy required for various kinds of movements, and so forth, we recognize that animals are limited in what they can do. What shall it profit a bluebird if he monopolize the attentions of a female and lose the energy required to copulate?

We can now begin to understand why Oster and Wilson are cautious in their assessment of the uses of optimality models. If the rigorous presentation of a model requires the kinds of insights just listed, we may well wonder where all the requisite information is to come from. Consider, for example, the choice of strategy set. How can we pin down the alternatives that are evolutionarily available to a population? Is it really possible to evaluate the possibility that a mutation or recombination might have produced an allele that, in conjunction with the rest of the genome and the cooperation of the environment, produced a particular trait? Oster and Wilson acknowledge the problem:

> Most of the models we have developed to describe caste structure have been economic in conception: they aimed at specifying the optimal allocation of scarce resources among a predetermined set of alternatives. These alternatives consisted mostly of guesses based on our knowledge of natural history. In some sense we had to anticipate the allowable strategic alternatives by reconstructing evolution in our imagination. (1978, 299)

This is a candid admission of difficulty. Yet Oster and Wilson's guesses are not blind nor are their imaginations untutored. Wilson's unparalleled knowledge of social insects is put to good use in developing ideas about the kinds of options that might have been available in their evolution. The result is a collection of analyses that is unusually rigorous in its formulation and defense of assumptions. One way to respond to the use of optimization models by sociobiologists and other adaptationists would be to treat Oster and Wilson as model modelers, beside whose work the defects of more casual studies (like Barash's discussion of mountain bluebirds) can readily be appreciated.

But the most prominent critics of optimization do not want to stop at this point. They are not simply suggesting that there is a set of rules, articulated and followed by Oster and Wilson, to which most sociobiologists and adaptationists usually fail to conform. Gould and Lewontin discern a deeper flaw in the adaptationist program. Their concern arises from a problem that is not explicitly addressed by Oster and Wilson, the problem of how one chooses properties of animals that are supposed to be optimally designed. Is it legitimate to seize upon any property of an organism that strikes one's fancy and ask how selection has worked to produce the optimal form of that property? Surely not. The characteristics of organisms are deeply connected with one another—recall Darwin's frequent references to the correlation of characteristics. In seizing upon a trait of an organism in the absence of any understanding of the developmental history, we have no basis for assuming that that trait was itself the focus of selection. Yet, in the flush of enthusiasm, we may set about constructing grand optimization stories, full of clearly stated assumptions that reflect our educated guesses—guesses that are only uneducated in their neglect of the developmental phenomena.

Consider, for example, the human chin. Humans develop the chins they do as the result of the action in two different developmental fields. The chin is not an organ; it is formed through the interaction of two different facial structures. But it is very easy to tell a story about the value of the chin. In the best sociobiological tradition, we consider the splendid male, and we ask for the optimal design for a male face. Surely the jutting chin is part of that optimal design. Well-chinned males announce their prowess to the world. They do not need to defend this delicate part of the face by withdrawing it. No. They are prepared to assert themselves, and they show it by advancing their vulnerable jaws.

Gould and Lewontin worry that, once started on the project of telling stories of this type, there will be no stopping until one has offered some account of the chin's presence that—temporarily, at least—stands unrefuted. They stigmatize the adaptationist program as follows:

> We would not object so strenuously to the adaptationist programme if its invocation, in any particular case, could lead in principle to its rejection for want of evidence. But if it could be dismissed after failing some explicit test, then alternatives would get their chance. Unfortunately, a common procedure among evolutionists does not allow such definable rejection for two reasons. First, the rejection of one adaptive story usually leads to its replacement by another, rather than to a suspicion that a different kind of explanation might be required. Since the range of adaptive stories is as wide as our minds are fertile, new stories can always be postulated. And if a story is not immediately available, one can always plead temporary ignorance and trust that it will be forthcoming. . . . Secondly, the criteria for acceptance of a story are so loose that they may pass without proper confirmation. Often, evolutionists use *consistency* with natural selection as the sole criterion and consider their work done when they concoct a plausible story. But plausible stories can always be told. The key to historical research lies in devising criteria to identify proper explanations among the set of plausible pathways to any modern result. (Gould and Lewontin 1979, 587–588)

In the light of our previous discussions, we can identify what is insightful and what is misleading in this critique. Consider first the charge that the standards for accepting adaptationist histories are too loose. We have already seen that it is difficult to achieve substantial confirmation for an adaptationist history, and we have explored some of the ways in which the problems may be overcome. Gould and Lewontin would be overstating their case should they deny that it is in principle possible to arrive at good grounds for believing that a particular trait has been produced through natural selection and that a certain benefit has been crucial to its emergence.

There is a deeper point behind their critique. The main aim is to urge biologists to exhibit in their practice what is always allowed in principle—namely, that there are alternative hypotheses to the claim that a trait has emerged under the action of selection. Gould and Lewontin stress the importance of pleiotropy, linkage, allometry, and developmental constraints. When we pick out some characteristic of an organism for evolutionary study, we ought not to commit ourselves immediately to the assumption that the characteristic has been fashioned directly by selection and begin casting about for the benefit that it conferred on ancestral bearers. Instead, we should try to understand the ways in which that characteristic develops in the growth of the organism, so that we can appreciate the genuine possibilities for the action of evolution. In this way we shall not rely on a priori judgments about which properties have been selected, but shall have a rational basis for the formulation of Darwinian histories. The route to the reconstruction of phylogeny may lie—at least in part—through the study of ontogeny.

This point should not be news. It figured in the discussion of the confirmation of adaptationist histories in chapter 2. There I suggested that the possibility of achieving strong evidence in favor of an adaptationist history turns critically on being able to assess the plausibility of rival hypotheses, and that this will typically require knowledge of the genetic and developmental basis of the characteristic in which one is interested. Gould and Lewontin campaign for attention to rival forms of evolutionary hypotheses, but they distort their point by suggesting that the unfalsifiability of adaptationist claims is an insuperable obstacle. If I am right, the correct position is that the successful pursuit of adaptationist hypotheses about traits of organisms already presupposes just that attention to rival possibilities that Gould and Lewontin urge upon their colleagues. Significant confirmation of adaptationist hypotheses is possible—but only if biologists are prepared to take seriously all the forms of evolutionary scenarios that they admit as possible and prepared to undertake the investigations necessary for articulating claims about allometry, pleiotropy, and so forth. (As John Beatty pointed out to me, the Gould-Lewontin point could be accommodated in two different ways. One might encourage all biologists to develop and consider alternative forms of evolutionary explanation.

Or one might foster a pluralistic *community*, in which some biologists pursue adaptationist explanations, other concentrate on developmental constraints, and so forth. For a lucid discussion, see Beatty 1985.)

Viewed in this way, the Gould-Lewontin critique escapes the most important objections that have been presented against it. There is no suggestion, as some have supposed, of jettisoning the concept of adaptation entirely or of refusing to allow any adaptationist explanation as correct. (It is worth noting, however, that Lewontin now espouses a more radical position than that maintained in the paper coauthored with Gould. See Lewontin 1983a, where it is charged that the traditional concept of adaptation is incoherent. I shall not consider the reasons for this latest departure.) Gould and Lewontin can allow for the possibility of sometimes confirming adaptationist histories—although they might also legitimately insist that there will probably be cases in which the evidence available to us is insufficient for us to be able to eliminate all but one of the plausible hypotheses about the evolutionary history of the emergence of a trait in a group of organisms. We should not be so devoted to the project of constructing Darwinian histories that we try to tell such histories even in cases where the methodological canons are unsatisfiable.

Second, it is not right to protest, as Ernst Mayr (1983) has done, that adaptationism is the only game in town and that we can never gain evidence for any rival evolutionary account. Accidents of history are, of course, hard to pin down; yet we know from general considerations that they can sometimes play an important role in evolution, so that a complete adaptationist treatment of the five kingdoms of organisms would be mistaken. Moreover, there are ways of investigating the possibilities of hypotheses that invoke the developmental constraints Gould and Lewontin emphasize; if those investigations are forsworn, then, ironically, the adaptationist histories suffer as well. (I think it is possible that when the conclusion is formulated in this way, Mayr might assent to it.)

Two examples of the adaptationist program may help to make this summary of the Gould-Lewontin critique more concrete. In their exposition of sociobiology (which they consider to be part of the science of "behavioral ecology"), J. R. Krebs and N. Davies acknowledge the dangers of adaptationist storytelling. Like Mayr, however, they do not see any alternative possibility for the evolutionary explanation of behavior. "The problem with nonadaptive explanations is that they are hypotheses of the last resort. Further scientific enquiry is stifled" (1981, 36).

Antiadaptationists are not intellectual Luddites, but they do take seriously the possibility that some hypotheses about the evolution of particular forms of behavior may not be subject to confirmation because of the practical impossibility of gaining evidence. Moreover, they identify certain ways of pursuing unstifled inquiry that behavioral ecologists sometimes seem to neglect. Consider Krebs and Davies on the application of evolutionary theory to animal behavior:

> No-one has studied the genetic basis for copulation time in the male dungfly but it seems reasonable to assume that this is a result of natural selection. In a sense all behavior must be coded for by genes; reduced to its simplest form behavior is nothing more than a series of nervous impulses and muscle contractions and the protein structure of nerve and muscle is coded for by genetic instruction. We expect copulation time, foraging behavior, and other forms of behavior to have been selected during evolution just like other traits such as colour. (1981, 12)

No one should doubt that there is an evolutionary explanation for the fact that dung flies have a particular coloration or that they copulate for a certain (mean) length of time. No one should underrate the role of selection in evolution. But it does not follow that the dung fly's color and the dung fly's copulation time have been "selected during evolution." It may well be true that certain extreme deviations from what are now the typical ranges of colors and copulation times have been selected *against*, the unfortunate organisms who tried these evolutionary novelties having perished without permanent trace. We are in no position, however, to know whether color or copulation time has been selected *for*—that is, whether flies with particular features have left more descendants because of some benefit conferred by the presence of those features—until we have just that understanding of the genetic and developmental basis that, as Krebs and Davies concede, we now lack. Without such understanding, we might be able to claim, on the basis of close agreement between a well-supported optimization model and the findings in nature, that the hypothesis that a particular copulation time has been selected because it confers a precise benefit has become fairly plausible. But to go further, we must take seriously just those alternative possibilities to which Gould and Lewontin draw attention.

For a second example of the constricting power of adaptationism, we can take Hans Kruuk's rightly respected study of the spotted hyena. Kruuk spent about forty months observing the behavior of hyenas in eastern Africa (the Serengeti Plain and the Ngorongoro Crater), and his painstaking, often technically imaginative work has served to overturn many once-popular myths about hyenas. In the course of his inquiry Kruuk was often led to ask about the adaptive significance of apparently puzzling traits. So, for example, pondering the fact that hyena females have external genitalia that resemble those of males, Kruuk concluded that these structures have evolved to aid recognition in the complex meeting ceremonies that are a distinctive part of hyena social life.

> . . . in the female, who is the dominant of the two sexes, the peculiar shape of the genitals has evolved, and Wickler . . . is probably right in saying that it is a mimic on the same structure in males. At present, it is impossible to think of any other purpose for this special feature than for use in the meeting ceremony. (Kruuk 1972, 229; see also 211 and Wilson 1975a, 29)

But as Gould has argued, there is no reason to think that there was selection for the shape of the hyena genitalia (1983, 152ff.). We can investigate the development of the external genitalia and explore the possibility that the striking feature of the female hyena is a by-product of selection for something quite different. The investigation reveals that female hyena fetuses have unusually high concentrations of two androgens. Gould therefore suggests that we take seriously the hypothesis that increased androgen concentration has evolved (possibly to promote greater size and female dominance) and that, as a by-product, female hyenas come to have external genitalia that look like those of males. When discussing the issues involved in this and similar cases, it is helpful to use Sober's distinction between selection *of* and selection *for*. The shape of the hyena genitals emerged from a selection process; there was selection *of* the genital shape. But there was no selection *for* genital shape. (See Sober 1984, 97–102.)

This is no isolated example. Like many other field biologists, Kruuk is fascinated by characteristics that seem to lack any positive adaptive value, and his book displays many thorough attempts to formulate precise adaptationist histories and to gather the observations pertinent to testing them. He is particularly honest in recording his puzzlement at the ways in which various prey of hyenas—wildebeest, zebras, and Thomson's gazelle—respond when they are in danger from a pack of predators (1972, 183–208). Zebra stallions engage in typically ineffectual aggression, gazelle perform their much-discussed "stotting" gait, and gazelle females sometimes try distraction, usually to no avail. Kruuk documents the rates at which hyenas are successful in hunting these animals, and he tries to consider various ways in which the antipredator strategies might be improved.

> Why do wildebeest run into water when chased? [The result is often death from drowning: PK.] Why do not all wildebeest females defend their calves or combine in defense? Why do not zebra females attack hyenas as the stallions do? Why do gazelle stot for so long when being chased before going into a fast run? Many more similar questions could easily be brought up. (1972, 207–208)

Kruuk's own response to these questions is to hope that they can be solved by undertaking the optimization analysis in the context of the full range of ecological variables. But he admits that forms of behavior that look like antipredator strategies still appear "mysterious" even when the broader context is considered, and he concludes that "many more observations are needed to gain insight into these functional aspects of ungulate behavior" (1972, 209). The prior choice of a behavior that, taken by itself, is to be shown to profit the organism exhibiting it goes unquestioned. Unlike more casual adaptationists, Kruuk does not inhabit a world where every prospect pleases, but he is haunted by the thought that every prospect ought to please, if only he understood it aright.

O Brave New World

The adaptationist lapses we have just reviewed pale in comparison with pop sociobiological practice. There are serious issues about how the community of evolutionary biologists should distribute resources among projects that seek adaptationist explanations for the presence of current traits and projects that explore alternative types of Darwinian history. For pop sociobiology these issues never arise. An integral part of the program is the identification of patterns of human—and sometimes nonhuman—behavior with the deliverances of casual optimization. We have already seen numerous examples. Recall the spur-of-the-moment thoughts about advantages reviewed in chapter 5. Recall Wilson's efforts to show that sexual differences in behavior conform to evolutionary expectations. More examples await us in the next chapter.

For those who want to climb Wilson's ladder, extreme adaptationism is an article of faith. Part of the method for fathoming human nature is to compare the results of optimality analysis (or what passes for optimality analysis) with a conclusion about what people actually do. When we find coincidence (more or less), pop sociobiologists discern the hand of selection and proceed to their conclusions about "genetic bases of behavior" and "whisperings within."

It should be no surprise that some of Wilson's followers equate evolution with selection and take it as a first principle that selection optimizes. Barash announces that "sociobiology has emerged from the recognition that behavior, even complex social behavior, has evolved and is adaptive" (1977, 8). The slide is not only licensed in principle but applied ruthlessly in practice. Shorebirds are subject to conflicting pressures. Clustering helps them to detect and avoid their predators but interferes with their efforts at obtaining food. Barash is nonetheless confident that "gregariousness in the case of each species of shorebird represents a unique and optimal compromise between the conflicting demands of predator avoidance on the one hand and avoidance of maladaptive foraging interference on the other" (1977, 119). No evidence is offered for this conclusion. Indeed, a refined analysis of the problem, using game-theoretic techniques, explicitly disavows any such evaluation. (See Pulliam and Caraco 1984, 138ff., especially 147; as in many other cases, the analysis given holds out exciting prospects for further development.) Barash's assertions about optimal compromise simply reflect Panglossian faith.

When we come to human beings, adaptationism hits its stride. All Barash's provocative conclusions about sexual relationships are obtained by comparing loose optimality models with data that is often anecdotal. Barash is explicit on the technique: "At this point, it is only possible to point out the similarity between the predictions of natural selection, confirmed time and again in all other species, and the reality of human behavior. We may then decide simply to marvel at the coincidence" (1979, 89). Of course, the suggestion is that we should not marvel because there is no coincidence. When human behavior fits optimality analysis, the shaping hand of selection is revealed.

Wilson is typically more cautious than Barash. Well aware that natural selection is not the only agent of evolutionary change, he takes explicit note of the fact (see 1975a, 20–25). Moreover, as his work with Oster makes clear, he is sensitive to the conditions that optimality analyses should meet. Yet, in the heat of the quest for human nature, restraint and caution vanish. Tendencies to homosexuality, sexual strategies, hostility to strangers, and a host of other forms of social behavior are to be viewed as optimal and thus as shaped by selection. Pages are spent in speculation about how those who submit themselves to bizarre and extreme demands of religion really are favored, somehow, by selection (1978, 185ff.). Flights of Panglossian fantasy are devoted to defending the implausible generalization that "when the gods are served, the Darwinian fitness of the tribe is the ultimate if unrecognized beneficiary" (1978, 184).

Wilson's search for adaptation everywhere takes some curious turns. According to the barroom stereotype, males are to be ruthless competitors, doing whatever they can to secure whatever it is that brings access to females. Yet some men face a difficult struggle and eventually give up.

> !Kung men, no less than men in advanced industrial societies, generally establish themselves by their mid-thirties or else accept a lesser status for life. There are some who never try to make it, live in rundown huts, and show little pride in themselves or their work. . . . The ability to slip into such roles, shaping one's personality to fit, may itself be adaptive. (1975a, 549)

Doubtless Lady Bracknell would have been comforted by this account of the obstinate persistence of the lower orders.

The idea that it is adaptive to retreat from competition is, of course, at odds with the more prominent idea of male competitiveness, and there is no serious attempt to demonstrate that males compete precisely where it would enhance their fitnesses to do so. There is no need to make any such attempt. Despite his prefatory remarks about rival agents of evolutionary change, Wilson's concrete discussions of patterns of human social behavior are dominated by an undefended answer to the question "What human behavioral traits are adaptive?" All of them.

Other pop sociobiologists share the urge to see optimality and selection here, there, and everywhere. Alexander asserts that "we must expect all functions of the organism to be reproductive, and maximally so" (1979, 25). The Panglossian claim is qualified a little but remains very strong.

> It seems ludicrous to suggest that all activities of humans derive from the reproductive strivings of individuals, or more properly their genes (except activities that because of environmental changes are temporarily maladaptive—in the biological sense of not being maximally reproductive). Unless there are flaws in the argument presented so far, however, we are compelled to examine this hypothesis. (1979, 26)

The examination fails to uncover flaws, and we are launched on an adaptationist tour of human social institutions.

The ultimate version of the adaptationist method is to allow optimality analyses to substitute for surveys of the ways human beings actually behave. There are numerous areas in which we are ignorant of patterns of human social behavior. Evolutionary expectations can turn this ignorance into knowledge. To see how powerful the technique can be, consider the following reflections on the relation between female orgasm and spontaneous abortion.

> A female rat exposed to the scent of a strange male will commonly resorb her litter if she is pregnant (Bruce 1966). This response may be considered adaptive since it terminates investment in a litter which would probably be killed by the new male had it been carried to term, while initiating the sequence leading to receptivity of the female to the new male. Throughout human history it has not been unusual to take women as spoils of war (Mead 1950). If the subsequent offspring were subject to infanticide, those women who aborted as a consequence of orgasm would be at a selective advantage. (Bernds and Barash 1979, 500)

Elsewhere the evidence for a correlation between orgasm and induced abortion is described by the authors as "circumstantial." No matter. Consider what we can learn from the adaptive story. There is a connection between orgasm and spontaneous abortion. Women were frequently taken as spoils of war. It was so likely a fate that it was important for them to be prepared. So they came to achieve orgasm in intercourse with their conquerors (or maybe even before—for the conquerors may sometimes have been a little slow in enjoying all their spoils). They spontaneously aborted, quickly bore thriving children, and lived happily ever after. And that, O Best Beloved, is the way the woman got her orgasm.

At its most extreme, pop sociobiology offers guesses about the ways people actually behave (and have actually behaved) based on "evolutionary expectations." The expectations are drawn from a truncated version of evolutionary theory in which the optimizing hand of selection is seen in every detail of human social life. We can only echo Miranda: "O brave new world, that has such people in it!"

Chapter 8

The Proper Study?

A Surfeit of Suspects

One of the best tricks of a good mystery lies in exploiting the reader's assumption that there is a short list of suspects, exactly one of whom is responsible for the crime. When Hercule Poirot assembles the suspects in the dining car of the Orient Express, we have well-grounded suspicions about each of them. Yet we are not prepared for Poirot's solution. All the suspects are guilty.

Many critics of pop sociobiology write as if there ought to be a single crucial flaw in the program. Criticism would be simpler if that were so. If we could expose one error underlying all the faulty analyses of human social behavior, then it would not be necessary to proceed, as I have done, by examining example after example. Unfortunately, sociobiology is a motley. Not only is there no single monolithic theory to be scrutinized, but the individual Darwinian histories offered by pop sociobiologists may be flawed in any of a number of different ways. There is a family of mistakes, and in distinct examples distinct members are implicated.

Wilson's ladder begins with an attempt to show that a particular form of behavior would maximize the fitness of the members of a particular group of animals. As we have discovered, the analysis that is intended to show how fitness is maximized may be carried out more or less rigorously. Assumptions may be formulated precisely. Constraints may be justified by reference to antecedent knowledge of the biology of the animals under study. Or we may be offered a story that is vague and qualitative, with conditions that are pulled out of thin air.

Already, at the first stage, the language employed to describe the behavior deserves our scrutiny. It is important to ask whether we have grounds for thinking that the animals whose behavior is described in the same terms are all doing the same thing. It is important also to wonder if evolutionary assumptions are being smuggled in via descriptions of behavior that identify a function for it. Finally, there are ample opportunities for sociobiologists to employ language that is vague or ambiguous, to slide back and forth between claims about groups and claims about individual animals, to extend the application of a concept that makes sense only in a narrow context.

At the next stage, when the deliverances of the optimality analysis are compared with the actual behavior, there are further opportunities for the argument to go astray. Are the conclusions of the model compared with precise observations? Are discrepancies taken seriously or are they dismissed? If hypotheses are introduced to explain anomalous findings, do those hypotheses resolve the difficulties? Does their introduction threaten to subvert examples in which the model has previously appeared to be successful? Does the analysis proceed from a one-sided view of animal interactions, neglecting to explain why conflicts should be resolved in a particular way?

Supposing that all is sound so far, we must ask next whether we are entitled to treat a promising optimality analysis as signaling a history of selection. We know that there are general problems in finding a coherent gene's-eye view of the optimizing power of selection. Can we attribute to the optimality analysis such power that it enables us to neglect our ignorance of the genetic details? Can we argue that the fit between the deliverances of the model and the findings is so precise, and the data are so surprising on any rival view, that we are justified in thinking there must be some genetic mechanism—we know not what—that satisfies the requirements of the analysis? Should we ignore the possibility of rival accounts, simply assuming that further knowledge of the proximate mechanisms and of the development of the behavior would not affect our interpretation of what the animals are doing? Those who perceive sociobiology, and large parts of evolutionary theory, as consisting in a misguided adaptationist program will contend that the answers to all these questions are invariably negative. I do not endorse their skepticism. Instead, I recommend that we think critically about individual analyses. Faced with a particular endeavor to construct an ambitious Darwinian history on the basis of an optimality analysis, we must raise the questions I have listed. In many cases we shall find that the optimality models are too imprecise and the consideration of rival hypotheses too scanty for us to be entitled to forget about the intricacies of proximate mechanisms, genetics, and development.

Even if sociobiological analyses scramble this high, they still do not attain the dizzying heights to which Wilson and his followers aspire. If we are to announce grand conclusions about human nature, then we cannot content ourselves with the simple truths that there has been a history of selection and that there have been genetic changes in our lineage so that people behave in particular ways in our typical environments. What we want to know, if we are to resolve the "major intellectual controversy of our generation," is what variations in behavior could be produced given different social environments. Pop sociobiologists can only parlay their adaptationist histories into answers to the main questions by adopting dubious reductionist tactics. They must try to show that the range of possible environments that might be expected to modify the behavior under study is smaller than we think, that it turns out that evolution has already adapted us so as to fix the behavior across that range, and that alterations would disrupt social institutions that are themselves expressions of our nature, thus forfeiting qualities that we naturally value. There is no general reason to believe that these tactics are based on sound assumptions; and in particular instances, as we have seen, they are highly suspect. A thorough study of human behavioral genetics would, of course, make up for the deficiencies and guesswork in the Wilsonian claims about modifying our behavior only at great cost. Indeed, the detour through evolutionary studies of behavior tempts the aspiring discoverer of human nature precisely because the relevant parts of human behavioral genetics are unavailable.

Early Wilsonian sociobiology achieves its impact by piling putative example upon example. Because there is no single locus of theoretical controversy, we are forced to examine the conclusions case by case. Sometimes the analyses are flawed in one way; sometimes a number of different errors combine. My aim has been to identify a surfeit of suspects. Each suspect is guilty some of the time, no suspect is guilty all of the time, and each grand conclusion about human nature involves at least one guilty suspect. Such is the predicament of Wilson's pop sociobiology.

Two Cheers for Homosexuality

I have indicated some of the main attractions of the program in the course of surveying the potential pitfalls. Having done so, I could bid farewell to the early Wilsonians, confident that the errors have been unmasked and that the task of identifying them in particular instances can be left as an exercise for those who have interests in specific topics. Or, at the risk of exhausting my own and the reader's strength and patience, I could begin a systematic overhauling of the entire collection of accounts of human nature offered by Wilson and his disciples. Rather than taking either extreme course, I shall devote this chapter to looking at a number of the most popular examples of human sociobiological analysis. By combining discussion of these with the previous critiques of Wilsonian analyses of gender differences, territorial defense, male tendencies for rape, recreational behavior, love of home, and so forth, I hope to scotch the impression that pop sociobiological research promises us new insights about ourselves.

Let us start with Wilson's account of homosexual behavior, an account apparently designed with the laudable aim of showing that homosexuality need not be "unnatural." The topic is introduced *en passant*, in a discussion of "castes."

> Castes in vertebrates, if such exist, should take the form of distinctive physiological or psychological types that recur repeatedly at predictable frequencies within societies. Some would probably be altruistic in behavior—homosexuals who perform distinctive services, celibate "maiden aunts" who substitute as nurses, self-sacrificing and reproductively less efficient soldiers and the like. (1975a, 311)

The focus seems to be on discovering vertebrate analogs for the sterile members of insect colonies. But Wilson is also inspired by Trivers's analysis of parent-offspring conflict and by the suggestion that a parent might maximize its inclusive fitness by manipulating the behavior of one of its offspring. A later passage develops the point.

> Under some circumstances parents can be expected to influence the offspring's behavior into its adult life. Altruistic behavior might be induced when it results in an increase in inclusive fitness through benefits bestowed on the parents and other relatives. The celibate monk, the maiden aunt, or the homosexual need not suffer genetically. In certain societies their behavior can redound to improved fitness of parents, siblings, and other relatives to an extent that selects for the genes that predisposed them to enter their way of life. (1975a, 343)

The picture begins to take shape. There is a "genetic predisposition" for homosexuality, and the relevant genes are favored because of inclusive-fitness effects.

Wilson attempts to assemble a "set of clues" (1978, 146) by juxtaposing the deliverances of evolutionary theory and behavioral genetics. He appeals to the data of twin studies to claim that there is "some evidence" for "some amount of predisposition to homosexuality" (1978, 145). Joining this claim with the suggestion that homosexuals might prove helpful to their kin, he campaigns for tolerance of homosexual behavior: "There is still another cost—and some of our members are already paying it in personal suffering—for the society that insists on conformity to a particular range of heterosexual practices" (1978, 148). We pay a price for trying "to circumvent our innate predispositions."

It is hard to disagree with Wilson when he pleads for the abolition of the prejudice and repression that have blighted the lives of so many homosexuals. Yet the proper argument for tolerance is surely the one given by John Stuart Mill long ago. We should restrain the liberties of people only when their actions affect the welfare or curtail the freedom of others. Wilson's line of reasoning generates the right result by offering a dubious hypothesis and making a palpable inferential error. Biology is supposed to tell us that homosexual behavior is natural—at least for some people. From this hypothesis we are to infer that homosexuality is to be tolerated, that we should not attempt to correct, punish, or coerce. Those with clear heads and repressive inclinations will laugh this idea to scorn. Even if "genes favoring homosexuality" can be maintained in a population "at a high equilibrium level by kin selection alone" (1975a, 555), the same can be said for many genes that predispose their bearers to conditions we would like to correct. The allele for sickle-cell anemia is maintained in African populations by natural selection. Sickle-cell anemia is a disease, for all that, and we do our best to alter the conditions of those who carry two copies of the deleterious allele. More generally, Wilson has been guilty of the common mistake of thinking that revealing something as natural suffices to present it as good, or at least tolerable. The thunderings of the Moral Majority can already be heard: "As natural as venereal disease, and as resolutely to be opposed!"

My present concern is not with Wilson's faulty ethical argument but with his dubious biology. The central message is that there "may well" be "some predisposition" for homosexuality and that this predisposition "may be" adaptive. Two questions confront us: What does the conclusion mean? How is it derived?

The most explicit account of Wilson's biological hypothesis is given in the following passage:

> Like many other human traits more confidently known to be under genetic influence, the hereditary predisposition toward homosexuality need not be absolute. Its expression depends on the family environment and early sexual experience of the child. What is inherited by an individual is the greater probability of acquiring homophilia under the conditions permitting its development. (1978, 145–146)

Despite Wilson's praiseworthy desire to encourage tolerance of homosexuals, this passage hardly offers three rousing cheers for homosexuality. The apparent comparison is with susceptibility to a disease ("acquiring homophilia"). The comparison does, however, help us to formulate Wilson's thesis.

Consider a very simple genetic model. Sexual preference is determined by a single locus. Individuals who carry the combination *AA* or *Aa* at this locus have a predisposition for heterosexuality; those bearing *aa* have a predisposition for homosexuality. The talk of predispositions is to be understood as follows. Associated with the allelic pair *aa* is a function that maps possible environments onto behavioral phenotypes. In a certain class of these environments—call it H_{aa}—*aa* individuals have a preference for engaging in sexual activity with members of their own sex. Similarly, the genotypes *AA* and *Aa* are associated with functions that map possible environments onto behavioral phenotypes, and there are classes H_{AA} and H_{Aa} in which *AA* and *Aa* individuals manifest homosexual behavior. One way to interpret Wilson's claim that *aa* individuals have a greater probability of "acquiring homophilia" is to claim that the class H_{aa} properly includes both H_{AA} and H_{Aa}. Intuitively, the idea is that individuals with a genetic propensity to homosexuality develop into homosexuals if they are reared in a certain class of environments, which includes all those environments in which individuals without genetic propensity become homosexuals and more environments besides. (See figure 1 for a graphical representation of this interpretation and an alternative that appears less representative of Wilson's intentions.)

Once the conclusion has been stated, it becomes obvious that it is not easy to support. We do not have the ability to breed pure lines of human beings and examine the sexual behavior of various strains when they are reared in different environments. So how are we to obtain evidence?

One source is human behavioral genetics. Behavioral geneticists vigorously debate the issue of whether there is a genetic propensity for homosexuality. They do so because the standard methods of studying the question do not yield clear answers. Wilson cites studies by Kallmann (Wilson 1975a, 555) and by Heston and Shields (Wilson 1978, 145). Both studies are admitted to "suffer from the usual defects that render most twin analyses less than conclusive" (1978, 145; this is to overpraise the Kallmann study; see Futuyma and Risch 1984, 162–163). Both are based on analyses of correlations in sexual behavior among siblings and twins. We find that twins who "originate from a single fertilized egg . . . are more similar in the extent to which they express heterosexual or homosexual behavior than is the case for fraternal twins, which [*sic*] originate from separate fertilized eggs"

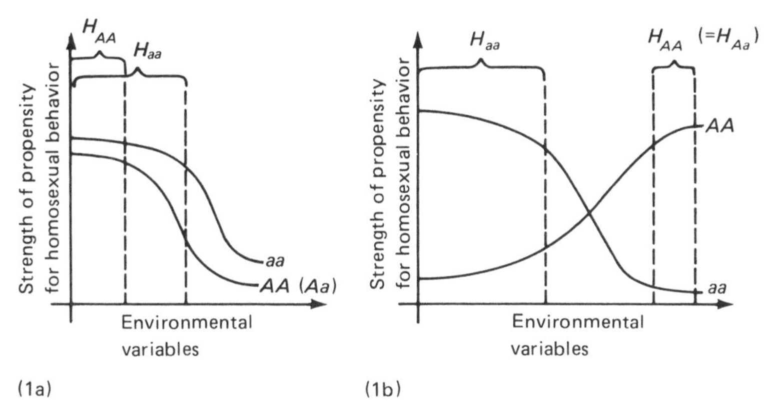

Figure 1
Both (1a) and (1b) represent ways in which H_{aa} might be larger than $H_{AA}(=H_{Aa})$. In both cases, if all values of the environmental variables are equally likely, *aa* individuals have a greater probability of developing into people with a strong propensity for homosexual behavior. The most natural interpretation of Wilson's claim is that he has something like (1a) in mind—*AA* (and *Aa*) individuals have a lower propensity for homosexual behavior given any developmental environment, and they only develop into homosexuals in environments in which *aa* individuals become homosexuals. (1b) is less representative of Wilson's apparent intentions because it does not support the idea of a genetic propensity for homosexual behavior: if (1b) depicted the actual situation, then there would be environments in which people lacking the alleged propensity developed a far stronger disposition to homosexual behavior (environments in H_{AA}).

(1978, 145). What does this show about predispositions to homosexuality?

Not much. Recall that the intended conclusion is that all the environments in which individuals without the genetic predisposition develop homosexual behavior are environments in which those with the predisposition develop the behavior, but there are environments in which the latter develop into homosexuals but the former do not. The rival account is that, irrespective of genotype, people develop very similar sexual preferences if reared in similar environments (see figure 2). How are we to support Wilson's conclusion and discredit the rival picture by showing that genetically identical individuals are more likely to have similar sexual preferences than siblings who are not genetically identical? Apparently, what is needed is a premise to the effect that the difference in the developmental environments for monozygotic twins is approximately the same as the difference in the developmental environments for any two siblings. If that premise were true, then one could argue that the rival picture is at odds with the observed correlations. Given the rival picture, there is no expecta-

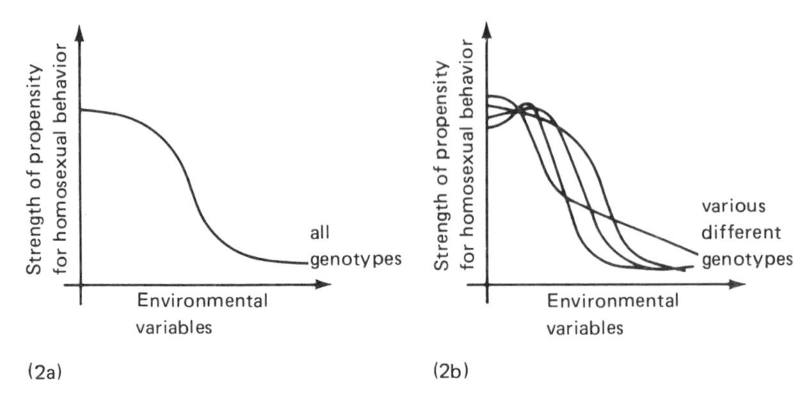

Figure 2
A rival picture of the possible relationships between genetic and environmental factors in the development of homosexual behavior. On the simple version (2a), all genotypes exhibit the same pattern of variation with the environment. Sophisticated proponents of the idea that sexual preferences are largely shaped by environmental factors will probably prefer something like (2b), in which differences in genotype cause small disturbances of a common environmental pattern. There are obviously vast numbers of ways of thinking about the roles of genes and environments in the development of sexual preferences. For example, a picture like (2b) may be an accurate representation of the relations among common genotypes, while something more like (1a) or (1b) may hold for certain rare genotypes.

tion that genetically identical individuals will be more likely to agree in sexual preference than people who are genetically dissimilar (see figure 3).

The premise is almost certainly false. As behavioral geneticists know all too well, monozygotic twins share environments that are far more similar than the environments of ordinary siblings or of fraternal twins. The similarity of the environment confounds the effects of the genetic identity. The grand (and unsurprising) conclusion is thus that individuals who are more alike genetically and who are reared in more similar environments are likely to behave more similarly than those who are genetically more diverse and who are reared in more varied environments.

Wilson's case does not rest on the deliverances of behavioral genetics alone, however. Most of his treatment of the subject is devoted to arguing that homosexuality may be favored by natural selection.

The tale begins with an important assumption. Wilson takes it for granted that there is a single type of sexual behavior—choice of a partner of the same sex—whose presence is to be understood. This is by no means as obvious as it may at first appear. Despite the advances that have been made in the study of human sexuality, we still

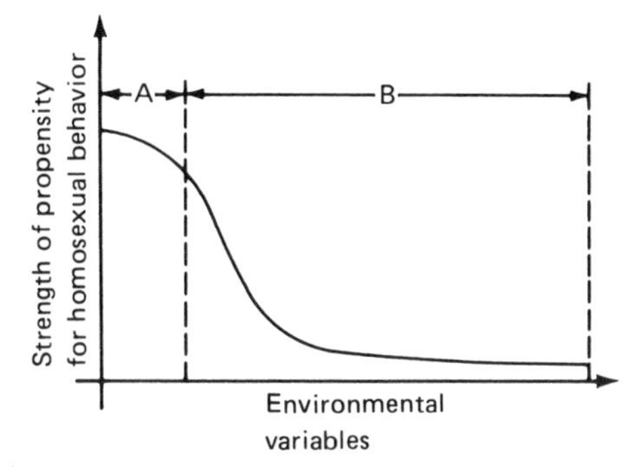

Figure 3
Suppose we take a simple version of the idea that environmental factors are critical in the development of homosexual behavior (as in figure 2a). The data reveal that monozygotic twins are far more likely than fraternal twins or ordinary siblings to share a sexual preference. If our picture is correct, then individuals who share sexual preference either both develop in environments in region A or both develop in environments in region B. Thus, monozygotic twins must be much more likely than other siblings to develop in similar environments. Hence, if we can assume that the environments of monozygotic twins are no more similar (in the important respects) than those of other siblings, then we can use the data to conclude that the simple picture is inaccurate. We can rehabilitate the picture by giving up the assumption. More sophisticated versions of the idea that environmental factors are crucial can also account for the data in the same way—and they have additional resources for explaining the observed correlations.

do not know whether our primary classification of sexual behavior should focus on the sexes of the participants or on some other features of the situation. We do not know if there are importantly different types of homosexuality and heterosexuality or whether some of the former are more similar to some of the latter than they are to other kinds of homosexuality. Lumping forms of behavior together may deceive us into thinking that there is a unitary phenomenon to be explained, whether in imperial Rome, ancient Greece, or the San Francisco of today. (We do know enough to recognize that there are important differences in patterns of homosexual behavior across cultures. See Money and Ehrhardt 1972, 227ff.)

Content to think of human homosexual behavior as a single form of behavior whose presence requires evolutionary explanation, Wilson quickly broadens the scope of his investigation.

> Homosexual behavior is common in other animals, from insects to mammals, but finds its fullest expression as an alternative to heterosexuality in the most intelligent primates, including rhesus macaques, baboons, and chimpanzees. In these animals the be-

> havior is a manifestation of true bisexuality latent within the brain. Males are capable of adopting a full female posture and of being mounted by other males, while females occasionally mount other females. (1978, 143–144)

Virtually any dog owner is familiar with the kinds of episodes Wilson cites. Some dogs—especially, in my experience, male lhasa apsos—have a pronounced urge to mount anything in easy reach. Tree stumps, human limbs, items of furniture, all prod them into a form of activity that they will happily force on any other dog (of appropriate size) that swims within their ken. Before we speculate on the plasticity of canine sexuality, on its spendid opportunism and thriving creativity, it is well to remind ourselves of the deliverances of an older tradition in the study of animal behavior. Primate watchers have offered a rival account of the familiar mounting behavior. After an aggressive interaction a defeated animal will often allow itself to be mounted by the victor. Perhaps the occasions of "homosexual behavior" to which Wilson refers can be understood in terms of the resolution of animal conflicts, actual or potential. Perhaps they have to do with something completely different. It is surely hasty to interpret them as signs of "true bisexuality latent within the brain."

At the first step of the ladder, then, we find a lumping together of forms of human behavior and an anthropomorphic description of the activities of other primates. The next step is to spin the adaptive story.

> How can genes predisposing their carriers toward homosexuality spread through the population if homosexuals have no children? One answer is that their close relatives could have had more children as the result of their presence. The homosexual members of primitive societies could have helped members of the same sex, either while hunting and gathering or in more domestic occupations at the dwelling sites. Freed from the special obligations of parental duties, they would have been in a position to operate with special efficiency in assisting close relatives. They might further have taken the roles of seers, shamans, artists, and keepers of tribal knowledge. If the relatives—sisters, brothers, nieces, nephews, and others—were benefited by higher survival and reproduction rates, the genes these individuals shared with the homosexual specialists would have increased at the expense of alternative genes. Inevitably, some of these genes would have been those that predisposed individuals toward homosexuality. (1978, 145)

The stereotype of the predatory pervert is to give way to that of the helpful homosexual. Every home should have one.

Unfortunately, Wilson's tale is just another story. There is no reason to think that the question with which Wilson begins is a realistic one. Do those who carry genes predisposing them to homosexual behavior (if such genes there be) have any fewer children than those who do not? Who knows? It might seem so—until we consider the possibility that homosexuals might have a greater propensity for sex than heterosexuals and that social pressure might channel their sexual energy into heterosexual behavior. Is there an inclusive-fitness effect of homosexual behavior that outweighs the loss in genetic representation that supposedly accrues from homosexual childlessness? Wilson offers us no model—only the dimmest suggestions of homosexuals helping their kin. No attempt is made to show that there are ecological conditions under which a family unit might have prospered from the presence within it of homosexual helpers (the prosperity resulting in enhanced inclusive fitness for those members who abandoned reproduction on their own). Are we to suppose that there is a similar explanation for the persistence of (hypothetical) genes that predispose parents to produce some sterile or asexual offspring? Is there some pleiotropic effect of the predisposing genes that associates homosexuality with artistic talent or with the ability to persuade others that one has second sight? How exactly were "seers, shamans, artists, and keepers of tribal knowledge" able to assist their relatives? Did they need to be homosexual to do so? The questions come flocking in. Since nobody knows the answers, we can be liberal in granting the right to tell stories—as long as we are clear that that is exactly what is being done.

Wilson begins his discussion of homosexuality with an important truth: "Nowhere has the sanctification of premature biological hypothesis inflicted more pain than in the treatment of homosexuals" (1978, 142). The remedy is not to replace a biological myth with some unfounded speculations. There is no good reason to think that there is a single type of behavior corresponding to what we call "homosexuality." There is no evidence for the existence of genetic predispositions for such behavior. There is no justification for viewing the behavior as selected because of inclusive-fitness effects.

By contrast, there is every reason to believe that bigoted repression of homosexuals should give way to tolerance. We do not need fantasies about genetic predispositions and adaptive behavior. We know that, given their genetic constitution and given their developmental environments, some people engage in homosexual behavior when they reach adulthood. Frequently their activities restrict no liberties and cause no harm to others. When that is so, elementary considerations of human liberty require that we should not interfere, either by force or by moral coercion.

"Mislike Me Not for My Complexion"

Portia's first suitor arrives from Morocco and, worried that his appearance tells against him, introduces himself with an allusion to his color: "Mislike me not for my complexion." The line signals an unpleasant feature of our lives, as familiar to us as it was to Shakespeare—interracial hostility.

Some people harbor hopes that racial dislikes can be eradicated if some of the causes of racial divisions are removed and if peoples of different ethnic heritage come to appreciate each other's backgrounds and traditions. They hold that we have no natural disposition to hate those who belong to different races. Their view is encapsulated in a song from *South Pacific*: "You've got to be taught before it's too late, before you are six or seven or eight, to hate all the people your relatives hate. You've got to be carefully taught, you've got to be carefully taught."

Barash quotes the lines, but his theme is not that of the sailor who sings the song. According to Barash, evolution may "incline us to a degree of racial bigotry" (1979, 154). Those who think that our social project must be to undo the harmful divisions and injustices imposed by our cultures, past and present, are cast as romantic idealists. "If sociobiology is correct, we've got to be carefully taught *not* to hate others who are different from ourselves, because it may be our biological predisposition to do so" (1979, 154). Barash's argument is that "the principles of kin-selected altruism" favor genes promoting antagonism toward those who look different from ourselves. The argument develops in three stages.

The first phase concerns not humans but ground squirrels. Barash reports on some experiments. Ground squirrels will sometimes "adopt" young who are not their offspring. It is thus possible to mimic the standard conditions beloved of human geneticists, producing pairs of siblings reared together, pairs of siblings reared apart, unrelated individuals reared together, and unrelated individuals reared apart. When the animals are brought together, kin are able to recognize one another and are prepared to treat one another with special attention.

> Biological siblings, whether reared together or apart, are "nicer" to each other than are sociobiological siblings. They showed less

> aggression and more huddling and mutual grooming. The ultimate benefit is clear enough: genes help themselves by being nice to themselves, even if they are enclosed in different bodies. (1979, 153)

The work is surely suggestive. There do appear to be mechanisms by which animals recognize their kin; and, *other things being equal*, we should expect that genes promoting cooperative exchanges would be favored if the inclusive-fitness effects outweigh the costs of engaging in the cooperative behavior. Other things are not always equal, however, and there are familiar kinds of activity in more complicated primate societies in which animals associate with troop members who are not their kin or in which they disperse to join a group of nonrelatives. Hence, while it would be wrong to deny the existence of inclusive-fitness effects in the social lives of mammals, the crucial questions concern the kinds of situations in which those effects are likely to dominate the other factors that affect fitness. There is no basis for assuming some blanket behavior of cooperating with relatives in all contexts. Even if we see the mutual grooming of kin among ground squirrels as a form of behavior that is favored by selection in accordance with Hamilton's inequality—and I hasten to note that the details of the fit have not been worked out—that does not mean that selection has favored some general tendency for animals to be "nice" to those who are likely to share their genes.

At the next phase we jump from careful experiments with ground squirrels to the elementary-school classroom and the summer camp.

> Experimenters with schoolchildren have shown that they can quickly divide a classroom into distinct camps, with real affiliation within each group and antagonism between them, simply by focusing on some identifiable trait that separates some and unites others. For example, blue and green eyes on the one hand versus brown eyes on the other. Or the purely arbitrary division into teams at summer camp "color wars." We seem always distressingly eager to make use of such distinctions, even when we are very young, and whereas the discrimination that develops may seem "inhuman," kin selection and inclusive fitness theory suggests that such behavior is all too human. Indeed, it even suggests an evolutionary tendency for racism. (1979, 153)

Barash provides no sources for his claims about the experiments with schoolchildren, so there is no way to assess the merits of the studies. However, unless he has misreported the findings, the research is irrelevant to his point. First, the kinds of behavior to which he alludes seem to have very little to do with *racism.* Some argument is needed to connect the competitive games of schoolchildren with the activities of organizations like the Ku Klux Klan. Second, the idea of an evolutionary tendency toward any form of propensity for discrimination is not supported by the mere possibility of indoctrinating children so that they discriminate against some of their peers. We need no elaborate experiments to show that that is possible. What is at issue is whether the discriminatory practices observed among young children develop only in environments where such discriminations are deliberately fostered. We know that we can kill a rose bush by stamping on it, failing to water it in dry weather, and neglecting to protect it from frosts. What we want to know is whether it will produce flowers if we treat it less harshly.

We are now ready for the denouement. Cooperative exchanges in related ground squirrels and socially inculcated discriminations in the classroom are supposed to point us to general conclusions about *Homo sapiens.*

> People of different races *are* different. Although we are all one species and quite capable of exchanging genes, the fact remains that members of any race seem likely to share more genes with each other than with individuals of a different race. Physical resemblance almost certainly has some correlation with genetic resemblance, and, accordingly, we can expect the principles of kin-selected altruism to operate on this fact. More to the point, we can expect the other side of the coin—antagonism—toward those who are different. (1979, 153)

In case the reader expects something more, let me note that this is it. Barash continues with some speculations about the evolutionary origins of gross racial differences, with a sketch of some highly controversial work on racial stereotypes (by the Dutch sociologist Hoetink), and with some expressions of regret that he should have been driven to such uncomfortable conclusions. Given the poverty of the argument, we can only wonder what does the driving.

Barash starts with a truism. Plainly people whom we count as belonging to different races are different. They differ in those superficial respects on which humans have based distinctions for centuries. They differ also in history, customs, habitat, and fortune. Because of the accidents of history, Caucasians have more common ancestors in the recent past than they have with members of other races (for example, blacks). As a result, the probability that I will share genes identical by descent with other Caucasians is higher than the probability that I will share genes identical by descent with blacks. What follows?

Not much. Let us take seriously the possibility that selection should have favored an allele in each race with the phenotypic effect of disposing the bearer to be "nice" to members of the same race. If the "principles of kin-selected altruism" are to operate, then we shall have to suppose that there is a disposition to perform actions helpful to others under the conditions of Hamilton's inequality. Suppose that the average coefficient of relatedness between members of the same race is r_s. Then the disposition should favor bringing benefits, *B*, to a member of the same race, at cost *C*, provided that $B/C > 1/r_s$. Equally, the principles of kin-selected altruism ought to dispose people to help members of different races provided that $B/C > 1/r_d$, where r_d is the average coefficient of relatedness of the members of different races. We should thus expect a willingness to help members of the same race under conditions that would not prompt help to members of other races. More exactly, when $1/r_s < B/C < 1/r_d$, the benefit should be dispensed to a member of the same race but not to a member of a different race.

Since r_s is larger than r_d (members of the same race have a higher coefficient of relatedness than members of different races), there will be some values of *B* and *C* that satisfy the condition. But now we come up against an obvious difficulty. The values of both r_s and r_d are very small. Hence, the behavior that can be expected is a tendency to aid members of the same race but not members of different races, only under very extreme circumstances. More exactly, this might be expected if we assumed that conditions were right for the attainment of an optimum and that there were no complicating factors based on histories of reciprocation with various kinds of people. Once we realize how very slight the differences in behavior toward members of the same race and members of different races are expected to be, it is easy to appreciate that differences in actual behavior might be obliterated beyond recognition.

So, when the appeal to "kin-selected altruism" is taken seriously, there is no reason to think that selection has favored some general tendency for humans to be "nice" to their kin, and even less reason to believe that people who belong to the same race as ourselves will bask in the radiant glow of our love. Even if we were to swallow this much of Barash's fable, a further step would be needed to reach the conclusion that we have been selected to be antagonistic to those who are different from ourselves. Recall the intricacies of our earlier discussions—based on admittedly oversimplified models—of hostile exchanges among animals. Barash is not one to confuse his readers with these mathematical niceties. Once assured that we have a genetic propensity to cooperate with those who look vaguely like ourselves, he concludes at once that we have a genetic propensity to be hostile to those who are different. Never mind the fact that hostile actions typically bring costs. Do not worry about the details of cost-benefit ratios and coefficients of relationship that would have to enter into any serious application of Hamilton's ideas about inclusive fitness. These complications should not be allowed to spoil a good story.

But the story is not good. It is nonsense, and dangerous nonsense to boot. We are asked to accept the idea of a genetic predisposition to racial hatred, on the basis of an argument that makes the following mistakes. First, it misreports an analysis of certain limited types of cooperative exchange in a nonhuman species, so as to suggest that the concept of inclusive fitness underwrites some general policy of being "nice" to kin. Second, it offers irrelevant and misleading information about the propensities of schoolchildren. Third, it waves its hands in the direction of an application of Hamilton's insight that breaks down as soon as we think seriously about the coefficients of relationship involved. Fourth, it leaps without argument, and without even the sketch of a theoretical analysis, to supposing that sympathy toward members of the same race will be mirrored by antagonism toward members of different races. If the issue of the persistence of racial distrust were not so important, there would be only one appropriate response to such "reasoning": derisive laughter.

Understanding Adolescence

Wilson has high praise for Barash's exposition of pop sociobiology. The passage I have just analyzed appears in a volume bearing the master's testimonial: "A delightful book. Forcefully, effectively argued" (Wilson, jacket blurb for Barash 1979). And that passage is not atypical. My earlier quotations from Barash and from Wilson's own popular treatment (1978) should make obvious at what expense these authors' arguments achieve their noted force and effectiveness. There is an excuse, say supporters: in the effort to reach a wide audience, pop sociobiology is sold short. To see Wilson's program at its best, we should trace the controversial conclusions to their roots in more technical pieces.

I do not accept the excuse. The popular presentations are where much of the action is. These are the pages on which pop sociobiologists advance their provocative claims about sexual behavior, human aggression, racial hatred, and all the rest. Errors in analysis and misleading reporting are not minor peccadilloes, when the human stakes are so high.

Nevertheless, even though the provocative popularizations have made pop sociobiology prominent on the contemporary intellectual scene, it is worth taking a look at some work that is not intended for popular consumption, in which technical ideas are extended in the direction of pop sociobiology. One obvious source is Trivers, who, unlike most of the other contributors to Wilson's program, has advanced some theoretical ideas that have been important to sociobiology. (Another is Dawkins. I have avoided taking examples of pop sociobiological claims and arguments from Dawkins's provocative book *The Selfish Gene*, because his presentations introduce a collection of issues that are extrinsic to the topics on which I concentrate—to wit, the issues that surround his radical ideas about genes as units of selection. When these issues are stripped away, many of Dawkins's arguments prove vulnerable to the same kinds of considerations that I raise in the case of other pop sociobiologists.)

In a seminal paper on parent-offspring conflict (1974) Trivers offers a more general version of the analysis presented at the end of chapter 3 and develops his model by offering an account of adolescent resistance to socialization. Consider a situation in which a young animal has the opportunity to aid one of its (full) siblings. If the aid is given, the fitness of the sibling will be increased by B, and the cost to the donor, in units of fitness, will be C. From the perspective of the young animal, the behavior will increase inclusive fitness provided $B > 2C$. From the perspective of the parent, however, the behavior will contribute to inclusive fitness if $B > C$. The expected number of grandoffspring will be increased if one sibling gains more than the other loses. Hence the different perspectives yield different conclusions about the desirability of the helping behavior.

Trivers concludes that we can expect human parents and their children to come into conflict during the process of socialization.

> Parents are expected to socialize their offspring to act more altruistically and less egoistically than the offspring would naturally act, and the offspring are expected to resist such socialization. If this argument is valid, then it is clearly a mistake to view socialization in humans (or in any sexually reproducing species) as only or even primarily a process of "enculturation," a process by which parents teach offspring their culture. (1974, 250)

There is something of a leap here. Even if we come to view socialization as a process in which parents maximize their own inclusive fitnesses at the expense of their offspring, it by no means follows that socialization does not also advance the interests of both parent and offspring by teaching the culture. I am more interested in the claim about conflict between parents and young, however, than in the further morals that Trivers draws from it. Should we expect the disagreement Trivers identifies?

From the very beginning some sociobiologists have been skeptical of the possibility of *any* kind of parent-offspring conflict. In an influential article Alexander argued that genes promoting resistance to parental demands would be selected against, because of a delayed cost: the mutant child who defies the parent is likely to have children who also prove defiant, and their resistance will detract from the mutant's inclusive fitness (1974, 340ff.). Subsequent attention to the genetic details has shown, however, that it is possible, given the right genetic conditions, for alleles predisposing a child to resist parental demands to increase in frequency (Parker and MacNair 1978; Stamps, Metcalf, and Krishnan 1978; Stamps and Metcalf 1980). Intuitively, the reason is that the harm done to unselfish cooperators is greater than the loss suffered by the selfish offspring who oppose their parents (and are, in turn, opposed by their own children).

We can thus assume that some scenario of the general type envisaged by Trivers is possible. There are a number of ways in which we can try to work out the details. The first question we must ask concerns the type of mechanism that is hypothesized. Here I shall consider two alternatives. One is to suppose that there is a *perceptive* mechanism that operates in the parent and a similar mechanism that operates in the child. Parents can tell which kinds of behavior in their children contribute to their (the parents') inclusive fitness. Children inherit the ability to discern which forms of behavior would boost their own inclusive fitness. On this alternative we should ask whether selection favors parental genes that pressure children into performing actions that promote parental inclusive fitness and whether selection favors genes in the children that stiffen resistance to parental pressure whenever the children are asked to act in ways that would detract from their own inclusive fitness. Conflict arises on just those occasions on which $\frac{1}{2}C < r_P B$ and $r_O B < C$ (where r_P is the coefficient of relatedness of the beneficiary to the parent, r_O is the coefficient of relatedness of the beneficiary to the offspring, and B and C are respectively the benefit to the recipient and the cost to the offspring; if the beneficiary is a full sibling of the offspring, then the inequalities reduce to the form that Trivers normally considers, $C < B < 2C$).

The second alternative is that the mechanism is *blind*. On this construal we suppose that conflict always arises when parents make demands. There are no mechanisms for discerning how inclusive fitness is to be maximized. Instead we assume that selection favors genes (expressed in the actions of the parents) that encourage cooperation among offspring. There are frequent occasions on which cooperative behavior among the children would enhance the inclusive fitness of parents. We also suppose that the encouragement is forthcoming on occasions when cooperation would detract from the inclusive fitness of the children, so that selection favors genes that express themselves in resistance on the part of the children. Because they act blindly, however, both parties sometimes make mistakes. Parents plead for cooperation even in circumstances where cooperation would detract from their reproductive success; children resist even when cooperation might help spread their genes.

Both scenarios have their difficulties. Consider first the assumption that parents and offspring are able to identify those actions that would promote the spread of their genes. (I hasten to point out that they may do this quite unconsciously.) The mechanism of identification will have to be attuned to the possibility of hidden costs and delayed benefits. Imagine two characters, Grudge and Goody. Grudges are sometimes unable to appreciate the possibility that cooperation with their siblings will bring important benefits in the long run. Their fitness calculators are improperly adjusted, and because they overlook long-term benefits, they resist parental urgings to cooperate even in situations in which it would promote their reproductive interests to do so. (The pertinent occasions are those on which the benefit B from an action to a sibling is greater than $2C$, but Grudges perceive B to be less than $2C$.) Goodies always do what Mommy and Daddy encourage them to do. As a result, they are sometimes exploited by their parents. On these occasions they do worse than Grudges. Quite evidently, Goodies will be favored over Grudges if $mB^* > nC^*$, where m is the expected number of occasions on which the Grudges miscalculate, B^* is the mean net benefit accruing to Goodies on such occasions, n is the expected number of occasions on which the Goodies are manipulated, and C^* is the mean net cost that they incur through being manipulated. The point is that if sizable long-term benefits can easily be overlooked, then it may be far better to settle for a bit of parental manipulation. If Mommy and Daddy know best, then the wise child should do what they say.

Articulating Trivers's scenario in terms of perceptive mechanisms of encouragement and resistance may set rather high demands on a mechanism for discerning possibilities of maximizing inclusive fitness. (Of course, there is no evidence for thinking that any such mechanism is physiologically, developmentally, or genetically possible.) Ironically, the hypothesis that people have *accurate* mechanisms for discerning what actions enhance their own inclusive fitness would raise interesting questions about the kinds of situations with which Trivers is concerned. What would be the point of efforts to socialize our children? On Trivers's hypothesis, parental indoctrination stems from the effort to maximize parental inclusive fitness. If both children and parents can accurately calculate how actions will affect their inclusive fitness (or if they behave as if they have performed an accurate calculation), then, in the style of pop sociobiology, we can argue that selection can be expected to favor parental behavior that encourages children to perform actions only when there is an incompatibility between the inclusive-fitness interests of parents and children. On the other occasions the children can be left to their own devices. They will recognize the course of action that maximizes their own inclusive fitness, and, since the case is one in which there is no conflict with the reproductive interests of the parents, they will act so as to maximize parental inclusive fitness. Hence there is no need for extensive socialization. All we should observe is conflict.

So, on our first alternative, where we assume a perceptive mechanism, we face a dilemma. If the mechanism works well, extensive socialization ought to prove unnecessary (and presumably should disappear as a waste of time and energy). If it does not work well, then it is likely that those who conform will be favored.

What happens when we look at the second alternative? Here the difficulty seems obvious: children who automatically resist parental advice seem likely to do badly; the meek should inherit the earth. In a population of rebels we would expect selection to favor conformists. They benefit on all those occasions on which the interests of parents and children coincide, and they lose only when their parents exploit them. Barring the possibility of massive parental exploitation, the cases of coincidence of interest can be expected to outnumber the situations in which conflict will arise.

Previous discussions should have alerted us to the idea that the superficially plausible scenario can sometimes dissolve under closer scrutiny. Hence it is worth taking a look to see if a more refined analysis endorses our qualitative assessment. Here is one way to approach the problem. Suppose that there are two kinds of parents—Exploiters and Paternalists—and two kinds of children—Conformists and Rebels. Parents and children periodically face situations in which the inclusive fitness of the parent can be increased if the child performs a particular kind of action. In some of these situations the action would increase the inclusive fitness of the child (call these *coincident* situations). In other situations the action would decrease the inclusive fitness of the child (call these *conflict* situations). When any of the situations arises, the parent requests that the child perform the appropriate action. If the child is a Rebel, there is a resultant battle, at cost to both parent and child. Half the time the parent wins and the action is performed; the rest of the time the child wins and the action is not performed. If the child is a Conformist, then it always performs the action. The difference between Exploiting parents and Paternal parents is that the former engage in more interactions in which there is a possibility of increasing the inclusive fitness of the parent if the child behaves appropriately. We can imagine that Exploiters act to replace some of the coincident situations in which they and their children would have found themselves by a greater number of conflict situations. They encounter less coincident situations than do Paternalists. However, the increase in the number of conflict situations more than makes up for the decrease in the number of coincident situations.

What combination of strategies would we expect to emerge under selection? Or, to pose a precise (and slightly different) question, which combinations are evolutionarily stable? When the analysis is developed (see technical discussion H), we find what we might have expected. Depending on how we adjust the costs and benefits, the opportunities to exploit, and so forth, we can reach almost any conclusion we want. If the benefits of coincident situations are sufficiently large, then it will pay parents to be paternal and children to conform. If the costs of battle are slight, and if conflict situations bring sizable returns, then it is possible for the combination Exploiter-Rebel to be evolutionarily stable. There are even ways of juggling the parameters to yield the result that none of the pure combinations is stable or (less plausibly) to secure the stability of Exploiter-Conformist or Paternalist-Rebel. So, when we probe Trivers's analysis a little, we see that there is no clear conclusion about what selection can be expected to favor or to preserve.

So far, I have omitted an important aspect of the problem. The beneficiary has been viewed as a purely passive part of the landscape. That perspective is entirely appropriate when we are considering some of the applications of Trivers's *general* idea of parent-offspring conflict. In discussions of weaning conflict, for example, the beneficiaries of an "altruistic" early relinquishing of the mother's breast are

Technical Discussion H

Suppose that in any coincident situation the child can act so as to increase the fitness of a sibling by an amount B_1 at a cost to itself of C_1, where $B_1 > 2C_1$. In any conflict situation the child can act so as to increase the fitness of a sibling by an amount B_2 at a cost to itself of C_2, where $C_2 < B_2 < 2C_2$. The cost of battle is C_3 for both parent and child. Exploiters engage in m_1 coincident situations with their children and m_2 conflict situations. For a Paternalist the number of coincident situations is n_1 and the number of conflict situations is n_2. In accordance with the description of the text, we suppose that $m_1 < n_1$, that $n_2 < m_2$ and that there is a number $k > 1$ such that $m_2 - n_2 = k(n_1 - m_1)$.

The payoff matrix for the game can be written as follows:

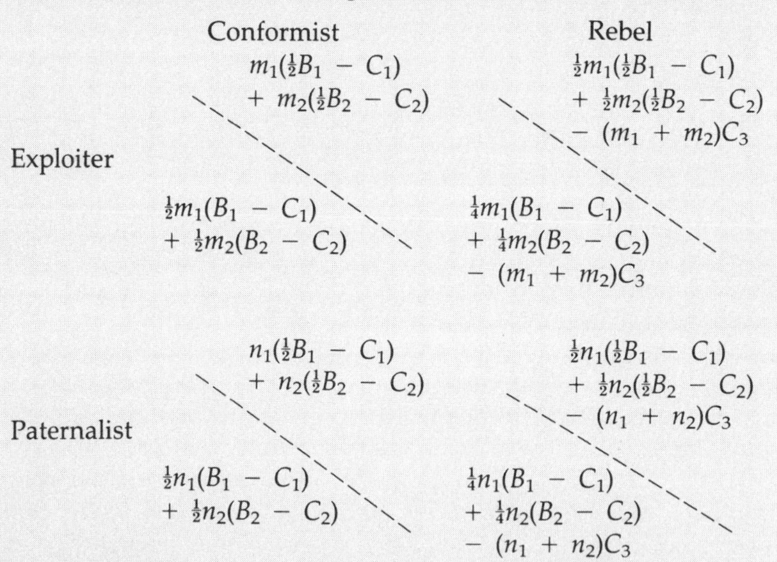

	Conformist	Rebel
Exploiter	$m_1(\frac{1}{2}B_1 - C_1) + m_2(\frac{1}{2}B_2 - C_2)$ $\frac{1}{2}m_1(B_1 - C_1) + \frac{1}{2}m_2(B_2 - C_2)$	$\frac{1}{2}m_1(\frac{1}{2}B_1 - C_1) + \frac{1}{2}m_2(\frac{1}{2}B_2 - C_2) - (m_1 + m_2)C_3$ $\frac{1}{4}m_1(B_1 - C_1) + \frac{1}{4}m_2(B_2 - C_2) - (m_1 + m_2)C_3$
Paternalist	$n_1(\frac{1}{2}B_1 - C_1) + n_2(\frac{1}{2}B_2 - C_2)$ $\frac{1}{2}n_1(B_1 - C_1) + \frac{1}{2}n_2(B_2 - C_2)$	$\frac{1}{2}n_1(\frac{1}{2}B_1 - C_1) + \frac{1}{2}n_2(\frac{1}{2}B_2 - C_2) - (n_1 + n_2)C_3$ $\frac{1}{4}n_1(B_1 - C_1) + \frac{1}{4}n_2(B_2 - C_2) - (n_1 + n_2)C_3$

In a population playing Paternalist-Conformist, a child playing Rebel cannot invade if

$$\tfrac{1}{2}n_1(\tfrac{1}{2}B_1 - C_1) + \tfrac{1}{2}n_2(\tfrac{1}{2}B_2 - C_2) + (n_1 + n_2)C_3 > 0.$$

A parent playing Exploiter cannot invade if

$$(n_1 - m_1)(B_1 - C_1) + (n_2 - m_2)(B_2 - C_2) > 0;$$

that is,

$$(B_1 - C_1)/(B_2 - C_2) > k.$$

It is easy to see that both conditions can be satisfied if B_1 is large and C_1, B_2, and C_2 are all much smaller.

In a population playing Exploiter-Rebel, a child playing Conformist cannot invade if

$$\tfrac{1}{2}m_1(\tfrac{1}{2}B_1 - C_1) + \tfrac{1}{2}m_2(\tfrac{1}{2}B_2 - C_2) + (m_1 + m_2)C_3 < 0.$$

A parent playing Paternalist cannot invade if

$$(m_1 - n_1)(B_1 - C_1) + (m_2 - n_2)(B_2 - C_2) + (n_1 + n_2 - m_1 - m_2)C_3 > 0.$$

We can satisfy these inequalities by making C_3 small, B_1 approximately equal to B_2, C_1 slightly less than $\tfrac{1}{2}B_1$, and C_2 slightly greater than $\tfrac{1}{2}B_2$, and by taking k to be large. Thus, let $B_1 = B_2 = 5$, $C_1 = 2$, $C_2 = 3$, $C_3 = 0$. The first inequality reduces to

$$m_2 > m_1,$$

which always holds by the conditions of the model. The second becomes

$$2k > 3.$$

It is possible for each pure combination to be unstable, as we can see from the following example. Let $C_1 = 2$, $C_2 = 4$, $C_3 = \tfrac{1}{4}$, $B_1 = B_2 = 5$, $m_1 = 10$, $m_2 = 50$, $n_1 = 20$, $n_2 = 10$. (It follows that $k = 4$.) The payoff matrix is as follows:

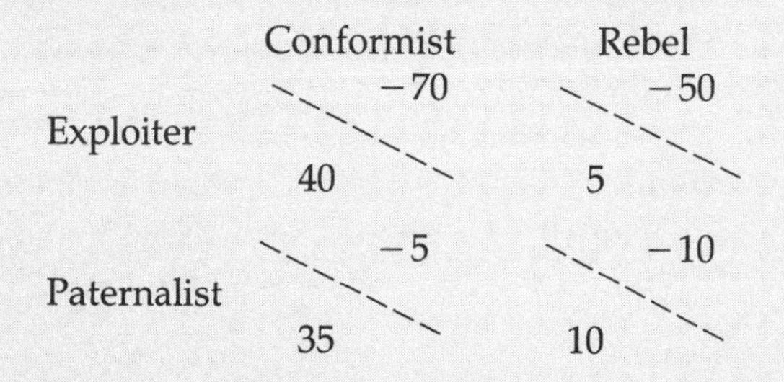

Under these circumstances, in a population playing Paternalist-Conformist, it will pay a parent to switch to Exploiter. If the population reaches Exploiter-Conformist, children playing Rebel will be at an advantage. Given the combination Exploiter-Rebel, parents playing Paternalist will be favored. Finally, if the combination Paternalist-Rebel occurs, children who play Conformist will be favored.

It is also not difficult to construct examples in which Exploiter-

Conformist or Paternalist-Rebel are stable. I leave these to the reader's ingenuity.

Although the analysis is obviously oversimplified in a number of ways, it is worth pointing out one important feature. The idea that Exploiters trade coincident situations for conflict situations may seem somewhat artificial. However, if we consider the more realistic game in which children increase their resistance as they find themselves exploited, we can see that something like a trade will be expected to occur. Parents who attempt to exploit their children may encounter resistance even in situations in which there is coincidence of interest. If this is so, then the apparently artificial idea of a trade (represented by the parameter k) may provide a simple way of capturing part of what goes on in a more realistic analysis.

The main point, however, is to show how sensitive evolutionary expectations may be when we isolate—and modify—crucial assumptions.

siblings yet unborn. When we are interested in the socialization of human children, however, the attitudes and behavior of the beneficiary need to be taken into account.

Imagine a family with two children, close in age, and two fair-minded parents. Periodically there are situations in which one child could help the other, enhancing the sibling's inclusive fitness and the inclusive fitness of both parents but incurring a net fitness cost. Imagine that there is a rough symmetry to these situations: if one child has the opportunity to help today, the other will have the chance tomorrow. Consider any such pair of situations. There are four possible outcomes, and we can represent the payoffs to the children as follows:

		Charles	
		Cooperate	Refuse
Andrew	Cooperate	$\frac{3}{2}(B - C)$ \ $\frac{3}{2}(B - C)$	$\frac{1}{2}B - C$ \ $B - \frac{1}{2}C$
	Refuse	$B - \frac{1}{2}C$ \ $\frac{1}{2}B - C$	0 \ 0

Here we suppose that B is the benefit brought to the sibling on each occasion and that C is the cost to the child performing the helpful action. Since we are assuming that the situation is one in which there is a conflict between parental inclusive fitness and the inclusive fitness of the child who would perform the action, we know that $C < B < 2C$. These inequalities ensure that the payoff matrix is that for Prisoner's Dilemma (see chapter 3) and that the condition employed by Axelrod in his discussion of the iterated Prisoner's Dilemma is satisfied. From this we can draw an interesting conclusion.

If the two siblings participate in a sequence of interactions throughout their childhood, then, in effect, they will be playing a sequence of Prisoner's Dilemma games against one another. We know from Axelrod's work that TIT FOR TAT is an ESS for iterated Prisoner's Dilemma. Hence, if we suppose that evolution has been able to reach the ESS, then we shall expect both the siblings to follow TIT FOR TAT. When TIT FOR TAT encounters itself, the result is a sequence of cooperative interchanges. It follows that the siblings can be expected to help one another and, in doing so, to maximize the inclusive fitness of their parents. Thus, when the situation is viewed from the perspective of a long history of occasions (assumed to be roughly symmetrical), each of which satisfies Trivers's conditions for conflict, an ESS for the entire history removes any conflict.

Quite evidently, my brief treatment of the problem involves a number of assumptions, some of which may be unrealistic. The point is, however, that Trivers's qualitative argument for expectations of resistance to parental socialization falls apart under analysis. There is just no reason to accept his conclusions about the likely course of evolution. We can give equally good—perhaps better—arguments for radically different conclusions.

Any attempt to articulate Trivers's reasoning in detail is likely to have "an obvious air of unreality" about it. The fault surely lies in the general problem with which Trivers began. We are invited to think of socialization as a process in which a single mechanism operates in the parent and a single mechanism operates in the child. This is implausible. The contexts in which parents attempt to shape the behavior of their children are highly diverse. To think of children's responses to these heterogeneous demands as primarily determined by some one mechanism is to present a caricature of the ways in which children behave. The natural picture of the situation represents parental pressure as motivated by the parents' desire to achieve certain social ends for themselves and for their children, and resistance as occurring when the offspring have secondary goals of their own. Committed sociobiologists will instantly reply that the parental goals are viewed as desirable because the attainment of them would maximize the inclusive fitness of the parents and that the secondary goals of the offspring are valued because achieving those goals would maximize the inclusive fitness of the child. Yet when we consider the range of cases in which parents and their children—particularly their adolescent children—fight most bitterly, it is surely hard to see any correlation whatever between maximization of inclusive fitness and the actions that are chosen. Trivers's story is not only underdeveloped theoretically. It also takes an impoverished view of the phenomena.

We know, in a vague way, that parents and children sometimes disagree. Trivers sketches an account of why they might sometimes be expected to do so. So far, we have at best something that is on a par with the groping efforts of early science. When we discover that a certain kind of disease tends to break out in low-lying, marshy areas, we do not content ourselves with the claim that it may have something to do with the water. We try, on the one hand, to develop a theoretical account of what might be going on. On the other hand, we attempt to describe in a more precise way the conditions under which the disease occurs.

In fact, we know much more about conflicts between parents and children than Trivers's tale accommodates. For example, battles between parent and child vary in both number and intensity during the course of the child's development. Moreover, rebellious behavior is directed not only against parents but against other adults and societal institutions as well. There are complicated relationships between such behavior and other attitudes and actions. How do these familiar facts enter into Trivers's simple calculus?

There is one passage in which the idea of some general tendency to conflict gives way to discussion of more particular disagreements.

> Parent-offspring conflict may extend to behavior that is not on the surface either altruistic or selfish but which has consequences that can be so classified. The amount of energy a child consumes during the day, and the way in which the child consumes this energy, are not matters of indifference to the parent when the parent is supplying that energy, and when the way in which the child consumes the energy affects its ability to act altruistically in the future. For example, when parent and child disagree over when the child should go to sleep, one expects in general the parent to favor early bedtime, since the parent anticipates that this will decrease the offspring's demands on parental resources the following day. Likewise, one expects the parent to favor seri-

> ous and useful expenditures of energy by the child (such as tending the family chickens, or studying) over frivolous and unnecessary expenditures (such as playing cards)—the former are either altruistic in themselves, or they prepare the offspring for future altruism. In short, we expect the offspring to perceive some behavior, that the parent favors, as being dull, unpleasant, moral, or any combination of these. One must at least entertain the assumption that the child would find such behavior more enjoyable if in fact the behavior maximized the offspring's inclusive fitness. (1974, 251)

It is worth scrutinizing the assumptions involved in the examples offered here.

Trivers seems to favor an evolutionary scenario in which selection has produced perceptive mechanisms that lead parent and child to conflict. Parents have been selected to value situations in which their children act so as to maximize the inclusive fitness of the parents. In consequence, parents find early bedtimes and periods devoted to homework highly desirable. (Let us not worry about the exact mechanisms that underlie these desires.) Selection also favors genes that express themselves in children as desires for states in which the children maximize their own inclusive fitness—by going to bed later or by playing cards. What basis is there for any of these assumptions? Do we have any good reason to think that early bedtimes really help parents maximize their inclusive fitness or that they diminish the inclusive fitness of their children? Do children really maximize their inclusive fitness by playing cards?

The peculiarities of the account are most evident in the assumptions about parental favoring of homework. How does parental insistence on the virtues of studying contribute to the inclusive fitness of the parents? Presumably by enhancing the inclusive fitness of the child. If it is indeed true that parents maximize their inclusive fitness by exhorting their children to study, then the apparent benefit lies in the increased probability of future academic success for the child (which is *assumed* to translate into an increase in the child's expected number of descendants). If the assumption about the benefit to the parents is correct, we seem to forgo the possibility of that diminution of the inclusive fitness of the child on which the whole story about conflict is based. We appear to be headed for a happy coincidence of interests. But not so fast! Perhaps the studying "prepares the offspring for future altruism." The parents reap a double dividend in that a successful child may also be in a position to help siblings. A tangled web is woven here. To make the story cohere, we have to suppose that the probability that the child will increase the number of its own descendants by studying is not high enough to make studying pay, nor is the inclusive fitness of the child enhanced when we take into account the help it *may* give to its relatives; however, the number of grandchildren for the parents *is* likely to be increased sufficiently to make the child's studying evolutionarily profitable for them. But why should we think that the expected benefit brought to siblings by an academically successful child will be of just the right size to make the course of studying in the evolutionary interests of the parents and not of the child itself? Trivers's scenario turns on delicate assumptions—which seem far more likely to be false than true.

Pop sociobiology is often dedicated to the enterprise of making close connections between our actions and the working of selection. Vulgar adaptationism sketches an account of how a behavior might contribute to fitness, offers some impressionistic data to suggest that the behavior is present, and so proceeds to conclusions about genetic bases for behavioral propensities. We have seen that Trivers's story fails both to work out a detailed analysis and to provide any clear connections to the facts of human behavior. It also makes the common adaptationist mistake of overlooking possible rival explanations. Confronted with the phenomena of conflicts about bedtime and about studying, the natural strategy is to suggest that the links between our actions and the working of selection are far less direct than Trivers takes them to be.

Our evolutionary heritage surely equips us with *something*. Perhaps humans have genes that predispose us, given the environments in which we typically live, to find certain situations desirable and to avoid others. Yet we also have extraordinary cognitive abilities, which we use to represent to ourselves many subtle features of the world around us. Furthermore, each of us is reared in a culture that provides us with a mass of information and misinformation, that shapes our appreciation of what is desirable and what is not. So, in our maturity, we make decisions. Those decisions are the products of many factors: our basic predispositions, our representations and reasonings, our interactions with the society in which we live. To suppose that they are automatically directed toward the maximization of our inclusive fitness is simply to forget that on any occasion of decision many pieces of the equipment with which our evolutionary history has left us combine. Even if we assume that each of them has been fashioned separately by selection and that *overall* each of them works to enhance our fitness, there is no basis for thinking that every combination of them in decision should lead us to a fitness-maximizing course of action.

Consider, from the general perspective of the last paragraph, a typical case of parent-offspring conflict. The parents would prefer that their child study. The child would prefer to go to the local arcade and play video games. How does the conflict arise? The parents tell us, sincerely, that they want their child's happiness. They believe that this happiness is more likely to be achieved if the lure of the arcade is resisted and if the homework is done. Perhaps the desire for the happiness of our children is indeed something to which evolution predisposes us. Yet it need have no close connection with inclusive fitness in the case at hand. There may be no correlation between the kind of success that studying would promote and the production of grandchildren—and the parents may even be aware that this is the case. They simply have a certain desire, and their decision represents an attempt to satisfy their desire in the light of what they believe.

The child also has desires, desires for recreation, for the company of friends, perhaps for future success as well. In the light of her beliefs she makes the decision that going to the arcade would do little harm to her chances of attaining her long-term goals and would bring pleasure in the short term. Again, we may see the hand of evolution in the predisposition for recreation, in the love of company, in the ability to consider one's future well-being. But the actual decision may have little to do with maximizing fitness.

So there is an obvious style of explanation that runs counter to the proposal introduced by Trivers. It requires no implausible assumptions. It invokes no underdeveloped theoretical model. It is simply discounted in the urge to see the shaping hand of evolution in every detail of our decision making. Trivers's general account of conflict between parent and offspring is a useful tool that can be applied *with caution* to the understanding of the evolution of behavior; it is misapplied in the speculative remarks about human parents and their children. The credentials of pop sociobiology prove no more impressive when grand claims about human behavior occur cheek by jowl with theoretical contributions to biology.

An Enchanting Castle?

Doubt may linger. Does my choice of cases merely pick on studies in which Wilson and his followers have been unusually unrigorous? I suggest that if the arguments reviewed here and in previous chapters break down, there is not much in the early Wilson version of pop sociobiology to become excited about. Many people have been attracted to the program because it seemed to promise genuine insights about important aspects of human behavior—our urges to defend family and territory, our intersexual relations, our fear of strangers, the existence of homosexuality, the persistence of conflicts among parents and children. If, under scrutiny, insight dissolves into speculation in all of these cases, then, even without extrapolating to other instances, we can claim that the Wilson program has failed.

I want to conclude my treatment of the program with a look at one of its most prized explanations. Champions of pop sociobiology point with pride to the possibility of explaining the human propensity for incest avoidance as an evolutionary adaptation (Ruse 1982). The story goes as follows. Not only do most cultures have taboos against sexual intercourse among close relatives (those whose coefficients of relationship are 0.25 or greater), but even when taboos are lacking, incest is typically infrequent. The low rate of incest can be understood in terms of a genetic propensity to refrain from sexual relations with those with whom one has been reared. Evolution would have favored the propensity because, in typical cases, those who surround the young human being are close relatives, and, owing to the well-known costs of inbreeding, breeding with a relative is likely to reduce an animal's fitness.

The most vigorous recent exponent of the sociobiological incest story has been van den Berghe (van den Berghe 1980, 1983; van den Berghe and Mesher 1980). There is certainly some evidence for some parts of the story. There are also anomalies, complications, and crucial questions about which we know very little. In the treatment of these difficulties we can discern the same relentless commitment to hewing a simple biological line that we have seen in earlier examples.

Studies of nonhumans have amassed considerable evidence to show that many different kinds of animals avoid inbreeding. There are detailed studies of patterns of outbreeding in Japanese quail (Bateson 1978, 1980, 1982b) that provide insight into the development of inbreeding avoidance. Observations of olive baboons and chimpanzees have revealed that the practice of resisting copulation with kin is very common in the higher primates (Packer 1979; Pusey 1980). However, there are also populations that show a high degree of inbreeding. The wolves of Isle Royale are a well-known example (Livingstone 1980; Mech 1970).

Investigations of the effects of inbreeding in humans support the conclusion that the offspring of matings between close relatives are likely to have a high frequency of birth defects (see, for example, Schull and Neel 1965). Behind the empirical results stands a simple theoretical argument. Close inbreeding increases the probability that two copies of a harmful recessive gene may find their way into a zygote.

The stage seems set for pop sociobiology. In a wide variety of animal groups a propensity not to mate with kin would contribute to fitness. The propensity is found among nonhumans and appears to be present in our closest evolutionary relatives. We should conclude that we share it and that it has a genetic basis. Even in the absence of cultural sanctions against incest, relatives should still refrain from copulation.

One important question concerns the frequency of incest in humans. As van den Berghe rightly notes (1983, 94), it is easy to go astray here. Suppose that we think of incest as involving heterosexual copulation. Then the number of cases of incest will be reduced. Alternatively, we might think of incest as involving that form of sexual activity in which the individual might be expected to engage with unrelated individuals. Kissing, petting, fondling, or other forms of close sexual contact between postpubescent siblings will then count as incest if it is the kind of activity in which people of comparable age engage with those to whom they are sexually attracted. On this conception the incidence of incest is likely to be considerably higher.

A further conceptual difficulty concerns the relatedness of the participants. Initially, we think of incest as involving related individuals. However, if we hope to test the existence in humans of a propensity to avoid sexual activity with those with whom one has been reared, it is important to consider the sexual interactions of stepsiblings.

Furthermore, it is important to recognize the differences among various types of incest and to formulate precisely the patterns of behavior that selection is supposed to have favored. Inspired by studies of humans and nonhumans, many sociobiologists, including van den Berghe, assume that selection has endowed us with a propensity to avoid copulating with those with whom we have been reared. It is not obvious that this propensity will account for the avoidance of father-daughter incest. Perhaps we can understand why daughters have a predisposition to resist copulating with their fathers, but the alleged propensity will not explain why fathers might refrain from copulating with their daughters. Given the obvious asymmetry in the situation, should we expect that fathers have no propensity to avoid incest with their daughters and thus that unwilling daughters are forced into incest?

Here the issue becomes very slippery. The coefficient of relationship between father and daughter is the same as that between brother and sister (0.5). Hence, any expectation that evolution should have favored a disposition in brothers to avoid copulating with their sisters ought to be mirrored in a similar expectation about fathers and daughters. But perhaps it will be suggested that there is a sexual asymmetry here. (See van den Berghe 1983, 97.) The fitness costs tell differently on males and females. Maybe evolution favors males who will tolerate incest and females who oppose it. Brothers do not have the power over their sisters that fathers enjoy over their daughters. Hence, father-daughter incest should be much more common than brother-sister incest.

Working out the details is rather complicated. Suffice it to say that it is not difficult to construct qualitative arguments for radically different conclusions. As long as we remain at the level of pop sociobiological discussions of the topic, we can argue either that father-daughter incest ought to be as rare as brother-sister incest or that it ought to be rampant. By appropriate choice of assumptions we can allow for any intermediate conclusion that might emerge from the data.

In the rest of my discussion I shall follow van den Berghe in focusing primarily on the example of sibling incest. It is well to remember, however, that a convincing account of incest avoidance between siblings will have to fit within a general explanation of why people do not copulate with their relatives. Factors that are cited in one case cannot simply be ignored when we turn to other examples.

Van den Berghe is interested in showing that human beings do not have sexual relations with those with whom they have been reared. We can consider four types of data. First, there are popular reports emphasizing that incest is far more common than many people previously believed. There are also two scholarly studies of situations in which children have been reared with unrelated individuals. In both situations, in the Israeli kibbutzim and in the practice of "minor marriages" in Taiwan, those who have been reared together show an apparent reluctance to engage in mutual sexual activity. Finally, there is a genetic study that appears to show high levels of inbreeding among the small bands of people from whom we descend. Van den Berghe discusses the first three. I shall take a brief look at them all.

Van den Berghe is anxious to rebut published accounts of widespread incest as "sensationalistic." He is justified in claiming that much of the popular literature on the topic employs a very broad definition of incest—often in the legitimate interest of dramatizing the plight of the many children who are victims of sexual abuse. His remedy, however, is to minimize the incidence of incest by choosing the narrowest possible definition. Despite his commitment to a hypothesis of inbreeding avoidance through rejection of close childhood associates, he automatically discounts matings between stepsiblings. Moreover, he supposes that incest requires full copulation, thereby eliminating the numerous cases in which young adolescents engage in just the same forms of close sexual contact with their siblings that others of their age group engage in with one another (kissing, fondling, and so forth). Narrowing the definition in this way serves van den Berghe's purposes as well as broadening it helps those who campaign for greater public sensitivity to the prevalence of incest. It enables him to hack away at the published statistics and to conclude that incest is relatively rare.

As he well knows, however, even after the published data have been tidied up, estimates of the frequency of incest must still be based on guesswork. "On the one hand, it is likely that incest is one of the more underreported of sex offenses. On the other hand, many children are not sexually molested" (van den Berghe 1983, 94). The trouble is that the extremes leave a vast space in which the true frequency of incest can be expected to lie.

There are two detailed studies of particular situations. Children reared together in the kibbutz showed little inclination to "marry or make love to each other" (van den Berghe 1983, 95; see also Shepher 1971 for details). Here we are supposed to see the effect of a rule proscribing sexual activity with childhood associates. That is a possible explanation. Yet there are confounding factors, which van den Berghe does not discuss. The medical director of a kibbutz child and family clinic in Tel Aviv writes of a recent sexual revolution in the kubbutzim, describing the period during which Shepher collected his data (the data used by pop sociobiologists) as "the puritanical era in the kibbutz" (Kaffman 1977, 208). He paints a picture of an environment in which young people were constantly reminded of the high ideals—including ideals of purity—of the adults around them. That picture suggests an alternative explanation.

Another important piece of pop sociobiological evidence is the study of "minor marriages" in Taiwan (see Wolf and Huang 1980). In many parts of China there has been a tradition of adopting a future daughter-in-law. Typically, the daughter of a relatively poor family is adopted by wealthier parents and later marries their son. There is much anecdotal evidence of resistance to such marriages, of a higher incidence of adultery and of divorce. Statistical data reveal that the reproductive success of minor marriages is nearly one third lower on the average than in other marriages. (What exactly this shows about the incidence of copulation or the intensity of sexual attraction is a nice question.)

Again there are complications. Wolf and Huang offer a fascinating account of the practice of minor marriages, drawn partly from registry data, partly from conversations with old people who recalled the past social activities and past gossip of their villages. There are many cases in which minor marriages seem to have been quite successful—for example, case #113: "Minor marriage. The couple produced twelve children and Mr. Li seems certain they are all the husband's" (157). There are three cases in which premarital sexual relations between the "brother" and his adopted sister hastened the marriage. Furthermore, Wolf and Huang make it clear that minor marriages were much despised. The adopted daughters resented the actions of their natural parents. They and their future spouses were often teased by friends and childhood associates, with the result that they frequently came to dislike their situation and each other (see 143–144). Finally, attention to the descriptions given by Wolf and Huang's informants indicates the very distinct possibility that wealthy parents sometimes adopted a daughter in the hope of securing a spouse for a "lame duck" son. Many of the men involved in minor marriages are not described in flattering terms.

Wolf and Huang are very cautious in their conclusions. They explicitly disavow the idea that the aversion to copulate with a childhood associate is "universal" and "absolute" (158). They also remark that the unwillingness to marry may have something to do with the minor-marriage arrangement itself (145; they cite the kibbutz data, however, as pointing toward a more general disposition to avoid copulating with childhood associates). Pop sociobiologists seem to hope that the conjunction of two cases that involve serious complications allows them to neglect the possible confounding factors.

Finally, there are some data that suggest a relatively high amount of inbreeding in our recent evolutionary past. By using the techniques of contemporary genetics, it is possible to compare the frequencies of alleles in a present population with what would be expected, given various levels of inbreeding in the past. Using this technique on a large number of loci in an Amerindian population, a group of geneticists have discovered indications of surprisingly high levels of inbreeding in the past. They conclude that if the results about Amerindians hold generally, then "the process of detribalization which has occurred since the advent of civilization some 2,000–4,000 years ago must have resulted in a marked relaxation of inbreeding and of its consequences for the selection against recessive alleles" (Spielman, Neel, and Li 1977, 369).

How strong is the human propensity to avoid copulating with those known intimately from childhood? How strong is it in contemporary members of our species? How strong has it been in the past?

We simply do not know. There are some suggestive, but by no means clear-cut, observations from Taiwan and from the kibbutzim. We have anecdotal evidence that incest has been dramatically underreported in the United States. Technical work in genetics indicates that some form of mating between close relatives may have been very common among our ancestors. Any serious study of incest should recognize the extent of our ignorance rather than rushing to pronounce on the behavioral rule that people must be following.

The question of what people actually do is only one part of the pop sociobiological picture. There are also claims about what it would maximize our inclusive fitness to do. Here we should not jump to conclusions. Other things being equal, mating with a sibling or another close relative may detract from the spread of one's genes. But other things are not always equal. If the attempt to find an unrelated mate brings with it a high probability of death or serious damage, then people may do better to stay home and copulate within the family. The case has been carefully studied (Bengtsson 1978; see also May 1979). It points in the direction of a refined analysis of the selective pressures to which our ancestors might have been subject. The results of that analysis are by no means obvious a priori.

I shall underscore the need for detail by considering one of van den Berghe's proposals. In his enthusiasm to provide a pop sociobiological explanation of everything connected with incest, van den Berghe takes on the cases in which incest has been culturally favored, even prescribed for certain individuals. Many of these examples involve marriage rules for royalty, and van den Berghe devotes himself to explaining these. Allegedly, royal incest has been favored in situations where it maximizes the inclusive fitnesses of the participants (the siblings). The situations are distinguished by the existence of polygynous mating for high-status individuals, especially for the king. Here is the argument.

> Under these conditions . . . royal incest is, in fact, a fitness-maximizing strategy. For the king, incest is a low-risk, moderate-gain strategy. If the king succeeds in producing a fit incestuous heir with an $r = .75$, that person will in turn become highly polygynous and produce many grandchildren with an $r = .375$. If the incestuous strategy is repeated over many generations, the king comes close to cloning himself. For commoners who are monogamous or small-scale polygynists, the risk of inbreeding depression militates against the incestuous strategy. For a large-scale polygynist, however, the risk is low. Should the strategy fail, the king's fitness loss is minimal (being largely limited to his

> sister's wasted effort), and he still has the opportunity of producing an outbred heir with any of his many other wives. (van den Berghe 1983, 100)

Van den Berghe continues by arguing that the sister's fitness is also maximized by taking the incestuous strategy. Incest is risky for her but has the possibility of yielding large evolutionary returns.

Questions flock in. Are we to suppose that incest is to be found *exactly* where it maximizes inclusive fitness for the participants (or for the more powerful of them)? How are we to explain the absence of incest in those royal families that can achieve polygyny either simultaneously or serially (for example, by divorce)? Are we to hypothesize some mechanism that enables people to inhibit the normal disposition to avoid inbreeding when considerations of inclusive fitness intimate that incest is a good idea? Are we to expect that humans attuned to possibilities of contraception will come to recognize that incest is no longer to be accompanied by loss in inclusive fitness, so that copulation between siblings may thrive? I shall waive these concerns and simply focus on van den Berghe's remarks about the king's inclusive fitness.

Van den Berghe summarizes his argument in a misleading way: "for the king, incest with a full sister represents a 50% increment in the first generation, at minimum risk" (1983, 100). Two commentators (Bateson and Dawkins) have pointed out that the summary appears to misunderstand the notion of inclusive fitness. If Rex I produces a child from a union with a sister and Rex II produces a child from a union with an unrelated female, and if we measure fitness in terms of expected number of offspring, then there is no reason to think that Rex I will have greater fitness than Rex II. For the inclusive fitness of Rex I is 0.75, the proportion of Rex I's genes expected to be present by virtue of descent in the incestuous progeny, while Rex II is to be attributed 0.5 through his child and 0.25 from the child his sister can be expected to produce. Other things being equal, Rex I monopolizes his sister's ovaries to no avail.

Van den Berghe does not intend that other things should be equal. His argument turns on the possibility of greater genetic representation in the second generation in a society in which high-ranking males have many wives. The suggestion that the king may clone himself is a red herring. As long as the king's descendants reproduce sexually, clones will not be a permanent feature of the lineage. The crucial issue is whether the king's incest increases the frequency of his genes in the second generation and beyond.

Suppose that Siegmund and Sieglinde are siblings. They have the choice of mating with one another or of mating with nonrelatives. If Sieglinde mates with a nonrelative, she will become one of several wives of a high-ranking male. Siegmund himself will have many wives. If the number of wives that Siegmund can have is unlimited, then, unless the costs of inbreeding are unusually severe, Siegmund will gain by adding Sieglinde—and any other female—to his list. This argument is far too strong, however, for it suggests that males can be expected to maximize their inclusive fitness by mating with sisters in any society that allows unrestricted polygyny (or by copulating with sisters in societies that do not). A more realistic version of the argument is achieved by supposing that the number of Siegmund's wives is fixed at some number N and that to mate with Sieglinde is to substitute his sister for an unrelated wife. If we adopt this interpretation, then we can begin to make sense of van den Berghe's numbers.

We can compute the expected genetic representation for Siegmund in the second generation by calculating the number of marriages contracted by his offspring, given the incest strategy and the nonincest strategy. Assuming that neither strategy affects the expected reproductive outcome of Siegmund's other marriages, we need to compare the products of his possible union with Sieglinde with the products of the marriage that would otherwise produce his heir. Let us suppose that N is sufficiently large that one of Siegmund's unions will produce a surviving male who is healthy enough at maturity to assume the throne. Then there are two main possibilities: (1) The union with Sieglinde produces no healthy surviving male. Siegmund's inclusive fitness is less than it would be had he followed the nonincest strategy, because of lowered potential for grandchildren via Sieglinde and because of the loss of Sieglinde's independent contribution. (2) The union with Sieglinde produces a healthy surviving male. Siegmund's inclusive fitness is increased because of the fact that a very closely related individual has the maximum number of wives but is decreased because of the loss of offspring due to the costs of incest and because of the loss of Sieglinde's independent contribution.

It should be obvious that the accounting is delicate. When we work out the details, using the best available estimates of the costs of incest in humans, we find that incest will not maximize the king's inclusive fitness unless he is allowed considerably more marriages than his junior brothers and his father's sister's sons. (See technical discussion I.) Once again a pop sociobiological discussion is flawed by the desire to seize on any qualitative argument that looks as if it may generate the sociobiologist's favorite results.

The truth of the matter is that we are ignorant of the prevalence of

Technical Discussion I

Assume that the number of healthy surviving male offspring in a nonincestuous union is expected to be k, that the expected number of healthy surviving female offspring is also k, that the expected number of healthy surviving males in an incestuous union is pk ($p < 1$), and that the expected number of healthy surviving females in an incestuous union is also pk. Suppose that in any union the expected number of zygotes produced is m and that half of these are male. Then the survival probability for an individual male zygote conceived in incest is $pk/\frac{1}{2}m$, and the probability that all the male zygotes from an incestuous union fail to survive (or to survive with sufficient health to inherit the throne) is $(1 - 2pk/m)^{\frac{1}{2}m}$. This is the probability that a situation of type (1) will occur. In a situation of type (1) the expected genetic representation of Siegmund in the second generation is measured by

$$0.75pk + \text{contributions from } N - 1 \text{ nonincestuous unions (one of which yields a male heir).}$$

Let us now suppose that the number of marriages achieved by a king's son is n_1 and that the number of marriages achieved by a king's sister's son is n_2 (where $n_2 < n_1 < N$). Then in a situation of type (1) the expected genetic representation to Siegmund if he pursues the nonincest strategy is

$$0.5kn_1 + 0.5k + 0.25kn_2 + 0.25k + \text{contributions from } N - 1 \text{ nonincestuous unions (one of which yields an heir).}$$

In a situation of type (2) the expected genetic representation of an incestuous Siegmund is measured by

$$0.75N + (pk - 1)0.75n_1 + 0.75pk + \text{contributions from } N - 1 \text{ nonincestuous unions.}$$

Hence the expected gain to Siegmund from engaging in incest is

$$(1 - 2pk/m)^{\frac{1}{2}m}\{-k[0.5n_1 + 0.25n_2 + 0.75(1 - p)]\} + [1 - (1 - 2pk/m^{\frac{1}{2}m}][0.25N + n_1(0.75pk - 0.5k - 0.25) - n_2(0.25k) - 0.75k(1 - p)].$$

Does incest pay? Obviously, much depends on the relative values of the parameters k, m, p, N, n_1, n_2. However, given some fairly reasonable assumptions, it turns out that the expected gain from incest is negative. So, for example, if we take $m = 10$, $k = 3$, $p = 0.66$, incest would bring Siegmund a fairly significant loss in expected genetic representation in the second generation, unless N is very large in relation to n_1 and n_2. Given these values, Siegmund gains by following the incest strategy only if

$$N > 1.5n_1 + 3.25n_2 + 3.25.$$

So, if monogamy is for everyone except the king, N has to be larger than 8; if $n_1 = 2$ and $n_2 = 1$, N has to be larger than 9; and if $n_1 = 3$ and $n_2 = 2$, N has to be larger than 14. I leave it to the reader to judge how likely it is that the cases for which van den Berghe gives his model accord with such mating patterns.

Plainly, van den Berghe could achieve a lower threshold for the incest strategy if p were larger. The value of p that I have chosen is drawn from the analysis in May 1979. (See also Bodmer and Cavalli-Sforza 1976.) However, even if we suppose that $p = 0.75$, which, given the available data on inbreeding depression, is far too high, matters are not much better. Now the crucial inequality is

$$N > 0.58n_1 + 3.16n_2 + 2.37.$$

This means that if $n_1 = n_2 = 1$, N must be 7 or more; if $n_1 = 2$ and $n_2 = 1$, N must again be 7 or more; and if $n_1 = 3$ and $n_2 = 2$, N must be 11 or more. These numerical examples illustrate the point that unless the distribution of wives is very sharply peaked, with the king having far more than his younger brothers and other members of his extended family, the incest strategy is unprofitable.

human incest, past or present, that we have only preliminary analyses of the relations between incest and human fitness, that we have no idea whether our hominid ancestors had the genetic variance needed for evolution to equip them with whatever dispositions maximize fitness, and that we are only beginning to formulate accounts that do justice to the possible roles that cultural factors might play. (I shall discuss the case of incest avoidance again in chapter 9, in the context of Lumsden and Wilson's theory of gene-culture coevolution. Attentive readers will already have noted that the pop sociobiological accounts considered so far do nothing to help explain why so many societies have incest taboos. Why do we need taboos to prohibit what we naturally resist?)

Into this mass of ignorance rush the pop sociobiologists, with fast answers about our dispositions to avoid copulating with relatives. Those answers overinterpret material (such as the careful studies of animal behavior and of resistance to minor marriages) that may some day be used in fashioning a comprehensive account of incest. For the present, the gap between the little we know and the answers we would like to have is bridged by pop sociobiological guesswork.

There are important unresolved difficulties even in the example that the early Wilson program treats most successfully. Robert May's critique of the attempts to show that avoidance of incest should be a "universal attribute of human behavior" concludes with an apt image: "piece after piece can too easily be added to build an enchanting castle that rises free, constrained to earth with too few anchorlines of fact" (1979, 194). Whether or not one finds the castle enchanting, the absence of anchors is all too obvious.

Chapter 9

Hypotheses Non Fingo

Genetic Agnosticism

"But hitherto I have not been able to discover the causes of those properties of gravity from phenomena, and I frame no hypotheses; for whatever is not deduced from the phenomena is to be called an hypothesis; and hypotheses, whether metaphysical or physical, whether of occult qualities or mechanical, have no place in experimental philosophy" (Newton [1687] 1962, 547). Newton's famous declaration, made with the apparent intent of avoiding fruitless controversies about the causes of gravity, might serve as the credo for one of the most influential movements in pop sociobiology. Wilson's program, with its provocative claims about human nature, captures the popular imagination. For many practicing biologists and social scientists, however, the important work in human sociobiology has been done by a group of biologists and anthropologists led by Richard Alexander. Those who work in this rival program emphasize the importance of the techniques that have been introduced into contemporary evolutionary theory. They proclaim that biology holds the key to understanding human nature and human culture. Yet they do not explicitly devote themselves to justifying claims about genetic limits on human behavior. Here they frame no hypotheses.

The counterpart to the major Wilsonian texts (Wilson 1975a, 1978) is Alexander's influential book *Darwinism and Human Affairs* (1979). The main claim of the work is no secret: contemporary evolutionary theory provides "the first simple, general theory of human nature with a high likelihood of widespread acceptance" (1979, 12). What is this "simple, general theory"? Alexander's answer is that human social behavior is to be understood in terms of the maximization of inclusive fitness. "It seems clear that *if we are to carry out any grand analysis of human social behavior, it will have to be in evolutionary terms, and we shall have to focus our attention almost entirely upon the precise manner in which both nepotistic and reciprocal transactions are conducted in the usual environments in which humans have evolved their social patterns*"

(1979, 56; italics in original). The end product of the analysis is not, however, to be some collection of statements about the impossibility of altering our social behavior by changing our social environment. Instead, we are simply to understand the ways in which our social behavior adjusts to different ecological conditions so that people who encounter those conditions maximize their inclusive fitness.

Attempts by critics (and enthusiasts) to draw conclusions about genetic limitations strike Alexander as "a profound misunderstanding of biology and evolution" (1979, 95). He continues,

> If there is one thing that natural selection has given to every species, it is the ability to adjust in different fashions to different developmental environments. This is what phenotypes are all about, and all organisms have phenotypes. If there is an organism most elaborately endowed with flexibility in the face of environmental variation, it is the human organism. (1979, 95)

We are to explain human social behavior by revealing how, in different situations, people modify their attitudes, practices, and institutions so as to maximize their inclusive fitness.

Jeffrey Kurland provides the most explicit declaration of the intent of this form of sociobiology. "Evolutionary biologists who study the evolution of sociality are concerned with the prediction or explanation of how behavior maps onto the environment, *not* how genes map onto behavior" (1979, 147). With respect to the issues of proximate causation—the mechanisms that underlie human social behavior and the genetic and developmental bases of those mechanisms—no hypotheses are to be framed. Indeed, old ideas about those issues may well be adequate (1979, 146). If, from time to time, some comment about "genes for altruism" should slip from a sociobiologist's mouth, that is only a *façon de parler*. "Such modeling procedures in no way commit one to strict determinism or reactionary politics" (1979, 147).

There is a marvelous moment in *Waiting for Godot* when Estragon, at last persuaded by Vladimir to say "We are happy," continues after a pause, "What do we do now, now that we are happy?" The predicament of the thoughtful reader is like Estragon's. Let us suppose that Alexander, Kurland, and others succeed in showing that in certain environments particular forms of social behavior maximize the inclusive fitness of those who engage in them. What exactly have we learned about human nature? The careful qualifications issued by Kurland and Alexander seem to block the most direct connection between premises about inclusive fitness and conclusions about fundamental features of human social behavior.

So it seemed to three reviewers of Alexander's book. In the post-script he reports that these readers "were disappointed that I had not more explicitly attacked the problem of what constitutes human nature, identifying its limits and explaining the consequences" (1979, 279). Alexander responds by admitting his "intuitive conservatism" about the project envisaged for him. Indeed, he offers a quick argument for thinking that the enterprise of identifying the limits of human behavior is self-defeating.

> As it concerns social behavior, human nature would seem to be represented by our learning capabilities and tendencies in different situations. The limits of human nature, then, could be identified by discovering those things that we cannot learn. But there is a paradox in this, for to understand human nature would then be to know how to change it—how to create situations that would enable or cause learning that could not previously occur. (1979, 279)

The argument is specious. Not only can we identify limits on our cognitive abilities without specifying the content of things that we cannot come to know (as, for example, when we find out that the answers to certain questions will forever be unobtainable or that our brains are too small to carry out certain kinds of computations), but we are perfectly able to recognize things that we cannot learn to *do*. Whatever else may be wrong with Wilson's pop sociobiology, it cannot be dismissed as founded on Alexander's "paradox." The poignancy of Wilson's account of human nature lies in its suggestion that, while we can appreciate our own aggressive tendencies and our predilection for ceding power to males, we can only change the resultant patterns of behavior and social institutions at considerable cost. Mere self-knowledge will not enable the rapacious aggressor to become a paragon of peacefulness, any more than mere theoretical knowledge will enable an incompetent like me to control skis on a downhill surface.

The three reviewers were not the only people to find the program envisaged by Alexander and his co-workers elusive. After confessing that he finds the particular studies of Chagnon, Irons, and Dickemann preferable to the grand philosophizing of much human sociobiology, Maynard Smith admits that he cannot tell what kind of explanation these workers are offering. Perhaps sociobiologists are asking whether, in a society with fixed rules, "the actions of different people in that society—rich and poor, young and old, male and female—are those which would be predicted if each individual is behaving, subject to the rules, in the way which would maximize his or her inclusive fitness" (Maynard Smith 1982b, 3). Maynard Smith continues,

> It may be that few sociobiologists think that inclusive fitness can be applied in such a direct manner. If so, it would help us if they told us what they do think. It cannot merely be that human behavior is influenced by kinship, and that kinship has something to do with genetic relationship, because surely, despite some very odd remarks by anthropologists, that is uncontroversial. (1982b, 3)

Maynard Smith has good reason to be puzzled. The professed agnosticism of Alexander, Kurland, and others where questions of proximate causation are concerned leaves us to wonder just what contributions they have made to a "grand analysis of human social behavior." Let us consider the alternatives.

Enriching the Conventional Wisdom . . .

Anthropologists often encounter forms of behavior that appear strange at first. There are societies in which a man supports his sister's children rather than his wife's. In other cultures there is a high frequency of female infanticide among the upper classes. Anthropology faces the challenge of explaining why these practices occur where they do; Alexander and his associates believe that evolutionary biology helps to meet the challenge.

Consider the case of infanticide. People's attitudes toward their infants differ in different conditions. Under many circumstances humans nurture their young, but there are situations in which parents are careless of their offspring (as with the unfortunate Ik; see Turnbull 1972) and societies in which some parents kill some of their children. Let us now suppose that we are given convincing reasons for believing that those who practice infanticide are really maximizing their inclusive fitness. How does this advance the project of anthropology?

We know what the anthropological explanation is *not* supposed to be. We are not entitled to infer that there are genetic differences among human populations, that selection has favored "nurturing" genes in most groups and "genes for committing infanticide" in a few groups where infanticide would maximize inclusive fitness. Chagnon, Flinn, and Melancon join the chorus: "We agree wholeheartedly that there are no such genes and that cultural differences in infanticide or any other human social practices are not likely to be the subject of genetic differences" (1979, 293). The variation in human behavior is a tribute to human plasticity. That much we knew.

One way of interpreting the Alexander program is to begin with a commonsense picture of human motivation that I shall call *folk psychology.* Folk psychology supposes that people typically act so as to maximize their chances of obtaining the things that they perceive as valuable, and it offers a catalog of the aspirations and goals that human beings are likely to acquire under a broad range of circumstances. So, for example, folk psychology suggests that people typically desire the welfare of their children, that they enjoy sexual activity, that they try to secure themselves against danger, and that, in various types of society, they compete for positions of wealth, power, and prestige. Dignifying this collection of truisms about ourselves with a title may seem to give it more than its due. Folk psychology is the motley of commonplaces that we employ every day to explain the actions of those with whom we interact. Nevertheless, folk psychology proves useful in historical explanation and in anthropological explanation. The successful historian makes past actions comprehensible to us by picking out those features of the historical situation that we would have overlooked and that, once seen, provide a purchase for our commonsense understanding of human motivation, so that the actions of past protagonists fall into place. The successful anthropologist can do something similar. When we read about the miserable conditions in which the Ik find themselves, we begin to understand how the desire for the welfare of others—even one's own children—can seem secondary to the overwhelming desire to stave off starvation. When we learn that in northern India the sons of high-class families were a source of wealth (through their ability to attract many wives and through the requirement that the bride's family pay a substantial dowry), while the daughters of such families were, at best, a drain on the family finances, our understanding of human greed enables us to appreciate why upper-class families might have killed many of their daughters at birth.

When it is put to work in the social sciences, folk psychology offers an account of the proximate mechanisms that underlie human actions. When the work is done well, we are given a perspective from which actions that initially seem strange or exotic appear as rational responses to the situations of the agents. Because it concentrates on the proximate mechanisms of actions, folk psychology is not a direct competitor with evolutionary biology. Indeed, we can envisage the possibility that evolutionary biology might enrich folk psychology by charting the evolutionary history that underlies the proximate mechanisms folk psychologists invoke. Here we discover one way in which Alexander's version of sociobiology might be articulated.

Clear-headed folk psychologists ought to recognize that there are evolutionary explanations for whatever mechanisms underlie our familiar human aspirations. If it is indeed true that, in almost any environment, humans will develop so as to have a disposition to nurture their children, then the presence of that disposition has an evolutionary explanation. (However, as the discussion in chapter 7 attempted to show, we should beware of leaping to the conclusion that there was natural selection for such a disposition.) We can foresee a discipline that tidies up the ramshackle collection of principles in folk psychology, traces them to underlying mechanisms, and provides evolutionary explanations for the presence of those mechanisms. Let us call this discipline *evolutionary folk psychology*—or EFP, for short.

The conservative interpretation of Alexander's program views it as a contribution to EFP. By showing how people maximize their inclusive fitness in various social situations, we obtain a deeper understanding of their behavior. Explanations given by folk psychology, which trace proximate mechanisms, are not abandoned as incorrect; rather, they are deepened through the incorporation of an evolutionary perspective. The behavior of those northern Indian patriarchs who killed their daughters is still to be understood in terms of the desire for wealth, power, and prestige, but we are supposed to gain a more profound understanding by recognizing the murders as contributions to inclusive fitness.

Some supporters of Alexander's sociobiological program seem to favor the idea that the evolutionary perspective supplements the conventional wisdom. Kurland remarks that "previous explanations about the *proximate* psychological or economic causes of human kinship behavior" may not be "incorrect or irrelevant" (1979, 146). When we look a little more closely, we find that the ecumenical attitude is hard to sustain. If Alexander's program is interpreted conservatively, then it appears to rest on confusion.

Proponents of EFP ought to believe that people do many things that do not maximize their inclusive fitness. They may decide to adopt a child, to devote considerable time to learning to play the violin, or to abstain from further reproduction. Careful scrutiny might possibly disclose unobvious ways in which these decisions really maximize the agents' inclusive fitnesses. Such discoveries are improbable; but their improbability in no way reflects on the credentials of

EFP. There is no embarrassment in the idea that the course of evolution should have endowed *Homo sapiens* with a number of different capacities; that when people endowed with these capacities grow up in novel environments, they acquire desires for various things that do not correlate with reproductive success; that when they form their plans on the basis of their desires, they are led to act in ways that do not enhance their inclusive fitnesses. By the same token, we do not increase our understanding when we find that complicated processes of decision lead agents to maximize their inclusive fitnesses. A priori expectations incline us to think that that might sometimes be the case and sometimes not. We learn little by recognizing how actual patterns of human behavior are distributed between the two categories.

For champions of EFP the crucial evolutionary question concerns not the pattern of behavior but the mechanisms underlying the behavior. When folk psychological explanations are fully articulated, they take the following form. At the first stage we try to understand the patterns of behavior that occur within a culture in terms of the aspirations of the protagonists and the opportunities afforded them (as well as the constraints imposed upon them) by the culture. The patriarchs of northern India are viewed as having certain desires (for wealth and power, for example) and as having certain opportunities for achieving those desires (by marrying off as many sons as possible). At the next stage we attempt to trace the emergence of the desires that lie behind the action. We hope to recognize the desire for wealth as stemming from a more general human disposition that, given the environment in which they grow up, leads the males in their maturity to favor increased riches over the survival of their female offspring. Only at the third step do evolutionary considerations enter the picture. What the evolutionary process bequeaths to the people who kill their daughters is the general disposition. If our project is to give an evolutionary explanation and if we are prepared to slide quickly to the idea that evolutionary explanations work by showing how inclusive fitness is maximized, then we need to show that a certain general disposition maximizes the inclusive fitness of those who have it. (In the interests of simplicity I have ignored the realistic possibility that the desires that prove efficacious stem from a combination of basic capacities and propensities that have been separately fashioned in the course of human evolution. It should be apparent that introducing further complications only strengthens the line of argument.)

When sociobiologists attempt to show that the patterns of human behavior observed in various unusual social situations maximize inclusive fitness, they show nothing that contributes directly to the advancement of EFP. To extend our understanding of female infanticide, we need an evolutionary explanation of the propensity that, in the context of northern Indian culture (and, possibly, in many other social environments), leads adults to desire wealth. The explanation would consider the evolutionary forces acting on our ancestors in the range of environments they encountered. The situation in northern India in the last millennium is entirely irrelevant.

Suspend for the moment all worries about adaptationism. Assume, equally implausibly, that human beings have a fixed propensity to be greedy. Then we can offer a simple (vulgar) account of what EFP suggests. Infanticide is the result of patriarchal greed. Patriarchs are greedy because humans have evolved the propensity for greed. They have evolved that propensity because greed pays. Now we confront the evolutionary question: Why does greed pay? To answer that question, the rampant adaptationist enterprise I am conceiving would try to show how greedy hominids were reproductively successful in ancestral environments. All kinds of things are wrong with this enterprise. But it does avoid the error of supposing that accounts of recent fitness maximization among the aristocracy of northern India are relevant to the evolutionary issues.

An admirer of Dr. Pangloss might suggest that the success of our noses in holding up spectacles is relevant to the issue of why our faces are designed as they are. Nobody should take the suggestion seriously. We recognize that our facial anatomy is the product of evolution and that, given the product, we have been able to put it to unanticipated uses. Similarly, we should recognize that human propensities, the results of our evolutionary past, are employed in unprecedented ways in our recent efforts at shaping our social arrangements. To link those efforts to the maximization of inclusive fitness is as irrelevant and misguided as descanting on the enhanced fitness of those whose sorry state of myopia is alleviated by their ability to balance spectacles on their noses.

I conclude that if EFP is a correct program for the understanding of human social arrangements, then the claims about maximization of inclusive fitness under special social arrangements do not contribute the evolutionary explanations that are really needed. On the conservative interpretation of Alexander's program, on which familiar proximate mechanisms are left intact, the evolutionary work is simply at the wrong level of generality. Perhaps a more radical construal will fare better.

. . . Or Subverting Our Self-Image?

Folk psychology might be inadequate. Moreover, ideas from evolutionary biology might help us to spot the inadequacy. Recognizing that a form of social behavior maximizes our inclusive fitness could challenge our prior conceptions of human motives and human decision making. We could discover that there is no way to understand the actions of members of a society without attributing to them motives and abilities to assess the evolutionary consequences of various strategies that are quite different from the motives and abilities that we normally take ourselves to possess.

Thanks to Freud, our twentieth-century ideas about human motivation and the springs of human behavior have already been extended beyond those that were available to our predecessors. The radical interpretation of Alexander's program promises further subversion of our self-image. Instead of viewing sociobiological explanations as evolutionary accounts of forms of behavior whose proximate mechanisms are presently well understood, the more ambitious thesis is that, by recognizing how the behavior patterns maximize inclusive fitness, we shall be led to identify hitherto unappreciated causes of human action. Not only does this construal avoid the muddle noted in the last section. It is also intrinsically more exciting. Even though an evolutionary explanation of the presence of familiar proximate mechanisms would be a contribution to EFP, it is not entirely obvious that it deserves star billing as a "grand analysis of human nature."

There are two ways of developing the radical interpretation. The first is to suppose that there is some general mechanism for calculating the expected payoffs of the available courses of action and to maintain that this always comes into play in the causation of our social behavior. We think of ourselves as acting in accordance with our conscious wishes and beliefs. However, the reflective decision making on which we pride ourselves is only a facade. Behind it lurks the general fitness calculator, quietly grinding out what is in our own best evolutionary interests. That is the engine that drives our actions. We become aware of its workings when we encounter situations in which people engage in social practices that maximize their inclusive fitness and in which their actions cannot be reconstructed by applying our everyday models of decision making.

The second possibility is less revolutionary. Perhaps our decision making proceeds in the general way that we normally suppose. People act so as to maximize their perceived chances of obtaining what they value. We represent various options to ourselves. We assess the extent to which these options would bring us what we want and the chances of realizing our goals. We choose that course of action that brings the greatest expected return. However, we are not always aware of our evaluations and assessments. By looking closely at the social practices of a range of populations, we find that, while the members of each group act so as to maximize their inclusive fitness, their actions cannot be understood unless we modify the collection of human interests and abilities. We are forced to suppose that the evolutionary process has fixed in us hitherto unnoticed proximate mechanisms that affect our decision making. There are motives of which we were not previously aware. There is a propensity not simply to love our kin, but to calculate exact coefficients of relationship and to value people accordingly. The form of human decision making agrees with our previous ideas, but the particular figures in the drama are different.

Pursued in either of these ways, human sociobiology can generate exciting conclusions. The popular imagination stirs to the new prospect that we are each equipped with an all-purpose inclusive-fitness calculator, that this calculator governs our decisions and actions, and that the conscious processes that seemed to play so critical a role are merely a false, imaginary glare. Similarly, if we discover that evolution has fixed in us basic desires of which we were formerly ignorant—perhaps a desire to be dominated that expresses itself in females, or an automatic aversion to strangers—then these unappreciated motives will form part of a new picture of human nature. On the radical interpretation, Alexander's program achieves the headline-grabbing power of pop sociobiology.

Indeed, it shares with Wilson's pop sociobiology a bold insistence on our coming to terms with ourselves. No task is more important than the fathoming of our nature. For, once armed with an understanding of how human patterns of behavior are directed toward the maximization of inclusive fitness, we shall begin to see how to achieve those social reforms that presently attract us. After elaborating his main claim about the human propensities for acting to maximize inclusive fitness, Alexander remarks,

> The single environment in which all that I have just said can become irrelevant, of course, is that in which the interactants have become consciously aware of this aspect of their natural history. Perhaps that is the most important point in this book. (1979, 131)

The theme recurs at the end of a discussion of the role of strife in hominid evolution:

> . . . what I am saying throughout this book may be right or wrong as an interpretation of human history. In any case, it does not imply a deterministic future. On the contrary, I have argued that the individuals and groups least bound by history are those who best understand it. (1979, 233)

The vision offers an optimistic contrast to Wilson's talk of "costs" that attend our attempts to resist our nature. Alexander seems to think that when our hidden motives and calculations are brought out into the open, their power over us will wither and we shall be able to organize our lives in ways that run counter to the demands of inclusive fitness. Unfortunately, the optimism seems to rest in large measure on the specious argument discussed in the first section of this chapter.

So we have identified a new version of pop sociobiology. The task of relating human social behavior to considerations of inclusive fitness is hailed as "the first priority" for any serious investigation of that behavior. Apparently we are to treat the social behavior of our own species using the ideas and techniques of contemporary evolutionary theory and employing the same rigor that is found in the best studies of nonhumans. The next step should be to draw the consequences for our understanding of the *proximate* causes of human decisions and actions.

> Is what is pleasurable, hence seems "good" and "right," that which, at least in environments past, maximized genetic representation? I suggest so, even to the degree that our pleasure depends upon a capability of judging reactions of others to each of our actions, including the use of planning, self-awareness, and conscience to establish our own personal guidelines to success in sociality. (Alexander 1979, 82)

Apparently we are to understand our real motives and abilities as adaptations that, in combination, lead us to actions that maximize our inclusive fitness. The hopeful conclusion is that once we have become aware of how we work, we shall be able to adjust our society in the direction of our ideals.

I shall henceforth ignore the theme that self-understanding can liberate us from the secret mechanisms that maximize inclusive fitness. My aim in the analysis that follows will be to see whether Alexander's pop sociobiology does offer interesting conclusions about how people maximize their inclusive fitness, whether it reveals previously unrecognized human motives and capacities, and whether it offers any evolutionary explanations for the springs of our behavior.

Predictive Potpourri

Although Alexander's version of pop sociobiology focuses on a small number of examples that it treats at length, there are many shorter speculations on a variety of aspects of human life and behavior. Addressing the possibility that homosexuals do not maximize their inclusive fitness, Alexander thinks it appropriate to ask if "the phenotypes of some homosexual individuals, or their developmental experiences, are such as to reduce their likelihood of success in ordinary sexual competition or increase their likelihood of success in some alternative kind of behavior" (1979, 204). He intimates that we can understand why we feel "alone in a crowd" if we remind ourselves that what matters in evolution is the relative fitnesses of individuals who compete with one another (1979, 17). One of his co-workers, Low, explains why men are "especially attracted to large-breasted but otherwise slender women" (Low 1979, 466). Allegedly the selection is driven by male preference for mates who will be good mothers.

In these and similar examples, such as the discussion of macho behavior in teenage males (Alexander 1979, 244) and the characterization of the circumstances under which females are most likely to achieve orgasm (Alexander and Noonan 1979, 451), we discover the same kinds of errors that we have seen in the pop sociobiology of Wilson and his followers. The speculations feature a yen for casual optimization and the parading of gossip as observational data. To take just one example, it seems legitimate to wonder why selection for big breasts has not proceeded apace in other mammalian species. We might also seek to know the correlation between breast size and ability to nourish young. Finally, those who are swept away by Low's account of what attracts men might enjoy a stroll through an art museum with a good Rubens collection.

Because the troubles of such cases should be familiar from previous discussions, I shall only deal with the major examples produced by Alexander and his co-workers. Alexander's own argument attempts to combine quantity with quality. He suggests that he has a monolithic account of human behavior, based on the idea that patterns of social behavior maximize the inclusive fitness of the participants (or, in the more cautious version, that they are the residues of forms of behavior that previously maximized inclusive fitness). He claims that this account gives rise to a large number of predictions and that the predictions are borne out when we examine the ethnographic record. He also offers detailed explanations of two cases that once seemed to him "the most provocative and outstanding apparent contradictions of an evolutionary view of human behavior" (1979, 168). These are the phenomena of the "avunculate," a social arrangement in which children are provided for by their mother's brother and not by their mother's husband, and the distinction of various types of cousins in the marriage rules of some societies.

In this section I shall take a quick look at the list of predictions with which Alexander hopes to dazzle his readers. Subsequent sections will take up the explanation of the avunculate, given qualitatively by Alexander and elaborated with more precision by Kurland, and also two other influential studies. We shall consider Chagnon's account of alliances in combat among the Yanomamo and Dickemann's explanation of patterns of female infanticide. With respect to all these examples we shall pose three main questions: Do the forms of behavior studied really maximize the inclusive fitness of those who engage in them? Even if they did, would that increase our understanding of familiar proximate mechanisms of human action? Would it offer us new suggestions about hitherto unappreciated motives and capacities that figure in human decision making?

Alexander introduces his list of twenty-five predictions with something of an apology: "While some of these predictions are trivial, taken individually, and some may be faulted as circular because our immersion in the human system of sociality already tells us that they are true, collectively they are significant, and the longer the list the more important it becomes" (1979, 156). The apology is justified but the confidence is misplaced. Militant adaptationists think that the details of animal morphology and physiology maximize inclusive fitness. This hypothesis by itself, however, does not tell us much about the structure and functioning of members of particular species. Similarly, the large claim that human beings act so as to maximize their inclusive fitness does not predict much about the details of human sociality. As we shall see, Alexander's list consists of a sequence of relatively banal facts about ourselves that are only loosely connected with his central claim.

The first "prediction" is supposed to derive from Hamilton. "When the abilities of potential recipients of nepotistic benefits to translate such benefits into reproduction are equal, then closer relatives will be favored over more distant relatives" (1979, 156). As Alexander goes on to note (157), this is not so much a prediction as a restatement of Hamilton's central insight. However, as I have emphasized earlier, there is no reason to think that the inclusive fitness of an action will be determined by simple considerations of relatedness alone. If humans behave so as to maximize their inclusive fitness, then in some situations they may do better to give benefits to more distant relatives (or to unrelated individuals) because of other factors that promote inclusive fitness—likelihood of future reciprocation, the benefits of joining an alliance, or whatever. So, in more exact form, the prediction tells us that when the abilities of potential recipients to translate benefits into reproduction are equal, then either the closer relative will be favored or there will be some other return in augmented fitness for the donor that outweighs the difference in relatedness. In this form the prediction is no observationally verifiable consequence of the hypothesis that humans act so as to maximize their inclusive fitness; rather, it is a simple corollary: we help our closer relatives except when it would detract from our inclusive fitness to do so.

Other predictions are more specific in concentrating on particular kinds of social situations but are stated so vaguely that they effectively allow the aspiring pop sociobiologist to cover all the possibilities. Consider the following.

> 4. Cooperativeness and competitiveness between particular sets of relatives, such as full siblings, may vary across essentially the entire spectrum of possibilities, depending upon their opportunities and needs to use the same resource (e.g., parental care or mates) and the value to each of having a cooperative individual available. (157)

To translate: siblings are expected to compete when they would maximize their inclusive fitness by disputing an important resource, and they are expected to cooperate when their inclusive fitness is maximized through mutual aid. How are we to compare this "prediction" with human social behavior? The short answer is that we cannot do so until we are offered a detailed account of the factors that affect the fitness of siblings in particular situations. While we remain at the level of generality favored by Alexander, nothing is tested and nothing is confirmed.

Consider next a suggestion about parental attitudes toward their children.

> 8. Older offspring are likely in many circumstances to be reproductively more valuable than younger offspring, and to serve their parents' interests better in the course of serving their own interests because of their typical age and dependency relations to their siblings (leading to primogeniture). Younger dependent offspring, on the other hand, may be given full attention with fewer reservations because of the diminished likelihood of additional dependent young (hence, may be "spoiled"). (158)

Here we have the first concrete expectations: people should favor primogeniture and they are likely to "spoil" younger children. I shall consider later whether evolutionary considerations add anything to our expectations about primogeniture. For the moment let us look at the reasoning that underlies the second prediction.

Imagine a human family with a number of children. First came first and Last came last, and there were a number in between. When First was young there were no other siblings with whom parental attention had to be shared. Hence First received all the energy and attention that the parents were ready to give. Now that Last has arrived, the parents no longer have to hold resources in reserve for children yet unborn. They can give their all to their brood. Of course, there is greater competition for energy and attention. Whether Last receives more attention at any given stage depends on whether the parents' optimal division of a greater resource pool among all their offspring assigns to Last more than the entire pool that was available when First was at the comparable stage. The accounting is not obvious. There is no evident reason why younger children should be "spoiled."

Finally, let us look at something even more specific.

> 14. Relatives by marriage are in a particularly favorable position to gain by cheating their affluent in-laws, since they do not gain directly by the distribution of any benefits to their spouse's relatives (hence the prominence of in-law jokes). (159)

Once again the optimality argument is suspect. Married people start with different attitudes toward members of their respective families. However, it is hard to see why people should be expected to cheat their affluent in-laws. If there are frequent situations in which two people can achieve substantially more by cooperating than by working separately, then it may be in the interests of sons-in-law to assist their fathers-in-law and conversely. Without detailed knowledge of the constraints and available strategies in the social situation, there is no justification for pronouncing a priori on expected human propensities for cheating their relatives through marriage. There is even less basis for the idea that the recognition of in-law treachery should express itself in the form of jokes. No doubt, if some other attitude toward in-laws were prevalent, Alexander would hail that too as evidence for some deep appreciation of our inclusive-fitness interests.

Enough of the predictions themselves. The claims advanced are connected with Alexander's central idea about fitness maximization only by the most casual assumptions about optimization. They are also almost always too vague to permit comparison with actual data about human behavior. Nevertheless, the full worthlessness of the list is only revealed when we consider Alexander's candid response to some cases in which studies of human sociality appear to threaten his general approach.

One obvious difficulty concerns adoption. Alexander discusses a cluster of cases described by George Murdock.

> "In Africa and elsewhere . . . it is common for the illegitimate children of a married woman by another man to be unquestioningly affiliated by patrilineal descent with her husband, their 'sociological father' " ([Murdock 1949] p. 15). One wishes to ask (a) whether such children are valuable to the sociological father, either as labor or as a source of bride-price, and (b) what other benefits may accrue to the man who accepts such children. Does he thereby retain a wife of value that he would otherwise lose? Does he maintain ties with an important group of affinal allies? Is the real father commonly a close relative such as a brother? Are such events reciprocal? How is the child actually treated by him? The cases to which Murdock refers are probably those in which a girl's mother selects for her a lover who fathers her first child. In such cases we need to know the status of the lover, the eventual distribution of inheritance, the extent of knowledge of paternity in all cases, the behavior of the real father toward the first child, and more. Without such kinds of information there is no way to analyze the apparent paradox. (1979, 166)

Before we fulminate against the invocation of ad hoc hypotheses and the zeal with which a favored hypothesis is defended, we should recognize that Alexander is quite right. The information for which he declares a need is required if we are to understand whether the men under discussion are or are not acting in the interests of their inclusive fitness. Perhaps on closer investigation we would arrive at a model that reveals them to be maximizing their inclusive fitness after all. However, an equal number of considerations are relevant in any other case in which we hope to discuss the relative fitnesses of forms of human behavior. We cannot invoke confounding variables just where it is convenient to do so. Appeals to the complexity of human social situations ring hollow when that complexity has been cheerfully ignored in deriving previous predictions. All those putative confirming instances of those vague general predictions correspond to complicated situations for which sophisticated analysis is needed if we are to discover whether, and in what way, the participants are maximizing their inclusive fitness. Just as the "apparent inconsistency" is no decisive objection to the thesis Alexander hopes to defend, so the softly focused appearance of consistency elsewhere provides no real support.

Moreover, the introduction of the evolutionary perspective brings with it no increase in predictive power. Folk psychology, unaugmented by any evolutionary ideas, will yield all the expectations Alexander claims for his own central hypothesis. Consider the most promising example, the institution of primogeniture. It is easy to devise an argument for thinking that elder children (even elder sons) are likely to inherit the family wealth. Parents typically desire the welfare of their children. If they should die before the younger children are mature, it is important that the goods they have acquired should be passed on to someone who will protect the surviving juveniles. Since siblings can be expected to defend one another, the best strategy seems to be to bequeath the family fortune to the eldest child (or the eldest son, in situations where men have greater powers and privileges than women).

In offering this as a rival story to that proposed by Alexander, do we beg the question? Folk psychology takes certain human propensities for granted, whereas Alexander's pop sociobiology achieves greater predictive power by beginning with the evolutionary underpinnings of such propensities. Yet how exactly is anything more achieved? Does the appeal to inclusive fitness offer a more refined account of what is going on, suggesting new mechanisms and offering more precise predictions? Hardly. Alexander's account offers no expectation that is more detailed than the one given in the last paragraph. So do we achieve greater depth by beginning with the points about inclusive fitness? By starting with considerations of inclusive fitness, do we understand the propensities that folk psychology takes for granted and thereby come to recognize the institution of primogeniture as something to be expected?

No doubt there is an evolutionary explanation for the commonplace facts that parents desire the welfare of their children, that siblings are typically disposed to help one another against outsiders, and so forth. Moreover, we can recognize the gross outlines of the explanation. Relatives enhance their inclusive fitness, *other things being equal*, by helping one another. So much is banal. But this has nothing to do with the claim that particular patterns of human social behavior maximize the inclusive fitness of those who engage in them. Specifically, it has nothing to do with the thesis that instituting primogeniture maximizes inclusive fitness. For suppose that thesis were true. It would be folly to continue by arguing as follows: "We now understand why there is a human disposition to desire the welfare of children. This disposition has been fixed in us as a proximate mechanism for leading us to maximize our inclusive fitness through instituting primogeniture." My imaginary argument sets the cart before the horse. Evolution has not equipped us with the dispositions to care for our children and aid our siblings because those dispositions serve as the means of achieving an allegedly fitness-maximizing institution of primogeniture. Those dispositions have evolved in us because of other factors, whose gross form we can appreciate by recognizing the elementary point that we share genes with children and siblings. Once in place, the dispositions lead us to behave in a variety of novel ways in a variety of novel contexts. Whether or not these forms of behavior maximize our inclusive fitness is irrelevant.

We cannot praise Alexander's hypothesis on the grounds that it leads us to more refined expectations than those offered by folk psychology, for it does no such thing. Nor can we regard it as deepening the understanding that folk psychology provides. For even if the claims that specific patterns of human behavior maximize the inclusive fitness of those who engage in them were true, those claims would not supply the evolutionary explanations we need. At most they would yield a more precise description of the phenomena that EFP ought to explain.

The attempt to bombard readers with a shower of predictions does nothing to exhibit the power of Alexander's pop sociobiology. Ironically, some writers whom Alexander despises (and despises with much justice) are fond of a similar style of argument. In *Scientific Creationism* Henry Morris draws up a table of "fourteen predictions" from his version of Flood Geology and announces triumphantly that the predictions are confirmed. Like Alexander's, the predictions are formulated in very vague terms, and it is easy for both authors to cope with worrisome exceptions. Both Alexander and Morris appear to think that some vague matching of trends in nature with trends loosely associated with their central ideas will help their cause. In both cases the right response is to press for details.

Since Alexander believes that only misguided Creationists and ignorant philosophers (1979, xvi) block the acceptance of Darwinism and the appreciation of its significance for human affairs, he should find this kinship of method somewhat disquieting. The policy of announcing large numbers of vaguely stated "general predictions" does nothing to support the hypothesis that human social behavior generally maximizes inclusive fitness. It does not enrich our understanding of the familiar mechanisms of human decision making. It does not point the way to any hitherto unrecognized springs of human behavior. Let us see if the more detailed treatment of a particular case fares better.

Absent Fathers and Kind Uncles

There are a number of societies in which it is not the custom for a man to use his goods and abilities to support the children of the woman (or women) to whom he is married. Instead, he directs his labor and its fruits toward the children of a sister. At first sight the arrangement appears inconsistent with the claim that men act to maximize their inclusive fitness. After all, a man is related to the children of a sister by 1/4 and to his own children by 1/2. By using his wealth to aid his nieces and nephews, he is pursuing a suboptimal strategy, for his goods might be put to better use in promoting the reproductive abilities of those to whom he is more closely related.

Once the apparent inconsistency is elaborated in this way, it is evident that it depends on an assumption. We suppose that the children of the man's wife (or wives) are his own. If this is not so, then the conclusion that aiding nephews and nieces is suboptimal will no longer follow. Indeed, if the probability of paternity is sufficiently low, a man's best strategy for maximizing his inclusive fitness may be to give aid to the children of a sister. The avunculate turns out to be an institution that maximizes inclusive fitness (at least for the men; for the moment we shall imagine that they are in control of how the benefits they acquire are to be dispensed).

Alexander outlines the reasoning I have just given and summarizes his conclusion as follows:

> . . . if a man lacks confidence of paternity, his nieces and nephews may be his closest relatives in the next generation. This alone does not mean that such relatives are the most appropriate targets of parental care, since, if his low confidence is unusual (i.e., other males are confident of their paternity), his nieces and nephews may be expected to have fathers willing to care for them. I repeat here that none of these calculations need be conscious to individuals who nevertheless behave as though they are. (1979, 171)

Let us postpone, for the moment, asking how the coincidence between actions and the decisions recommended by inclusive-fitness calculations is actually achieved. Let us begin with the details of the fitness-maximization story.

We imagine a society in which there is widespread low confidence of paternity. For concreteness, suppose that men live apart from their wives and that the separation persists because of some aspect of the social situation (for example, the need for young men to put in long periods of service as soldiers). Conditions of this sort are present in some of the cases for which Alexander designs his analysis. Let us also assume that the fitness of a child is directly proportional to the amount of aid received from a man. In other words, if M_1 gives more aid to C_1 than M_2 gives to C_2, then C_1 will have greater fitness than C_2.

We now imagine that there is a tradition of men aiding the children of sisters rather than the children of wives. Consider the possibilities for the most wealthy man in the society. Under the prevailing system his wife's (or wives') children are the recipients of benefits from someone less wealthy than himself. Would it be in the interests of his inclusive fitness to alter the traditional arrangement and give some of his wealth to support his wife's (wives') children? We can simplify the analysis by supposing that his only choice is whether to contribute the difference between his wealth and the wealth of his wife's brother to his sister's children or to his wife's children. Which strategy maximizes his inclusive fitness?

If the probability that a wife's offspring are her husband's is p, then the expected fitness payoff if the male gives his support to his wife's children is $\frac{1}{2}Bp$ (where B measures the benefit in units of fitness). The expected fitness payoff from helping his sister's children is

$$\tfrac{1}{2}B[\tfrac{1}{2}(\text{probability that ``sister'' is a full sibling}) + \tfrac{1}{4}(\text{probability that ``sister'' is a half sibling})].$$

We can reasonably suppose that the probability that the woman known as "sister" is either a full sibling or a half sibling is 1. The probability that she is a full sibling is the probability that she and the man have the same father.

Kurland takes this probability to be p^2 (1979, 150–151; the result is a "correction," due to Greene 1978, of the value assigned in Kurland 1976). This is not quite right, however. The probability that their mother's husband sired both the male and "sister" is p^2. The probability that they share a common father is greater. If their mother copulated with n males during some relevant time period (in addition to copulating with her husband), then, if each male had some probability q_i of fathering any one of her children, the probability that two children share the same father is $p^2 + \Sigma\, q_i^2$. Under conditions of promiscuity, where n is sufficiently large, we can assume that each q_i is small in comparison with p and that the relevant probability is approximated by p^2.

We can now state the condition under which the wealthy man will do better to buck the system and direct his surplus wealth to his wife's children. We require that

$$\tfrac{1}{2}Bp > \tfrac{1}{2}B[\tfrac{1}{2}p^2 + \tfrac{1}{4}(1-p^2)];$$

that is,

$$p^2 - 4p + 1 < 0.$$

As Kurland notes, this inequality signals a *paternity threshold* ($p = 0.268$, the root of $p^2 - 4p + 1 = 0$ that lies between 0 and 1) such that if the probability p is above the threshold, the man will do better by investing his resources in his wife's children. Below the threshold he does better to abide by the rules of the avunculate.

So far, my analysis differs from that of Kurland and Alexander in focusing only on the wealthiest man in the society. In this way I sidestep the obvious worry that a poor man who is married to the sister of a rich man would benefit from the avunculate even if his wife's children were actually his own. However, the treatment I have offered can easily be extended. Suppose that the n wealthiest men in the society maximize their inclusive fitness by each supporting the children of a sister. Consider now the predicament of the $n+1$st man (in the ordering determined by wealth). Either his wife is the sister of a man who has more wealth or she is the sister of a poorer man. (We can treat women without brothers as if they had brothers with zero wealth.) In the former case his wife is already accommodated by the system. (We can imagine that she is part of the household of her brother.) In the latter case the man's predicament is just that of the wealthiest man in the original situation. Hence, if the fitness-maximizing strategy for the n wealthiest men is to direct benefits to sisters' children, then that will also be the fitness-maximizing strategy for the $n+1$st man. We thus have an inductive argument for concluding that below the paternity threshold all men will maximize their inclusive fitness by continuing the tradition. In a similar way the original establishing of the tradition can be viewed as a fitness-maximizing strategy.

Kurland refines his analysis by considering optimal strategies in situations in which men can divide their resources (1979, 177–180). He also considers, but does not articulate in detail, another refinement:

> Paternity certainty is something that a woman can give a man. In the hypergamous marriage system of the Nayar, a woman might increase her reproductive success by guaranteeing paternity to a high-status man who would therefore be willing to, and capable of, investing in the child. (163)

Exploring the possibility of female strategies leads to some interesting conclusions.

Imagine a population in which the avunculate exists. We suppose that the probability that a woman's child is fathered by her husband, p, is below the paternity threshold. Now a woman decides to play a new strategy: she announces to the wealthiest man that she is willing to submit herself to the supervision of his female relatives, and thus to guarantee his paternity, if he, in return, will marry her and direct his resources to their children. Call this the *Calpurnia* strategy. Evidently, it is in the reproductive interests of any woman except the sister of the wealthiest man to play Calpurnia. For by doing so, she secures more wealth for her offspring than she would otherwise have done and thus—by the assumption that the fitness of the offspring increases as the benefits given to them increase—produces offspring who are maximally fit. It is also in the interests of the wealthiest man to take up the offer. For by doing so, he achieves a payoff of $\frac{1}{2}B$, which is greater than the $\frac{1}{2}B[\frac{1}{2}p^2 + \frac{1}{4}(1-p^2)]$ that he would otherwise obtain. Hence, in a population practicing the avunculate, we would expect a woman to maximize her fitness by playing Calpurnia and the wealthiest man to maximize his fitness by cooperating with this woman. Calpurnia can invade.

Now imagine a society in which the n wealthiest men do not practice the avunculate but mate with women who play Calpurnia. Consider any unmated female except the sister of the $n+1$st male. It will be in the reproductive interests of this woman to make a Calpurnia offer to the $n+1$st man (assuming that wealthier men are all "saturated" with wives). For, given that the n wealthiest men are directing their resources toward the children of their wives, the woman cannot expect to achieve any greater level of support for her offspring. Hence, if the n wealthiest men abandon the avunculate to direct their resources toward the children of wives who play Calpurnia, the $n+1$st man should have the opportunity to do the same, and in the interests of his fitness he should seize it. By induction we can conclude that the society can be expected to abandon the avunculate in favor of an alternative arrangement in which wives submit to supervision and husbands support the children of wives. That is what would be expected if individuals maximized their inclusive fitness.

Once we have begun to explore rival strategies, we can see that they are legion. If p falls below the paternity threshold, then there must be considerably more copulation outside marriages than inside. It is important that such copulation should be broadly spread. If each man has only three or four regular sexual partners, then it will be possible for many men to identify women (not necessarily their wives) whose children stand a sufficiently high probability of being their offspring to make it worth investing in them. Similarly, if we take seriously the idea that women act to maximize their inclusive fitness, then we should consider the possibility that they will simply abandon their fickle ways and that this turn to chastity will be detected by their fitness-maximizing spouses.

The case illustrates a general point that I have emphasized before. Claims about optimization, evolutionarily stable strategies, or maximization of inclusive fitness must be advanced against a background of constraints. The constraints cannot simply be invented. Careful work is needed to discover the alternatives that are really available to the people whose behavior is under study. The point of the Calpurnia example (and of the other strategies that I have indicated more briefly) is that, given starting assumptions that are *no less plausible* than those favored by Alexander and Kurland, we can argue that the avunculate does *not* maximize the inclusive fitnesses of the men and women who engage in it.

Indeed, if we were to discover that the ethnographic record corresponds to Kurland's model—that is, that the avunculate exists exactly where the society is below the paternity threshold—then we should have to explain how the society has managed to resist invasion by women playing Calpurnia. In fact, the available data do not provide convincing support for Kurland's model. As he himself admits, "Testing this model by means of the ethnographic literature is difficult" (1979, 157). The problems are not of his own making. Traditional anthropology has been preoccupied with other questions, and the measurements relevant to Kurland's model have not been made. Indeed, even if anthropologists now turned with gusto to the task of testing the hypothesis that the avunculate exists just where the society is below the paternity threshold, they would have to overcome obvious difficulties. Except in societies in which all copulations take place in public (or in which there are detailed genetic data), it will be hard to achieve reliable estimates of the crucial parameters, the probabilities for each man-woman pair that a child of the woman was fathered by the man. (Recall that the crucial probability in determining whether "sister" is a full sister may involve other terms beside p^2.)

The best we might hope to show, then, is that when extramarital intercourse is relatively common, societies practice the avunculate and that this corresponds, in a general way, to the deliverances of Kurland's model. Alexander believes that even this much may not be forthcoming: "The avunculate is more widespread than is dramatically lowered confidence of paternity" (1979, 175). He advances an alternative scenario to accommodate some of the difficult cases:

> . . . in the case of the avunculocal extended family, even prepubertal boys may leave their parental homes, going to live with a maternal uncle in another village. If, in such cases, these boys marry their uncle's daughter (MoBrDa), this system amounts to prepubertal adoption (and supervision and guidance) of a daughter's husband, on the one hand, and assurance of a wife for a prepubertal son on the other. (1979, 175)

Before we marvel at the manifold ways in which inclusive fitness can be maximized, we would do well to note a small difficulty with the proposed analysis. Like other human sociobiologists, Alexander is impressed with the idea that people have a propensity to avoid copulating with those with whom they have associated most closely prior to puberty (1979, 79). If the analysis of human avoidance of incest is correct, then it seems that the strategy sketched by Alexander may well not maximize inclusive fitness for the participants. According to the results from the kibbutz and from Taiwan (discussed in the previous chapter), there is a significant probability that the adopted son will fail to consummate the marriage, and a lowered expectation of offspring from it. Hence it is far from obvious that the beneficent uncle maximizes his chances of transferring his genes into the second generation by directing his resources toward his nephew.

Let us take stock. We have a problematic model that fails to explain why the avunculate does not suffer invasion from women playing Calpurnia and, more generally, appears to depend on undefended assumptions about the alternative strategies that are available to members of the pertinent societies. We have an admittedly inadequate ethnographic record, unable to furnish data that are sufficiently precise to enable us to compare the deliverances of the model with actual human behavior. We have examples of the avunculate that fall outside the scope of the model, and we have attempts to explain these cases that are seemingly in tension with other favored pieces of sociobiological research. How then does the study contribute to our understanding of human nature?

At this point is is helpful to pose our other two main questions. Do the practices discussed by Kurland and Alexander point toward a different account of human decision making from that encapsulated in folk psychology? Or does their study deepen our understanding of the mechanisms that folk psychology would invoke to explain the presence and persistence of the avunculate? I shall argue that the answer to both questions is "No."

In those societies in which marital infidelity goes hand in hand with the practice of the avunculate, we need only appeal to relatively commonplace human desires. Few people are going to deny that we usually develop a disposition to help our kin and to do what we can to promote their welfare. Typically, this disposition leads us to give aid to our own children but not to assume a primary role in caring for the children of siblings. It is obvious, however, that in the societies where Kurland's model of the avunculate works, there will be very good reason to doubt that a woman's children were fathered by the woman's husband. Hence, the man's natural disposition to help his offspring cannot be confidently expressed in the provision of benefits to these children. In a situation of uncertainty the man turns to those with whom he has some assurance of kinship. There is no need to hypothesize some unconscious mechanism of inclusive-fitness calculation or some hitherto unsuspected motives that redirect the man's efforts. Banal points about common human aspirations make the practice comprehensible. Perhaps this is the reason why Kurland is at pains to note that his model may be compatible with traditional accounts of proximate mechanisms. It is.

Similarly, in the case of the uncles who "adopt husbands" for their unmarried daughters, there is no need to look further than some familiar points about human motivation. As Alexander himself suggests, we might expect this practice to occur in societies that favor polygyny, for in these cases the family of the adoptee will want to make sure of a wife for their son (again the old concern for the welfare of offspring appears), and the adoptive family will want to increase the chances that their daughter will be well treated by her husband. The arrangement promises a neat solution to both of these problems—although, as I have remarked, it may suffer the difficulties that are supposed to attend arranged marriages between people who have been reared together. Perhaps those who make the arrangement are not sufficiently attuned to the genetic consequences for these difficulties to override the familiar human motives to which I have appealed. Perhaps adoption after puberty would genuinely maximize inclusive fitness, but this option is valued less because the adoptive family wants greater influence on the attitudes and actions of their daughter's husband.

Our explanations in terms of ordinary human motives are vague. Moreover, we should expect that in different cases a host of different extra factors will come into play. In a very general way we can understand the avunculate as arising from the participants' attempts to satisfy perfectly ordinary human desires and as being maintained in the same way. More detailed investigation *might* reveal to us that the practice can only be understood if we postulate hidden mechanisms that really underlie our decisions. For example, we might discover that there is a cluster of societies in which the paternity probability—the parameter p —takes values in a small interval around the paternity threshold, that those societies above the paternity threshold adopt the system of supporting the children of wives, that those below adopt the avunculate, and that those at the paternity threshold are somehow unstable. Evidence of this sort would reveal that humans have a much more fine-grained ability to respond to the dictates of inclusive fitness than is suggested in our everyday recognition of propensities to care for kin. We do not have anything remotely resembling such data, however. From what we know about the relevant societies there is no reason to believe that there are mechanisms that adjust human behavior to the optima defined by Kurland's model (even granting that those are the real optima).

Vague as it is, the explanation by appeal to everyday human motives proves superior to an account that would ascribe to us some unknown mechanism for adjusting our behavior to maximize inclusive fitness. It allows us to understand why Calpurnias do not typically subvert the avunculate. When we think in terms of maximization of inclusive fitness, there is a genuine puzzle: women really would enhance their inclusive fitness by making Calpurnia bargains with men. But when we think in terms of proximate mechanisms, the puzzle dissolves. Women might have to make personal sacrifices in order to provide their husbands with confidence of paternity. They might have to forfeit the company of friends and kin, to subject themselves to the scrutiny of suspicious people, and so forth. Common desires override the propensity to advance the interests of future children. More exactly, we may expect that a woman will sometimes be attracted by the prospect of marriage with a very wealthy and powerful man—when she can improve her own situation and that of her children at a relatively small cost to personal freedom—but that she will be reluctant to make such a bargain when the expected improvements are slight in comparison with the personal losses. Calpurnia may be the best female strategy from the perspective of inclusive-fitness maximization, but it is easy to understand why women do not play it.

Folk psychology triumphs over the appeal to inclusive fitness. The conservative interpretation (and defense) of Alexander's pop sociobiology would have it that the desires and dispositions to which folk psychology appeals are themselves to be understood in terms of our evolutionary history. Very true. Yet the reply is the same: the proposals of Kurland and Alexander contribute nothing to the evolutionary explanation of that which needs evolutionary explanation. Suppose it were true that the avunculate maximizes the inclusive fitness of those who practise it. How would that fact bear on the evolution of the dispositions, such as the desire for the welfare of one's close relatives, that underlie the practice of the avunculate? It would be absurd to suppose that the evolutionary explanation of the presence of these dispositions is that they have been fixed in us as a means to maximize our inclusive fitness when paternity is uncertain.

When Alexander's program is interpreted conservatively, the same error emerges again and again. Evolutionary folk psychology proposes to trace social arrangements to proximate mechanisms and to seek an evolutionary explanation of the presence of the mechanisms. Alexander and his followers (construed conservatively) seem to propose an evolutionary account of the patterns of behavior themselves. Their discussions are askew because they shed no light on what really needs evolutionary explanation—to wit, the proximate mechanisms.

Mayhem, Murder, and Marriage

Pop sociobiology's discussion of the avunculate proves a disappointment. We are left with some interesting anthropological description, the story of the conditions under which men choose to help their sisters' children; the rest is the idle addition of concepts from evolutionary biology. The result is typical. Other major examples of anthropological research inspired with pop sociobiological zeal reach the same terminus. Let us look at one that has been most thoroughly documented: Chagnon's study of the Yanomamo Indians of South America.

Chagnon began his research on social organization among the Yanomamo before the banner days of pop sociobiology. In recent years he has offered sociobiological analyses of data that were obtained before he adopted the hypothesis that human social behavior should be studied from the perspective of the maximization of inclusive fitness. Chagnon makes the point that he can hardly be accused of looking for data that fit a preconceived hypothesis, and he adds that "the definitiveness of [the] conclusions would have been enhanced" had he been aware of the idea of inclusive fitness (Chagnon and Bugos 1979, 217).

One example of his sociobiological analysis is a detailed study of an ax fight between two groups of Yanomamo, filmed by Chagnon and his colleague Asch in 1971. According to Chagnon and Bugos, the network of alliances formed in a conflict of this kind is especially significant.

> Ambiguity and metaphorical aspects of kinship, it would seem, should be minimized as the actors elect to follow particular courses of action exercising choices that have obvious costs and benefits. If we are interested in examining individual human behavior with an eye toward understanding the extent to which that behavior is "tracking" biologically relevant dimensions of kinship relationships, it seems that crisis or conflict situations involving potential hazard to the actors are a reasonable place to begin looking. (1979, 215)

Chagnon's story of the events leading up to the fight is fascinating, and it draws on his detailed understanding of Yanomamo social structure. (See Chagnon 1968, 1974, 1977. For a different view of the Yanomamo, which lacks the depth of immersion in the culture so clear in Chagnon's writings but which considers a broader range of Yanomamo groups, see Smole 1976; Smole finds much more variability in hostile behavior and a much more sexually egalitarian society.) For our purposes, we can strip the account to its skeleton. A group of Yanomamo that had recently left their original home village and gone away to found a new village, returned for a visit to their old home. A few of the residents wanted the visitors to move back, but the majority of the people in the host village quickly became weary of the visit and hoped that their guests would leave.

> Matters came to a head when one of the men from among the visitors (Mohesiwa 1246) ran into a party of women in the garden and demanded that one of the women (Sinabimi 1744) give him a share of the plantains she was carrying back to the village for her own family. She refused to give him plantains, punctuating her refusal with an insult. Mohesiwa was incensed, and beat Sinabimi with a piece of wood. (Chagnon and Bugos 1979, 219; the four-digit numerals are used by Chagnon and his co-workers as convenient, unambiguous designators of individual Yanomamo)

Sinabimi complained, and one of her kinsmen, Uuwa, took up her cause.

The first phase of the fight was a struggle with clubs between Uuwa and Mohesiwa. Each man attracted some supporters, but after some menacing glares and insults the conflict seemed to be over. However,

Sinabimi's husband, Yoinakuwa, and his brother Kebowa armed themselves with a machete and an ax. They attacked Mohesiwa's house and, after a struggle, dragged him out into the open. Mohesiwa's younger brother, Tourawa, came to his defense, eventually attacking Kebowa with an ax. In accordance with the customs of Yanomamo combat, both Kebowa and Tourawa used the blunt sides of their axes at this stage. After Tourawa turned the sharp side of his ax up, threatening Kebowa, someone seized Tourawa. Kebowa was then able to inflict a severe blow, striking Tourawa in the back with the blunt edge of the ax. The fight was then halted by some of the older men.

Chagnon and Bugos are interested in the alliances that developed, or were revealed, in this incident. Their extensive knowledge of kinship among the Yanomamo enables them to compute the coefficients of relationship of all the individuals who took part (see the table in technical discussion J). So they are in a position to test their major hypothesis.

> One would predict, from kin selection theory, that the supporters of Mohesiwa will be more closely related to him and to each other than to Mohesiwa's opponent, Uuwa, and his supporters. Conversely, we also expect that Uuwa's supporters would be more closely related to him and among themselves than they are to Mohesiwa or his supporters. (Chagnon and Bugos 1979, 223)

Two questions arise. What exactly would we predict from "kin selection theory"? Do we need kin selection theory to predict what Chagnon and Bugos actually predict?

Zealous pop sociobiologists might argue as follows: If hostilities develop to a sufficiently high level, then each person around will be forced into the conflict, and the issue will be which side to join. Humans are equipped with mechanisms that enable them to maximize their inclusive fitness. Inclusive fitness is maximized by helping kin, or by helping closer kin against more distant kin. Hence we shall expect people to fight on the side of those to whom they are, on average, most closely related. The trouble with this argument is that, even if we grant the hypothesis about our fitness-maximizing equipment, a false assumption is needed to derive the conclusion. It is simply not true that in the context of a Yanomamo ax fight, inclusive fitness will invariably be maximized by helping closer kin. There are intricate social arrangements among the Yanomamo involving "contracts" for the exchange of women, coalitions among aspirants for power, and so forth. If we are serious about the maximization of

Technical Discussion J: How to Choose Your Allies

Using Chagnon's data, we can present the relationships among the principal participants and their defenders as follows. (The members of the two groups are listed in the left-hand column, with four-digit numbers serving as designators of individuals. The second column gives the present home of each—either the new village, I, or the original village, M. The remaining columns list coefficients of relationship to the four principal participants.)

Mohesiwa band	**Village**	**Mohesiwa**	**Tourawa**	**Uuwa**	**Kebowa**
0029	I	.2656	.2656	.1796	.0312
0067	M	.5156	.5156	.1796	.0312
0259	M	.5156	.5156	.2968	.0626
0336	M	.2656	.2656	.0626	—
0517	M	.1718	.1718	.1250	—
0714	M	.1250	.1250	.2500	.0626
0723	I	.1250	.1250	.2500	.0626
1278	M	.5156	.5156	.1796	.0312
1312	M	.5156	.5156	.1796	.0312
1335	I	.2656	.2656	.2968	.0626
1568	M	.1954	.1954	.1562	.0312
1837 (Tourawa)	M	.5156	—	.1796	.0312
1929	M	.5156	.5156	.0626	—
2194	M	.2656	.2656	.0626	—
2505	M	.0390	.0390	.0390	.1562
2513	I	.1094	.1094	.1718	.0468
Uuwa band					
0390	M	.0390	.0390	.0938	.0626
0789	M	.1094	.1094	.1718	.0468
0910 (Kebowa)	M	.0312	.0312	.0626	—
0950	M	.0156	.0156	.0312	.0312
1062	M	.0860	.0860	.3126	.1250
1109	M	.0938	.0938	.0938	.0626
1744	M	.1250	.1250	.2500	.0626
1827	M	.1796	.1796	—	.0626
1897 (Uuwa)	M	.1796	.1796	—	.0626
2209	I	.0312	.0312	.0626	.5000
2248	M	.0312	.0312	.0626	.5000

Mohesiwa's supporters

Almost all are close relatives or people who come from I. The only close relative of Uuwa who is from M is 0714. She is the sister of 0723, a resident of I.

Uuwa's supporters
All are more closely related to him than they are to Mohesiwa. In no case is there a strikingly close relationship. Uuwa's supporters share the common property that they are almost all people from M who do not have any close relationship with Mohesiwa. The one exception is 2209. Recall that the party originally injured, Sinabimi, was married to Yoinakuwa, and that Kebowa, Yoinakuwa's brother, was largely responsible for escalating the conflict. 2209 is sister to both Yoinakuwa and Kebowa. She deserts her fellow villagers to join the fight on the side of her brothers (and her sister-in-law).

Kebowa's supporters
Two of Kebowa's champions are close relatives (2209 and 2248). The rest are neither closely related to him nor closely related to his opponent Tourawa. Four of the eight (0789, 1109, 1744, 1897) are more closely related to Tourawa than to Kebowa. The mean relatedness of the eight to Tourawa is .0859; the mean relatedness of the eight to Kebowa is .0567.

Tourawa's supporters
With one exception, the support for Tourawa can be understood in the same way as the support for Mohesiwa. The exception is 2505. According to Chagnon and Bugos, 2505 is related to Kebowa by .1562, making 2505 a closer relative of Kebowa's than any of his other defenders except his brother and sister. Yet 2505 fights *against* Uuwa and Kebowa, on behalf of Mohesiwa and Tourawa, to both of whom he is only distantly related.

A look at the genealogy given by Chagnon and Bugos makes the ascription of a high coefficient of relationship between 2505 and Kebowa rather mysterious (see their Figure 8.2, 1979, 238). Perhaps Chagnon and Bugos are drawing on their knowledge of connections that are not revealed in the family tree that they supply. So let us assume that the figure of .1562 is correct. In this case, 2505 acts against both considerations of kinship and village loyalty in siding with Mohesiwa and Tourawa against Uuwa and Kebowa.

There is one connection in the genealogy that may shed some light on the role of 2505. His father-in-law is 1335, one of the fighting visitors. So perhaps we can explain his behavior in terms of a decision to defend his wife's family and to oppose Kebowa (who is a somewhat distant relation).

inclusive fitness, then such considerations will have to figure in the calculations. Hence we have no reason to expect a *perfect* match between ties of kinship and patterns of mutual aid.

Tempering ambition a little, pop sociobiologists must settle for a weaker argument. Inclusive fitness is likely to be maximized by helping closer kin against more distant kin, both because of the straightforward advantages in spreading copies of genes and because contracts and coalitions are themselves often (though not always) founded on kinship ties. So we might expect that people who fight on the same side will, on the average, be more closely related to one another than they are to the opposition. There may of course be exceptions, anomalous folk who, because of some particular social relationship to somebody who plays a role in the conflict, find it in the interests of their inclusive fitness to fight with more distant kin against closer kin. How many such cases should we anticipate? That obviously depends on the specifics of the situation. Probably not too many, so that overall the network of alliances will reflect the coefficients of relationship.

On one interpretation, the sociobiologists of the last paragraph are Chagnon and Bugos. Much depends on how seriously we take the quotation marks in the following passage.

> For the example under discussion here, the ax fight, the major variable appears to be closeness of genetic relatedness: individuals seemed primarily to "decide" to aid others on the basis of the degree of relatedness obtaining between themselves and other participants in the fight. (1979, 222)

The strong hypothesis is that people have an unappreciated ability to assess coefficients of relatedness, an ability that issues in a fine discrimination among kin of varying degree. The ax fight reveals this capacity at work.

A weaker hypothesis, which might also be attributed to Chagnon and Bugos, is to suppose that they share Kurland's ecumenical position with regard to traditional ideas about proximate causes. Perhaps people fight on the side of those they like and against those they dislike, or those they like less. Given a propensity to like close kin, especially those with whom one has been reared, and also to be well disposed toward those with whom one has a history of reciprocal exchanges, it may be that the resultant behavior will coincide with the action recommended by inclusive-fitness maximization. However, there is no unconscious computation of coefficients of relatedness. Ties of amity are formed in complicated ways, and these ties express themselves in our reactions to situations of conflict. Because many of these ties are founded in our propensities to like brothers and sisters, we expect that those who fight on the same side will be more closely related to one another than to their opponents. On the weaker hypothesis, appeals to evolutionary theory are idle.

Once we have tempered our expectations, we see that both the weak and the strong hypothesis lead to similar—and similarly vague—predictions. So we might think that it will be hard to decide between the two hypotheses. Indeed it is.

When we look closely at the data presented by Chagnon and Bugos, we discover that the picture is much more confusing than a simple appeal to maximization of inclusive fitness would have suggested (see technical discussion J). True, the average relatedness among Mohesiwa's supporters (.2124) is significantly higher than the average relatedness of a pair, one of whom supported Mohesiwa and one of whom supported Uuwa (.0633). But the average relatedness among Uuwa's supporters is not impressively high (.0883). Recognizing the point should prepare us for some of the anomalies of the situation.

The two initial antagonists, Mohesiwa and Uuwa, are quite closely related (.1796). Five out of the sixteen people who rally to Mohesiwa's defense are related more distantly to him than is his antagonist, and eight of the ten who come to help Uuwa are more distantly related to him than he is to Mohesiwa. Moreover, Mohesiwa attracts four defenders who are less closely related to him (or to Tourawa) than they are to Uuwa. Why do these people join Mohesiwa's band?

There is a relatively obvious answer. Recall the initial situation. Mohesiwa and some of his fellow villagers had returned to visit the village in which they had originally lived. The fight was preceded by tension between visitors and hosts. When hostilities broke out, Uuwa, described by Chagnon and Bugos as a representative of "local authority," attacked Mohesiwa. We can imagine that the visitors felt threatened at this point and closed ranks. Perhaps Mohesiwa's supporters consist of people from the new village and close relatives in the old village.

This suggestion nearly makes sense of the data. Almost all Mohesiwa's supporters are close relatives—siblings, uncles, close cousins—or people who come from the new village. In three cases visitors from the new village who are more closely related to Uuwa than to Mohesiwa fight with Mohesiwa, presumably on the grounds of solidarity against the hosts. The remaining close relative of Uuwa who joins Mohesiwa's band is the sister of one of the visitors who rally to Mohesiwa's aid.

As technical discussion J shows, the same approach will account for the whole pattern of alliances. Two variables are important: village loyalty and ties to close kin. When they come into conflict, ties between siblings and other close relatives seem to take precedence over village loyalty. There is no reason to think that more remote kinship connections figure in the story. (The approach can be refined by recognizing that the composition of villages is itself largely determined by considerations of kinship, reciprocation, and personal friendship. Since I wish merely to show that folk psychology will provide *an* account of the proximate causes that explains the data, I shall not attempt to determine the *best* such account.)

The approach looks ever more plausible when we consider Kebowa's champions. Two of these are close relatives—his brother and sister. The remaining eight are close neither to Kebowa nor to his antagonist Tourawa. They are fairly distant relatives and should be expected, according to our hypothesis about proximate causes, to line up on the side of village loyalty. That is exactly what they do. All eight fight with Kebowa, even though four of the eight are more closely related to Tourawa and even though the mean relatedness of the eight to Tourawa is greater than their mean relatedness to Kebowa.

The treatment of this problem offered by Chagnon and Bugos is instructive. They remark,

> . . . as in the case of Mohesiwa's supporters mentioned above, some of the supporters of Kebowa are more closely related to his opponent than they are to their own champion. However, it is clear that despite the relatively large number of loops connecting Kebowa's supporters to his opponent, they are in fact more closely related to their champion—even though the number of relationship loops is small. (230–231)

Alas! When push comes to shove, it seems that fitness maximization has to give. If the data are to be interpreted as defending the strong hypothesis that we have a mechanism for keeping track of kinship and adjusting our behavior accordingly, then it seems that the mechanism is imperfect. It can be fooled if there are a lot of relatively weak connections between ourselves and some of our kin. Once this concession has been made, why do we need to think of the Yanomamo as motivated by any more than the most simple and straightforward appreciation of kinship? Seven members of Uuwa's band are related to Kebowa by less than .1. Is there any reason to think that they fight on his side because of their desire to defend a distant relative? Why should we prefer this hypothesis to the simple suggestion that they, like other host villagers, perceive their guests as overstaying their welcome and therefore choose to fight on the side of the home champion?

I conclude that a close look at the data provides no reason for thinking that the combatants have any ability, conscious or unconscious, to make anything more than a relatively coarse division of one another into kin and nonkin. A *preliminary* explanation of the patterns of alliance is that people choose to defend close kin (that is, those who are counted as kin in the coarse division) and choose sides on the basis of village loyalty where factors of kinship do not prescribe a course of action. We can develop and refine this explanation by articulating the relationships that had developed in the period preceding the fight, by tracing the formation of bonds among those who originally left to found a new village, by noting the conflicts with ties of kinship, and so forth. My tentative suggestions only show that folk psychology, with its strategy of appealing to proximate mechanisms, promises to provide a far more illuminating account of the alliances in the fight than does the invocation of inclusive fitness.

Chagnon and Bugos give no detailed account of how the participants are maximizing their inclusive fitness. They admit that "relatedness alone cannot account for all the bonds of attraction or tactics of recruitment" (237). So to what end are the data on coefficients of relationship assembled? There is no striking correlation that might incline us to think that some subtle mechanism for gauging relationships was at work. Nor does the study cast any light on the evolutionary origins of those human dispositions (for example, the disposition to aid siblings) to which a folk psychological account would appeal. The introduction of ideas from evolutionary theory again proves idle. We are offered a fascinating description of a particular incident, a description that undermines the anthropological position that ties of biological kinship *never* matter. So we achieve what Maynard Smith prophesied we would get—the refutation of "some very odd remarks by anthropologists." And that is all.

Enough of mayhem. Let us now take up marriage and murder. In another extremely influential study, Dickemann offers a pop sociobiological explanation of female infanticide in certain kinds of society. Dickemann characterizes the basic situation as follows:

> In certain stratified societies, characterized by intense competition for scarce resources, male reproductive success shows extreme variance, men of high rank acquiring access to a disproportion of females through polygyny, and in addition enjoying greater health and earlier entry into reproduction, while those at the bottom are disproportionately excluded from repro-

> duction through delayed marriage, heavy mortalities, and the imposition of celibate roles, while their RS [reproductive success: PK] is further reduced through heavy mortalities among their progeny. Further, the relative probable RS of males and females at different levels of the social hierarchy is culturally exacerbated through female preferential infanticide, suicide, and celibacy (including prohibitions against widow remarriage or against sexual intercourse by mothers of reproductive offspring), all reaching greatest intensity at the top of the social pyramid. Middle- and upper-class marriage systems are hypergynous, with competition between families (mostly patrilineal in these societies) for higher-status grooms possessing greater access to scarce resources. (Dickemann 1979, 323)

Among the upper classes the family of the bride pays money to the family of the groom, while in the lower classes grooms pay to obtain brides. In this situation upper-class families are supposed to maximize their reproductive success by practicing female infanticide. For upper-class daughters can, at best, be part of the household of an upper-class man, to whom money must be paid. Upper-class sons receive many wives and all the dowry money that accompanies the wives. While sons are profitable, daughters are a drain on the family finances. It pays to kill them at birth. (Strictly speaking, the most that this line of reasoning shows is that daughters are an *economic* loss. How this translates into the "currency" of reproduction is a matter for investigation, as we shall see.)

Dickemann's study largely consists in a wide-ranging description of practices in many different societies that conform to the basic conditions of upper-class polygyny and hypergyny (females marry men of equal or higher caste). She undertakes to show that female infanticide is prevalent among the upper classes in these societies and that there is a "striking fit" (367) between human social behavior and the predictions of sociobiology. We can admire the descriptive anthropology. But we should begin by asking just what sociobiology predicts.

The apparent source of Dickemann's biological ideas is the work of Trivers and Willard (1973) on adjustment of sex ratios. Among polygynous mammals, Trivers and Willard argue, females in good reproductive condition should be expected to adjust the sex ratios of their offspring so as to produce sons, and those in poor reproductive condition should produce daughters. The intuitive idea is that well-endowed males can win big in the mating game, while even poor-quality females have the opportunity to mate. The proposal is important and interesting, and it has been developed by a number of subsequent workers (see Charnov 1983 for a lucid account of the general problem of optimal sex ratios). There are some obvious difficulties, however, in simply appealing to the Trivers-Willard model in the context of human infanticide.

One crucial flaw in the analogy is that a physiological variable, reproductive condition, gives way to a socioeconomic factor, wealth and status. The transition raises a host of questions about the relations among biological and economic variables, questions that require detailed exploration. It cannot be taken for granted that economic gains will translate into reproductive gains.

In the second place, the adjustment of sex ratio by infanticide introduces complications into the Trivers-Willard argument as sketched above. Suppose that a man and woman adjust the sex ratio in their offspring by killing all their daughters. *If nothing else changes,* they achieve the adjustment by decreasing the total number of their offspring. By contrast, a mother who can adjust the sex ratio at conception does not decrease the number of her progeny. She simply produces a larger number of the favored sex (and a smaller number of the undervalued sex) than she would otherwise have done. This is the point on which the Trivers-Willard argument turns. One sex is a better vehicle for transmitting the mother's genes into subsequent generations, and she maximizes her fitness by producing *more* of that sex.

The crucial issue is whether parents can manage to produce more sons by killing their daughters. Consider the position of the mother immediately after giving birth. There will be some interval of time (greater than nine months) before she gives birth again. If that interval is decreased by the death of the newborn child, then it is possible that murdering an infant daughter will increase the number of sons produced by the mother. If that interval is the same whether or not the child lives, then the expected number of sons is independent of the practice of killing daughters. As the child-rearing practices of Victorian England clearly show, the institution of wet nurses makes it possible for women to breed at maximal speed. Hence there is a rival strategy that produces the same number of expected sons as does female infanticide: all that is required is that the babies be fed by people other than their mothers.

Dickemann does not attempt to show that parents who kill their daughters are able to produce their next child more quickly and so achieve a greater number of sons. She seems more concerned with the economic costs of daughters (see, for example, 339–340). In any case, as the argument of the last paragraph reveals, if there are no *fitness* costs in rearing daughters, then parents who save their daughters and hire outsiders to nurse them can be expected to have greater reproductive success than parents who practice female infanticide. (They will have the same expected number of sons and some extra daughters.) So the argument that female infanticide maximizes inclusive fitness for upper-class parents, in the appropriate kinds of social systems, has to depend on considerations of economics, which are assumed to translate into gains and losses in reproduction.

What are these considerations? Dickemann does not offer any explicit account. Although her study is billed as "a preliminary model," that is a misnomer. The only genuine model is that of Trivers and Willard, and that model deals with a critically different situation. Moreover, it is no trivial matter to fill in what is missing. When we try to supply the details, it is far from clear that Dickemann's expectations are sustained.

There are two possible sources of trouble with daughters. They cost money to rear and they cost money to marry. We might initially think (as Dickemann seems to: 326, 334, 339–340) that the primary trouble is the cost of the dowry. When the analysis is pursued, however, we find that economic loss leads to reproductive loss only if the dowry is unrealistically high. (See technical discussion K for details; I do not maintain that the analysis given is the only way to articulate Dickemann's "model," but there is no obvious way to reach her conclusions by modifying the main assumptions that I have made.) Intuitively, the problem with infanticide is that those who practice it forfeit an important way of spreading their genes. Even though upper-class daughters only marry once, if they marry upper-class men then they have upper-class sons. These upper-class sons—who are each expected to have several wives—may make an important contribution to long-term genetic representation.

Of course, marriage costs could always be avoided by allowing upper-class daughters to marry "down." An upper-class patriarch who preserved his daughters and married them to middle-class men would apparently achieve greater genetic representation in future generations than a rival who destroyed all daughters at birth. Dickemann's discussion assumes that this strategy is not available. It is hard to understand how the constraint precluding marriage of daughters to middle-class men is itself to be understood in terms of inclusive-fitness maximization.

Difficulties also arise if we consider the suggestion that the costs of rearing daughters make female infanticide an optimal reproductive strategy. Suppose that upper-class families only bestow on daughters the minimum investment required to enable the daughters to reach the age of marriage as healthy young women. Unless the cost of this

Technical Discussion K: When to Murder Your Children

As noted in the text, constructing a model that actually fits the social situations Dickemann describes turns out to require a large number of assumptions. Although the analysis I shall offer attempts to capture the main features reported in her data, it should be borne in mind that different initial suppositions might easily yield different expectations. Nevertheless, some analysis is better than no analysis. Given the qualitative remarks offered by Dickemann, we are in no position to judge which strategies would maximize inclusive fitness.

Assume that there are three classes, upper, middle, and lower. The society is taken to be highly pyramidal: 5 percent of the population is upper-class, 25 percent is middle-class, and 70 percent is lower-class. When a marriage takes place, the economic transactions meet the following conditions:

- If Groom is upper-class, Bride's family pays P_1 economic units.
- If Groom is middle-class and Bride is lower-class, Bride's family pays P_2 economic units.
- If Groom is middle-class and Bride is not lower-class, there is no payment.
- If Groom is lower-class, Groom's family pays P_3 economic units.

Each upper-class nuclear family enjoys W_1 units of wealth; each middle-class nuclear family, W_2 units of wealth; and each lower-class family, W_3 units of wealth. For an upper-class male, the number of wives that he can attract is determined by the wealth of his parents. We suppose that this number is hW, where W is the wealth of the family at the time at which the sons mate. Finally, let C be the minimal cost of rearing a child from birth to adulthood. We can assume that this basic cost does not vary across the classes and that it is the same for males and females. (It is possible to refine the analysis I shall give by considering rearing costs that are specific to class and sex. Thus, the total rearing cost for an upper-class female might be $C + C_{1F}$, that for a lower-class male $C + C_{3M}$, and so forth. I shall briefly consider this refinement later.)

Each wife can bear m children, and the sex ratio at birth is 1 to 1. In the initial situation an upper-class male will have the potential to rear hW_1m children. The expected number of marriages contracted by his children, if all the children are allowed to survive, is $hW_1m(\frac{1}{2}hW+\frac{1}{2})$, where W is the wealth of the family at the time the marriages take place.

Case 1

Rearing costs are negligible. Net gains and losses during the rearing period are also negligible. The principal differences for those who pursue different rearing strategies come in the costs of dowries.

Under these circumstances, if Sultan is an upper-class male who starts in the initial situation, then at the time of the marriage of his sons, his family wealth remains at W_1. As noted in the text, there is no loss if Sultan now marries off his daughters to middle-class men who do not demand dowries. If he pursues this strategy, the wealth passed on to his sons—the wealth that determines their ability to attract wives for their sons—is undiminished by the rearing of daughters. So Sultan loses no wives for his grandsons by rearing daughters. But he achieves a reproductive gain through his surviving daughters. (Some of Sultan's genes find their way into the bodies of middle-class granddaughters. The granddaughters enjoy a fairly high probability of marrying "up." So Sultan gets some upper-class great-grandsons who reap the advantages of polygyny. Moreover, all these genetic gains come free.) Suppose, however, that there is a constraint. Daughters are not allowed to marry "down." Hence, the consequence of rearing daughters is that Sultan has to pay dowries and reduce the family fortune. We simplify the life cycle by assuming that all the children marry simultaneously and that Sultan dies immediately after the marriages. His residual income is distributed as follows: each son except the eldest receives the dowry money from his brides, and the eldest receives his dowry money and the remainder of Sultan's estate. Given this assumption about roles of inheritance, Sultan's rearing behavior makes no difference to the wealth of the younger sons (hence, no difference to the number of wives that they can attract for their sons). However, the wealth passed on to the eldest son is diminished if Sultan chooses to marry his daughters to upper-class men. The wealth of the eldest son is determined as follows:

given infanticide, $W_1 + hW_1P_1$;
given marriage "up," $W_1 + hW_1P_1 - \frac{1}{2}hW_1mP_1$.

Suppose that Sultan practices infanticide. Eldest son (Sultan II) has hW_1 wives. Assume that Sultan II also practices female infanticide. Then he will have $\frac{1}{2}hW_1m$ sons, each of whom will contract $h(W_1 + hW_1P_1)$ marriages. So, from Sultan's (posthumous) perspective, the total number of marriages expected for his grandchildren is

$$\tfrac{1}{2}hW_1m(hW_1 + h^2W_1P_1) + \text{contributions from younger sons.}$$

Alternatively, suppose Sultan rears his daughters to marry upper-class males. Now eldest son (Sultan II) has wives but decreased initial wealth. Assume he repeats Sultan's practice of allowing daughters to survive. This time the total number of marriages that Sultan can expect among his grandchildren is

$$\tfrac{1}{2}hW_1m(hW_1 + h^2W_1P_1 - \tfrac{1}{2}h^2W_1mP_1) + \tfrac{1}{2}hW_1m$$
$$+ \text{ contribution from younger sons}$$
$$+ \text{ contributions from daughters.}$$

Because the number of younger sons is the same on both strategies and because they receive the same amount of initial wealth, we can ignore their contributions for purposes of comparison. The contributions from daughters are as follows: Each daughter marries an upper-class male. Hence, for each daughter, there will be $\frac{1}{2}m$ sons, each of whom contracts $\frac{1}{2}hW_1$ marriages, and there will be $\frac{1}{2}m$ daughters, each of whom contracts one marriage. Sultan has $\frac{1}{2}hW_1m$ daughters. Hence the contributions from daughters are

$$\tfrac{1}{2}hW_1m(\tfrac{1}{2}hW_1m + \tfrac{1}{2}m).$$

Infanticide is preferable if

$$\tfrac{1}{2}hW_1m(hW_1 + h^2W_1P_1) > \tfrac{1}{2}hW_1m(hW_1 + h^2W_1P_1 - \tfrac{1}{2}h^2W_1mP_1)$$
$$+ \tfrac{1}{2}hW_1m + \tfrac{1}{2}hW_1m(\tfrac{1}{2}hW_1 + \tfrac{1}{2}m),$$

which simplifies to

$$hW_1m(hP_1 - 1) > 2 + m.$$

The significance of this inequality is not hard to fathom. hP_1 is the number of extra wives that are available to a man because of his father's acquisition of one dowry. A necessary but not sufficient condition for infanticide to be preferable is that $hP_1 > 1$.

Recall that the number of wives attracted by a son of a family of wealth W is hW. Hence, the income, *per son*, for an upper-class family of wealth W is hWP_1. For infanticide to be preferable, this has to be greater than W. So the wealth of upper-class families would grow at an amazing rate in a society in which infanticide is preferable. If each upper-class family has 5 sons, then the influx of wealth into the upper class from the marriages of one generation would be five times the antecedent wealth of the upper class. Even making very generous assumptions about the resources available to the middle class, it is not hard to see that this situation is economically unfeasible. However, if upper-class families have less than 5 sons, then it is hard to see how upper-class patriarchs could have more than 2 wives. Hence, in this case, $hW_1 = 2$, so that $P_1 > \frac{1}{2}W_1$. Once again, there are economic difficulties for the middle class. Specifically, it is hard to see how a middle-class family could afford to pay a dowry on more than one daughter.

Conclusion: For female infanticide to be preferable because of the costs of marrying off daughters, the bridegroom price has to be so high that the system cannot function.

Case 2

Rearing costs are not negligible. The comparison of fitnesses for upper-class families who kill or save their daughters now proves complicated. We can simplify matters by supposing that the practice of "marrying up" is prevalent in the lower and middle classes and considering the economics of the arrangement.

Let us focus on the plight of a lower-class family with one son and one daughter. We assume that they begin with a small amount of wealth W_3, that they can gain a modest amount during the period of rearing, and that the costs of rearing are C. At marrying time their wealth is as follows:

given infanticide of one child, $W_3 + G - C$;
given infanticide of both children, $W_3 + G - 2C$.

If infanticide is practiced, then it should be preferable to kill the male. For if

$$W_3 + G - C > P_2,$$

then the family can afford to marry the daughter into the middle class, thereby obtaining greater probability of surviving grandchildren; and even if this inequality does not obtain, the probability that a surviving daughter will marry is greater than the probability that a surviving son will marry. Male infanticide should clearly be favored if

$$W_3 + G - C > P_2 \text{ and } W_3 + G - 2C < P_3;$$

for under these circumstances the couple has no chance of marrying off their son, but they do have the option of marrying their daughter into the middle class. If rearing costs are significant and if P_2 and P_3 are of similar size, then it seems likely that this inequality will be satisfied.

Fairly elementary economic analysis enables us to compare the size of P_2 and P_3. Both can be expected to vary in accordance with supply and demand. Hence we may suppose

P_2/P_3 = [(number of potential lower-class brides for middle-class males)
× (number of potential lower-class brides for lower-class males)]
/ [(number of middle-class grooms)
× (number of potential lower-class grooms)].

Since 1/4 of the population is middle-class and 1/20 upper-class, let us suppose that $\frac{1}{4}n_2 + \frac{1}{20}n_1 < 1$ (where n_1 is the mean number of wives for an upper-class man, n_2 the mean number for a middle-class man). Some women are left over for lower-class males to marry. We can compute the ratio P_2/P_3 as

$$\left(1 - \frac{n_1}{20}\right)\left(1 - \frac{(n_1 + 5n_2)}{20}\right) \Big/ \frac{1}{4}\left(\frac{7}{10}\right).$$

A sufficient condition for lower-class male infanticide to be a likely fitness-maximizing strategy is $P_2/P_3 \leq 1$, that is,

$$(20 - n_1)[20 - (n_1 + 5n_2)] \leq 70.$$

Assuming middle-class monogamy, the condition will be satisfied if $n_1 \geq 9$. In the situations described by Dickemann, where upper-class males fill their households with "large numbers of concubines" (336), these women are removed from the pool for lower castes, and it thus seems quite probable that lower-class couples would maximize their fitness through killing sons. Dickemann notes that her model "appears superficially symmetrical" (325), but, as she concedes, the practice of male infanticide among lower classes appears both less common and less intense (325, 341). If the analysis that I have offered is correct, then upper-class female infanticide can *only* be expected where the costs of rearing are significant (if then); and if the costs of rearing are significant, then the presence of pronounced polygyny at the top should lead to lower-class male infanticide as a fitness-maximizing strategy.

Finally, let us consider the forgotten middle class. We can write the wealth of a family at the time of marrying as

$$W_2 + G_2 - (c+d)C,$$

where G_2 is the economic gain during the rearing period, c is the number of daughters raised, and d is the number of sons raised. The numbers c and d should be chosen so that all daughters and all sons marry. This requires that

$$W_2 + G_2 - (c+d)C + dP_2 > cP_1,$$

if we suppose that sons marry lower-class women and daughters marry upper-class men. Evidently, if C is large, p_2 is small (as would be expected from the fact that P_2 has to be paid by peasant families), and if P_1 is much higher, then it is likely that a family with a number of sons and daughters will not be able to rear them all and marry the daughters to upper-class males. Hence, in a system in which middle-class families marry their daughters into the upper class (and, by hypothesis, maximize their fitness by doing so), we shall have to suppose that there is some middle-class infanticide.

For example, let $W_2 = 100$, $G_2 = 100$, $C = 50$, $P_2 = 5$, $P_1 = 40$. Then a family with two sons and two daughters cannot rear all the children and marry the daughters to upper-class males. Similarly, a family with two daughters and one son cannot rear all the children and marry both daughters "up." The latter family can, however, achieve upward marriage for both daughters by killing the son. Assuming that the increased production of grandsons through daughters who marry "up" makes the strategy of marrying daughters into the upper class preferable, we should expect to find male infanticide in the middle class in such cases. Although the numbers here are plainly arbitrary, it should be clear that the general predicament is relatively probable:

to make infanticide a fitness-maximizing strategy for the upper class, the costs of rearing have to be high with respect to upper-class wealth; hence, we must expect that C is of the order of magnitude of W_2.

Summary
We have briefly looked at two types of case. If the costs of rearing are negligible (more exactly, negligible to the upper class), then the analysis of case 1 suggests that female infanticide will not be an inclusive-fitness-maximizing strategy for upper-class patriarchs. If the costs of rearing are not negligible, then the analysis of case 2 shows that in a society in which women marry "up," there are very strong pressures for infanticide in the lower *and* middle classes. (Note: I have not considered the question of whether, under these conditions, female infanticide turns out to maximize upper-class fitness.) We can resist the conclusion by denying that the rearing costs are the same across classes. However, the crucial rearing costs are the costs of rearing daughters. Since middle-class daughters and upper-class daughters compete for the same husbands, it is hard to defend a distinction of the rearing costs. Hence, we should expect that pressure for upper-class female infanticide will be reflected in pressure for middle-class male infanticide.

Plainly, the problem is extremely complex, and it may well be that the analyses I have offered are inaccurate. The point of the exercise is not to argue that we know that Dickemann's expectations are wrong, but to show that we have absolutely no basis for thinking that her expectations are correct.

investment is extremely high, the loss that it brings to the family in terms of decreased ability for male descendants to attract large numbers of wives will be negligible in comparison with the genetic contributions that are made by the surviving daughters. As a result, the strategy of killing daughters can only be favored if the costs of rearing are very high. On the other hand, if the minimal costs of rearing are set very high, there will be intense pressure on the lower and middle classes to practice infanticide. Parents in those classes face the same minimal costs of rearing, and the costs represent a greater proportion of their resources. Hence it seems that pressure for female infanticide among the upper classes should be combined with even greater pressure for infanticide lower down. (See technical discussion K for details.)

One way to escape this conclusion is to propose that the costs of rearing affect different classes differently. Perhaps upper-class daughters must be given special cultivation if they are to mature as marriageable women. (The general idea will already have occurred to fans of Jane Austen.) Whether or not this is so, the proposal does not avoid the predicament. If middle-class women and upper-class women both compete for upper-class husbands, it appears that the costly refinements lavished upon the daughters of the upper classes will also have to be undertaken by the middle classes. Again, the pressure on middle-class families will be greater. (And again, Jane Austen has taught us as much.)

I draw the obvious conclusion. Whether female infanticide is a strategy that maximizes inclusive fitness for upper-class parents is an entirely open question. There is no model that resolves the issue in the way Dickemann suggests. When we plunge into the details, it proves very difficult to discover an analysis that yields anything like Dickemann's conclusions. Under these circumstances it seems a little premature to claim that the case provides a "striking fit" of predictions to findings.

When we turn our attention to the findings themselves, however, we can begin to provide an explanation by applying the commonplaces of folk psychology to the historical contexts in which the relevant people have made their decisions. The societies discussed by Dickemann have a cultural tradition in which middle-class families attempt to secure alliances with upper-class families by supplying upper-class sons with wives. Families strive to achieve wealth and power. Upper-class males exploit the middle classes by obtaining both brides and dowries from them. Given the system, upper-class females are a financial loss. Even though they might ultimately contribute to the spread of the parental genes, it seems that the short-term costs of rearing them and marrying them off are enough to seal their fate.

Dickemann is aware of the possibility that the stratified social systems she describes might seem to signal a history of power struggles. Rich and powerful men come to enjoy the possibility of having many wives and concubines. Their social inferiors gain protection, influence, and short-term advancement by supplying them with women. Competition for the role of "friend of the mighty" leads to the institution of dowries, so that, in their desire for social position, the middle classes increase the wealth of the very rich. Once in place, the system of dowries traps the daughters of the upper class. Their marriages too must be paid for; and their families, who have no need of their daughters for social advancement, are unwilling to foot the bill. They rationalize their greed by constructing elaborate rules about marriage, and they destroy their own daughters at birth.

Dickemann alludes to this alternative explanation in the following passage:

> Central to hypergynous marriage systems is the dowry, or bridegroom price, characteristic of high-ranking North Indian groups. That female infanticide was rationalized as a consequence of competitive demand for large dowries and other wedding expenses from the bride's father, and the small number of available sons at the top of the hierarchy, is clear from the testimony of native leaders. (334)

Why should we think that the testimony offers a rationalization? Why believe that considerations of inclusive-fitness maximization are at all relevant? As we have alredy seen, there is no obvious basis for the supposition that female infanticide is an inclusive-fitness-maximizing strategy. Even if it were, the upper-class families who practiced it might well have reached it coincidentally. There are familiar proximate mechanisms that would lead them to the strategy. Perhaps we should accept the obvious. *Radix mortis filiarum est cupiditas.*

Dickemann's account of infanticide in China underscores the point. Again the explicit motive for infanticide among the gentry is "concern over dowries" (341). Yet female infanticide did not figure only among the gentry. Ironically, Dickemann's best documentation is for infanticide among poor peasants, even though, as she admits, peasant women "played important agricultural and laboring roles" (342). Although there is considerable evidence that lower-class Chinese males had a hard time finding wives, male infanticide was not widely practiced. When we recall that population-control programs in modern China have foundered owing to the common preference for male children, it is hard to resist the suggestion that systematic undervaluation of women has a role to play in the complex story of female infanticide.

Two more passages will show how the invocation of biological ideas is an intrusion on Dickemann's informative survey of infanticide in various societies. According to Dickemann, "The intensity of hypergyny is revealed in many accounts of well-to-do families whose crippled, insane, or idiot sons could contract marriages with healthy girls of poorer families (344). There are probably exercises in apolegetics designed to show how such marriages really contributed to the reproductive success of the parents who used their daughters in this way. The less contorted story is that the long-term interests of the genes are being sacrificed in the cause of status, influence, and perhaps money in the immediate future. Similarly, Dickemann takes an optimistic view of the reproductive successes of prostitutes and concubines. "While the concubine's status within her mate's household was often no better than that of a servant, her increased security and wealth as compared to those of alternative roles must surely have advanced her reproductive success on the average" (345). There is simply no basis for this Panglossian vision of a world in which even the exploited somehow maximize their inclusive fitness. The idea that an unmarried slave might increase her wealth while taking a few months off to reproduce and that the resultant offspring would be more likely to survive and reproduce than those of an impoverished woman married to a peasant is simply speculation.

We can now evaluate Dickemann's interpretation of what she has shown.

> Recognizing the enormous amount of testing which remains to be done, nevertheless in this provisional exercise the close fit of familiar human social structures to that theoretical structure which would be predicted by sociobiology is striking. Behind the surface complexity of human cultural forms, a general mammalian model, maximizing reproductive success through male competition and the manipulation of sex ratios, is clearly visible. (367)

The "general mammalian model" concerns physiological condition, not socioeconomic status. When we try to mimic the deliverances of the Trivers-Willard model, paying attention to crucial differences, it appears most likely that Dickemann's expectations will be confounded. Not even those who seem to have the power to control the situation maximize their inclusive fitness by engaging in female infanticide. Considerably more strain is required to relate the strategies of the less powerful to their reproductive interests. What we have is a fascinating—and disturbing—account of practices in a number of different societies, overlaid with some irrelevant and misleading apparatus from evolutionary biology.

The general moral of this chapter can be summarized by reference to the practices Dickemann discusses. In order to understand the high frequency of infanticide in the societies of northern India and China, we need to trace the history of cultural institutions, recognizing how institutions affect the dipositions of those who grow up in societies dominated by them and how, in turn, those dispositions modify existing institutions. In tracing this history, we shall suppose that human beings have propensities that lead them, when they grow up in certain social and physical environments, to acquire as adults the desires and aspirations with which we are familiar. Here there are some genuine evolutionary questions. Ultimately we shall want an evolutionary explanation of basic human propensities. The point is that the evolutionary questions arise relatively late in the inquiry. Only after we have traced a complex social arrangement to the fundamental proximate mechanisms that have been at work in its historical development does it become relevant to inquire how the inclusive fitness of our ancestors might have been maximized by their having those proximate mechanisms. (Of course, it is possible that the evolutionary explanation of the proximate mechanisms might not identify them as maximizing inclusive fitness, but let us simplify matters by waiving legitimate worries about adaptationism.)

Anthropology, when pursued in the way recommended by Alexander and exemplified by the studies of Chagnon and Dickemann, introduces evolutionary considerations in the wrong place. There is much striving to show that the end products of a complicated process really maximize human inclusive fitness. Usually the striving is in vain. The products should not be expected to maximize inclusive fitness. Yet, even if they were to do so, the proper focus of evolutionary attention is on the mechanisms that drive the process. In the haste to see fitness maximization everywhere, those mechanisms—and the historical process in which they figure—are ignored.

The heart of the dispute is not a difference about styles of anthropological explanation. The revolutionary vision of Alexander's pop sociobiology is that our actions are those that would have been prescribed for us by a being that actually calculated how to maximize its own inclusive fitness and then acted on the calculation. We have found that vision to be profoundly misguided. There is no reason to suppose that a being of the kind hypothesized lurks inside us, no reason to suppose that there are hidden proximate mechanisms that help us to imitate its behavior, no reason to think that reference to its calculations will help us to understand the proximate mechanisms with which we are familiar, no reason to clutter up descriptive anthropology with irrelevant incantations about inclusive fitness. As we began with Newton, let us end with Laplace: we have no need of that hypothesis.

Chapter 10

The Emperor's New Equations

Beginning Again

Faint heart never won fair theory. Many of the most celebrated accomplishments of science have been achieved because an investigator has recognized important problems and has struggled to overcome them. Wilson's recent work in human sociobiology might be seen as an example of the phenomenon. The objections raised by critics are to be taken seriously and to be met. The program begun in 1975 is to emerge with new strength and even greater promise.

In *Genes, Mind, and Culture* (1981) Lumsden and Wilson introduce their enterprise by connecting it explicitly with Wilson's previous work in sociobiology and with the reactions to that work.

> Why has gene-culture coevolution been so poorly explored? The principal reason is the remarkable fact that sociobiology has not taken into proper account either the human mind or the diversity of cultures. Thus in the great circuit that runs from the DNA blueprint through all the steps of epigenesis to culture and back again, the central piece—the development of the individual mind—has been largely ignored. This omission, and not intrinsic epistemological difficulties or imagined political dangers, is the root cause of the confusion and controversy that have swirled around sociobiology. (ix)

Later in the same work they describe their new theory as "an extension of sociobiology" (343), viewing it as a response to the criticism that traditional sociobiological studies leave "the genetic foundation and the epigenetic mechanisms . . . largely unexplained" (349). Thus they view earlier sociobiology as using the "first principles of population biology" to "deduce" conclusions about how animals and humans can be expected to behave, but as leaving a lacuna that is especially worrying in the human case.

> In the case of insects and most nonhuman vertebrates, this vague conception of the linkage between genes and behavior perhaps

> leaves fewer mysteries. . . . But in order to have a real evolutionary theory of mind and culture, one must begin with genes and the mechanisms that the genes actually prescribe. In human beings the genes do not specify social behavior. They generate organic processes, which we have called epigenetic rules, that feed on culture to assemble the mind and channel its operation. (349)

In *Promethean Fire* (1983a), their popular exposition of the theory of gene-culture coevolution, Lumsden and Wilson offer a more personal history of their joint study, explaining how critics of early Wilsonian sociobiology made a genuine point about the inability of the early program to deal with mind and culture, how Lumsden's arrival broke "the impasse," how the pooling of their talents made it possible for them to cross "the no-man's land" and to introduce biology into the social sciences (44–50).

What was the problem that allegedly brought the early debates about sociobiology to a stalemate? Apparently it was the problem of explaining how the human capacity for complex representations and complex decision making might combine with an existing collection of social institutions to defeat or to divert the dictates of natural selection. Claims about the possibility of maximizing our fitness by performing actions that are prevalent among human societies need not be interpreted as signaling the presence of any direct genetic control of those actions, as long as one can offer the explanation of a complicated animal with a complicated brain that attempts to manipulate its social environment in its perceived interests. Many of Wilson's opponents exploited just this possibility, arguing that, whatever its merits in explaining the behavior of nonhuman animals, sociobiology could not challenge the thesis that the genes set only minimal limits on human behavior. The theory of gene-culture coevolution was designed to elaborate Wilson's earlier metaphor. Our genes are supposed to hold human culture on a leash (Lumsden and Wilson 1981, 13).

The issue comes into focus when we read Lumsden and Wilson's formulation of human sociobiology: "The central tenet of human sociobiology is that social behaviors are shaped by natural selection" (1981, 99). Casual reading might take this for a truism. Of course all human behavior is ultimately shaped by evolution (whose main agent is natural selection), in the sense that evolution has equipped us with cognitive capacities and elementary motives that figure in our everyday decision making. Lumsden and Wilson's use of the plural, "behaviors," gives the game away, however. They intend to advance a stronger principle, the principle that individual aspects of human behavior are shaped by selection. Evolution does not just supply us with an all-purpose decision-making tool, with which we go to work whatever the social and biological environment we find ourselves in. Instead, our forms of behavior are restricted in particular ways, and these restrictions limit, in their turn, the forms of culture that we devise. The principle emerges starkly in the concluding section of *Genes, Mind, and Culture*:

> Only with difficulty can individual development be deflected from the narrow channels along which the great majority of human beings travel. In most conceivable environments, and in the absence of a forceful attempt to produce other responses, these behaviors will persist as the norms of culture in most or all societies. . . . although they cannot escape the inborn rules of epigenesis, and indeed would attempt to do so at the risk of losing the very essence of humanness, societies can employ knowledge of the rules to guide individual behavior and cultural evolution to the ends upon which they agree. (357–358, 360)

Ah, the old stories. There is nothing like them.

If my earlier arguments are correct, then Wilson's pop sociobiology will not be vindicated even by a successful defense against the criticism that Lumsden and Wilson strive to rebut. For, as we have discovered, Wilson's ladder falls apart at each rung, and careful attention to the top step is not going to redeem the speculative optimization practiced lower down. Nevertheless, it is worth looking closely at the theory that Lumsden and Wilson propose. For it may be that their new attempt to resolve one difficulty of the old program sidesteps some of our previous objections.

Scouting the Territory

There is an obvious picture of human evolution that brings out the general ways in which genes and culture can be expected to coevolve. Let us suppose that we have a human population that, at time t_0, encounters an environment E_0. For our present purposes we may suppose that the environment can be separated into two parts, the biological environment B_0 and the cultural environment C_0. We shall represent the totality of genes in the population—the "gene pool"—as G_0. Our task is to understand how *G* and *C* interact and how they change over time as a result of their interaction. (We may also be interested in charting the impact on *B*.)

Stage 1
The individuals in the population develop. Given their genotypes and the environment E, they acquire a set of phenotypes P_1. Because of these phenotypes there are important differences in survival probabilities. Hence, at mating time t_1 not all of the young who were formed at t_0 are still alive. There is a new gene pool, G_1. Also, as a result of the phenotypes of the developing individuals and of the environment in which they find themselves, there are actions of the individuals on the environment. So at t_1 we have a new cultural and biological environment, C_1 and B_1.

Stage 2
At t_1 the human beings mate. Because of phenotypic differences and because of the characteristics of the new environment, some individuals are more successful than others. Again, as a result of struggles to obtain mates, cultural institutions or the biological environment may be modified to C_2, B_2.

Stage 3
At t_2 the next generation is born. The new gene pool is G_2, which may differ from G_0 as a result of the differential reproductive success of the individuals of the previous generation. The individuals who make up the new generation develop in the social environment C_2 and the biological environment B_2. They repeat the process of stage 1.

I take it that this three-stage picture represents our pretheoretical ideas about the interaction of genes and culture. We expect culture to play a role both in determining the phenotypic characteristics of human beings and in determining which phenotypic characteristics promote reproductive success. Conversely, we expect culture to be altered by the actions of those human beings who develop to maturity, and we suppose that the alterations will reflect phenotypic properties of these people, to wit, their cognitive capacities and their preferences. Finally, we assume that there will be a process of cultural transmission from generation to generation and that this process will be distinct from the process of reproduction.

Quite evidently, the picture I have sketched is simplified. Discrete generations are a convenient fiction that allows us ease in analysis. Similarly, we avoid all sorts of complications by assuming that there is a single social environment that all members of the population encounter. Nevertheless, even though the picture already departs from important aspects of the actual situation, it is a useful beginning. How can we make it more precise?

Any attempt to analyze all, or part, of the three-stage process can reasonably be called a theory, or partial theory, of gene-culture coevolution. Since the middle 1970s there have been a number of suggestions for theories of gene-culture coevolution. Some authors have concentrated in depth on a single part of the picture. For example, in an important monograph Cavalli-Sforza and Feldman (1981) pose the problem of how to understand the transmission of items of culture. In effect, they choose to ignore the links between genes and phenotypes and to inquire what distributions of cultural traits will be expected under conditions in which various channels of transmission operate with particular probabilities of success. (Thus, in their simplest models, Cavalli-Sforza and Feldman examine situations in which children acquire cultural traits from their parents, and they develop analogs of the usual theorems of population genetics. See especially chapter 2.) An alternative approach is to attempt to work with the general picture and to link it to particular anthropological examples (see, for example, Durham 1976, 1979). Lumsden and Wilson offer one particular way of articulating the general picture. Other authors propose different theories, often directed at different problems (beside those already mentioned, see Richerson and Boyd 1978 and Plotkin and Odling-Smee 1981).

To evaluate the theory of Lumsden and Wilson, we need to draw up an inventory of questions to which a full theory of gene-culture coevolution should provide answers. First there are questions of *representation*. We are concerned with the dynamics of a complicated system over time, and we need an adequate way of representing the state of the system at each particular time. So, for example, we must pick out those aspects of the social environment that are causally relevant to the processes whose dynamics we want to fathom. We must also decide how to talk about the phenotypic characteristics whose interaction with the social environment will determine both the relative reproductive success of the individuals who bear them and the changes that occur in culture. Hence we arrive at the following general questions:

1. What properties of culture should we select as our main causal variables?
2. What phenotypic characteristics are relevant (a) to the determination of individual fitness by the social environment and (b) to the modification of the social environment?

More straightforwardly, and less precisely, we need to know how to talk about culture and about minds.

The next task is to understand the developmental processes. To articulate stage 1, we need to know how the social environment influences the assignment of a phenotype to a genotype. In elaborating stage 2, we need to know how the social environment affects the fitness of phenotypes. So we achieve two more general questions.

3. To what extent and in what ways do social factors modify behavioral phenotypes?
4. To what extent and in what ways do social factors modify the fitnesses of behavioral phenotypes?

Finally we have to deal with the processes of transmission across the generations. One of these, the process of reproduction, is already well understood. We can use the insights of population genetics to understand how the relative fitnesses will be reflected in changes in gene frequency. The novel problem concerns the modification of the social environment in response to the behavior of individuals. The question is

5. How is the social environment at a later time determined by the social environment and the behavioral phenotypes present at an earlier time?

This is the question that seems to concern Cavalli-Sforza and Feldman. Their models can be viewed as attempts to provide answers for important special cases.

I think it is obvious that we cannot expect simple and general answers to these questions. Research on gene-culture coevolution will surely be advanced if it parallels work in the general theory of evolution—that is, if aspiring theorists attempt to offer precise analyses that can be applied to a broad range of important special cases. Hence, it may be appropriate to begin by making some further simplifying assumptions. Perhaps we can select a small set of causally relevant social variables. Perhaps we may reasonably restrict our attention to phenotypes that are modified in very simple ways by our set of social variables. So, for example, we may concentrate on human propensities to perform certain kinds of behavior, we may consider propensities that are simple functions of genotype and the social variables, and we may then try to identify the pattern of behavior that results when individuals develop in various social environments. (Notice that the social environment plays a double role here: it not only affects the behavioral dispositions that develop but also determines the way in which those dispositions are manifested in actual behavior.) At the next step we may attempt to give a model for determining relative fitnesses, and thus apply standard population genetics to compute the changes in gene frequency that occur. Combining our conclusions with an analysis of the dynamics of changes in the social variables, we may be able to give a model of the coevolution of our hypothetical situation.

I interpret Lumsden and Wilson as doing something of this kind—although it will become clear that they are not always aware of the issues that they are addressing and of the simplifications that are being made. How should we assess this kind of enterprise? First let us note that the general picture of gene-culture coevolution is neutral on the controversial questions about the role of the genes in our social behavior. That picture can be elaborated in ways that make human behavior appear very sensitive to the social environment, and it can be articulated in a fashion that leaves little behavioral plasticity. If Lumsden and Wilson are going to support the conclusions at which earlier pop sociobiology aimed, if they are to show that there are rules that we can escape only at the cost of losing our essential humanness, then there will have to be evidence for the particular elaboration of the general picture that they provide.

This means that assumptions about the appropriate social variables, about the modifiability of behavioral phenotypes, about the transmission of cultural traits, about the ways in which social variables determine relative fitnesses, will have to be confirmed either as part of the total theory that is set forth or independently of it. Those assumptions cannot simply bask in the light of the general picture. They must obtain support from somewhere; and there are, potentially, two sources. Perhaps we can articulate the theory of gene-culture coevolution by drawing on our knowledge of parts of the process. Maybe we can appeal to the existing social sciences to furnish evidence for conclusions about the modifiability of behavior or about the ways in which cultural traits are transmitted. This source of evidence cannot, however, be expected to resolve the truly controversial questions. The very fact that the theory of gene-culture coevolution is needed to settle issues about behavioral plasticity and about the aptness of the metaphor of the leash tells against the idea that the corresponding parts of the theory can be imported from thriving sciences. Instead, we must suppose that the simplifying assumptions of the theory of gene-culture coevolution gain their support from their capacity, taken together, to predict or to explain phenomena that would otherwise have been unexpected or incomprehensible.

The road to success lies in framing assumptions about human behavior and its relations to genes and to culture that simplify the vast problem of answering the general questions we have listed. Armed with some ideas about how to proceed in special cases, the aspiring theorist can draw up models that can be applied and tested in those cases. So, for example, one might test and confirm a hypothesis about the limits set by our genes on a certain form of behavior by showing that, in combination with other hypotheses (perhaps hypotheses that received independent support), it could account for patterns of human behavior that were inexplicable on any available rival hypothesis. There is no reason *in principle* why Lumsden and Wilson should not obtain the kinds of conclusions they want from the kind of theory they intend to construct. Practice, as we shall see, is another matter.

Atoms of Culture

The most prominent—and most criticized—feature of Lumsden and Wilson's theory stems from their atomistic conception of culture. The authors invite us to imagine "an array of transmissible behaviors, mentifacts, and artifacts, which we propose to call *culturgens*" (1981, 7). Culturgens are to be the units of culture. Conversely, a culture is to be specified in terms of the culturgens that make it up. A culture is an aggregate of culturgens.

We shall examine the official definition of "culturgen" later. For the moment, the term can be introduced by means of examples. Culturgens are a varied lot. Tools, taboos, food items, forms of behavior, dispositions to behavior, dreams, works of art, scientific theories, all count as culturgens. The liberality of usage buttresses the thesis that cultures are aggregates of culturgens. To confound those who complain that a culture is not simply reducible to a set of culturgens but rather is determined by how and how often the culturgens are employed, one need only remind them that patterns of behavior are themselves culturgens. We shall consider later whether the liberality that comes in so handy here brings other difficulties in its train.

Introducing the idea of a culturgen allows Lumsden and Wilson to answer the first question on our inventory. The state of a culture at any one time is to be represented by the probability distribution of the use of the various alternatives within a family of culturgens. If the cultural phenomenon in which we are interested is incest avoidance, then the state of the culture is to be represented by the probability that an individual in the culture will adopt the culturgen "engage in incest." If the cultural phenomenon in which we are interested is choice of flavor among ice cream consumers, and if the only available flavors are vanilla and chocolate, then the state of the culture is given by the probabilities of "choosing chocolate" and "choosing vanilla." Differences in the state of society at various times are to be traced by examining differences in culturgen use.

This idea also enables Lumsden and Wilson to answer our second question. The phenotypic properties on which they focus are propensities to employ various culturgens. Thus the propensity to engage in incest or to choose vanilla ice cream is taken to be the behavioral characteristic whose fitness, in the context of the social environment, will be used to determine the course of coevolution. Similarly, the alterations in cultural state from time to time will be understood as the result of individual propensities for the relevant culturgens.

To answer our third question, Lumsden and Wilson need to give an account of the way the social environment modifies the behavioral phenotype. They approach the problem by introducing the idea of *epigenetic rules*: "The rules comprise the restraints that the genes place on development (hence the expression 'epigenetic'), and they affect the probability of using one culturgen as opposed to another" (1981, 7). The epigenetic rule associated with a genotype represents the dependence of behavioral propensities on the genotype by showing how the behavioral phenotype varies with the social environment. Given Lumsden and Wilson's attenuated conception of cultural state, they have few options for representing the dependence. The behavioral phenotypes are taken to be the probabilities of choosing among a set of rival culturgens. (Thus, if there are just two competing culturgens—chocolate and vanilla, say—the behavioral phenotype is the probability of choosing one of them; of course, the probability of choosing the other is then fixed.) The social environments are represented by the surrounding frequencies of culturgen choices. The epigenetic rule for a genotype tells us the probability of choosing a culturgen when all those about are choosing it, when nobody is using it, and at all intermediate frequencies. In the simple case of ice cream choice, the epigenetic rule for a genotype shows how the probability that a person with that genotype will choose one culturgen (say vanilla) depends on the frequency of users of that culturgen (vanilla choosers).

Similarly, in response to our fourth question, Lumsden and Wilson can only allow for a dependence of the fitnesses of using culturgens on the frequencies with which culturgens are used in the society under consideration. In fact, they typically proceed by supposing that the fitness of using a particular culturgen is independent of the social environment.

Finally, they attempt to solve the problem of how the social environment develops in response to individual propensities by understanding how the choices of individuals will be reflected at the societal level. If we conceive of the propensities of individuals to adopt culturgens (or to switch from using one culturgen to using another) as determined by their genotypes and by the frequency of culturgen use in the society in which they develop, then it is possible to explore the ways in which patterns of culturgen use will change over time. Full analysis of the problem would involve consideration of the way in which natural selection acts on the underlying genotypes. Even without considering selection, however, we can "capture in a crude manner the essential properties of genes, mind, and culture" by investigating the problem that Lumsden and Wilson call "gene-culture translation." The idea is to specify an epigenetic rule and to inquire how the societal pattern of use of the pertinent culturgens is expected to change with time. If, for example, we specify the way that choice of ice cream flavor depends on the frequencies of choosing chocolate in the surrounding culture, we can ask whether the probability that all members of the culture will opt for chocolate increases with time and whether it will tend to a limit.

Obviously, Lumsden and Wilson are omitting plenty of factors that we might antecedently think of as important to the evolution of a lot of social characteristics. In a large population we shall expect that for each individual there will be an important subpopulation (roughly, the people the target individual knows) such that the frequency distribution of culturgen use within this subpopulation is what is primarily relevant to the individual's choices of culturgens. Moreover, we shall expect that the choices of particular surrounding individuals (parents, teachers, peers) will have special influence on the individual's decisions. Their influence will also vary at different stages in the individual's development. Finally, the decision to choose a culturgen at a particular time will surely depend crucially on the past experiences of the decision maker. People learn that they are bad at using certain tools or that particular courses of action land them in trouble, and the deliverances of friendly or unkind experience rightly have more weight with us than raw peer pressure.

Considerations like these remind us that the attenuated conception of culture embraced by Lumsden and Wilson will have to be enriched before we can do justice to some aspects of gene-culture coevolution. It does not follow that there are no cases to which their preferred formulation can profitably be applied. After all, conventional population genetics began with simplified models. Lumsden and Wilson may remind us that there are some societies (for example, bands of gatherer-hunters) in which everybody knows everybody. There are some forms of behavior—perhaps incest or foraging—on which all adults are equally authoritative and with respect to which past experience makes little difference. So we can hope to achieve results about some forms of behavior in some societies and thereby to confirm hypotheses about the forms of particular epigenetic rules.

Lumsden and Wilson believe that they have found actual cases where their general framework is elaborated by advancing hypotheses about the forms of epigenetic rules and where the hypotheses are tested and confirmed. Responding to a critical review by Lewontin (1983b), they write, "To develop and test the general concept of genes and culture exemplified by the incest case, we constructed a series of quantitative translation models that trace the coevolutionary circuit" (Lumsden and Wilson 1983b, 7). After citing their three favorite cases (incest avoidance, fissioning of villages among the Yanomamo, and changes in women's formal fashions), Lumsden and Wilson announce that in the case they like best of all, they have successfully confirmed their hypothesis: "We cite experimental data indicating a relatively strong bias against incest and show that the worldwide cultural pattern is at least roughly in accord with the pattern derived from our model" (7). This is surely the right type of argument for Lumsden and Wilson to offer. They are entitled to claim a triumph for their theory if they can show that it yields recognizably correct conclusions about patterns of human social behavior. Provided, of course, that it is not possible to do equally well without the theory.

There are three sources of evidence to which Lumsden and Wilson can appeal. They can attempt to use the theory of gene-culture coevolution to account for the distribution of pervasive (and, preferably, puzzling) cultural institutions. They might, for example, attempt to show how it is possible to understand the common repudiation of incest, which is often expressed in the form of an explicit social sanction. A second source would be the patterns of behavior that we observe across societies. Instead of focusing on attitudes toward incest, they could endeavor to explain the pattern of incest avoidance, showing us why we should expect incest to be infrequent in almost all societies and almost entirely absent in many. Finally, at their most ambitious, they could use the full resources of their theory to chart the course of some significant cultural change and to show how it can profitably be understood in terms of hitherto unrecognized genes, epigenetic rules, relative fitnesses, and so forth. A number of successes on these three fronts would give us reason to take the concept of culturgen and its associated principles seriously.

A Reductionist Revival

There are two easy global criticisms of the Lumsden-Wilson enterprise that have figured in early responses to it. Critics have gleefully pointed out the problems of defining the central term "culturgen," and they have chided Lumsden and Wilson for adopting a reductionist approach to culture. Before we examine the detailed elaboration of the theory and the attempts to test it, it is worth seeing if there is any substance to these general charges.

Lumsden and Wilson define "culturgen" in an appendix to their introductory statement of the theory:

> A culturgen is a relatively homogeneous set of artifacts, behaviors, or mentifacts (mental constructs having little or no direct correspondence with reality) that either share without exception one or more attribute states selected for their functional importance or at least share a consistently recurrent range of such attribute states within a given polythetic set. (1981, 27; the entire definition is italicized in the original)

The definition is not worth much. With a little ingenuity in the construction of sets of attributes, one can group together just about any behaviors one likes and hail the result as a single culturgen. Lumsden and Wilson want flexibility, and hence they do not insist that all the elements of a culturgen should share a single common property. Without limits on the construction of polythetic sets (sets of attributes of which most but not all must apply to each of the entities included in a culturgen), the way is open to gerrymander culturgens as one likes. Moreover, the astute reader will notice that the Lumsden-Wilson definition allows for relations of inclusion among culturgens, so that the universe of culturgens has a structure that will constrain the choices made by an individual. I shall return to this point in a moment.

The fact that a definition is not worth much is itself not worth much. Gone are the days when righteous philosophers could puff indignantly at the unhygienic practice of not defining the key terms of a new theory. We know all too well that rigorous and exact definitions come relatively late in the development of sciences, that they are often preceded by a period in which theoretical language is put to work with highly flawed views about the objects to which it is intended to apply. During this period scientists may yearn for relief from the obscurity into which their linguistic practices lead them. They struggle along, however, leaning heavily on examples in which the intended interpretation of the theoretical language is manifested, and trying, in a piecemeal way, to make their statements more exact, so that they will at last know what they are talking about. A recent example of such struggles is the history of the concept of the gene in our century (Carlson 1966; Kitcher 1982b), and Lumsden and Wilson rightly advert to this example in forestalling criticisms of their own definition (1981, 30). (However, they do not get the conceptual history quite right, for they assume that the classical concept of the gene was well defined!) This and numerous other instances—"atom," "electron," "species," "energy"—testify to the fact that new sciences need not undertake the difficult task of providing neat definitions for the new terms they introduce. Understanding can be achieved initially by focusing on central cases, and later investigators can try to resolve the worries about how exactly the usage of the term is to be extended.

Lumsden and Wilson correctly take this pragmatic line: "The best research strategy for gene-culture coevolutionary theory . . . would seem to be the same as that employed in biology and ethnography: start with examples in which the units are most sharply and readily definable, establish them as paradigms, and then proceed into more complex phenomena entailing less easily defined units" (1981, 30). In this way they can shrug off one line of criticism. They have a similarly straightforward response to the objection that their approach presupposes a mistaken atomistic conception of culture. As Lumsden and Wilson explicitly proclaim in response to their critics, they see no reason why reductionist approaches should not work as well in the social sciences as they do in physics (1983b, 33).

Yet there are important points lurking behind the criticisms, and we need to bring these points out into the open. However Lumsden and Wilson choose to define "culturgen" in the ultimate version of their theory, the form of the theory they propose dictates certain constraints on culturgens. Their goal is to understand how individual propensities will be translated into patterns of behavior within a society. A significant portion of the theoretical machinery is designed to link cultural states with the behavioral dispositions of individuals. The appropriate form for describing the behavioral dispositions ascribes to a person a propensity for choosing a culturgen; the appropriate form for describing the state of a society attributes a frequency distribution of culturgen use. These two general ideas, which are central to the Lumsden-Wilson articulation of our pretheoretical picture of gene-culture coevolution, place a pair of constraints on the concept of a culturgen. Culturgens are things to which particular human beings can have attitudes, things that can be chosen, adopted, or used by individuals. Moreover, they are the elements of cultural states. Cultures are distinguished from one another by their patterns of culturgen use. Is it possible to satisfy both of these constraints simultaneously?

Here lies the common root of two apparently different concerns about the terms in which Lumsden and Wilson frame their analysis. When we move beyond both the simple charge that "culturgen" has not been given a respectable definition and the blanket accusation that the approach to culture involves a misguided reductionism, there is a serious worry that may express itself in either of two ways. Lumsden and Wilson hope to do justice to human culture by employing the notion of a culturgen. If they are to achieve a sufficiently broad characterization of cultural states, then perhaps the concept will be broken loose from its relationship to individuals and will become ambiguous. Alternatively, if they preserve an unequivocal usage, then they will sell culture short.

Consider an example that is very closely connected with the authors' favorite case. Many societies have taboos against brother-sister incest, either formulated in items of cultural lore or expressed through almost universal public disdain or repugnance. If Lumsden and Wilson hope to apply their theory to the understanding of social institutions, such as incest taboos, they will have to show how to characterize a culture state that consists in having such a taboo, in terms of individual attitudes toward culturgens. To see that there is an important difficulty here, let us compare three distinct, hypothetical societies. Among the Shunsib there is an almost universal avoidance of brother-sister incest. Shunsib people do not, however, express any distaste for the actions of others. Indeed, it is considered poor form to prescribe the kind of behavior that is to be expected. So there is no public criticism of incest. It is simply avoided. By contrast, among the Moralmaj exactly the same pattern of incest avoidance is accompanied by personal sanctions against those brothers and sisters who are suspected of sexual relations. Moralmaj adults tell their children not to engage in incest, but there is no public expression of disgust. Attitudes are expressed toward individuals by individuals. Finally, among the Tabuit, who, coincidentally, have just the same pattern of incest avoidance, brother-sister incest is forbidden by an explicit taboo, inscribed in the book of laws. If a pair of Tabuit are convicted of incest, there are public punishments. Among the Tabuit, whispered criticisms are not enough.

How would the Lumsden-Wilson approach differentiate these three societies? Plainly the case is rigged so that the distinctions cannot be drawn by attending to patterns of culturgen use if the culturgens in question are those of engaging in incest with a sibling or avoiding incest. With respect to these culturgens the patterns of use are the same in all three societies. Yet there are surely three distinct cultural states.

The state of the Moralmaj can apparently be separated from that of the Shunsib by focusing on a different culturgen pair. If we consider the culturgens "criticize those who engage in incest" and "do not criticize those who engage in incest," then we can claim that there is a different pattern of use of these culturgens in the two societies. This tactic raises some very obvious questions. Does the activity of criticism make sense apart from a social context in which individuals are attributed with the right to appraise the conduct of others? How does the propensity for the culturgen of criticizing incest relate to the propensity for the culturgen of avoiding incest? These questions point toward issues that will occupy us later. For the moment, let us suppose that the choice of a different pair of culturgens will enable Lumsden and Wilson to distinguish the two cultural states and hence the two societies.

The case of the Tabuit seems less tractable. For the property that separates the Tabuit from the Moralmaj is a group property: within the culture there is a public system of punishing the incestuous. Of course, the individual attitudes of the Tabuit may be exactly those of the Moralmaj. There may be the same propensities to disapprove of incest. There may even be the same propensities to favor a public system of punishing incest. For the presence of a public system of punishment is compatible with a very broad range of attitudes toward that system. It would be foolish to suggest that the public institution is identical with some pattern of propensities for having it.

A sophisticated reductionist should acknowledge the problem. If the Lumsden-Wilson approach is to be carried through, there will have to be some way of identifying the property of having a public institution of punishing the incestuous with some (very complicated) pattern of individual behavior. It is quite possible that the reductionist thesis is false: although all societies that punish incest do so through the actions and attitudes of individuals, there need be no common pattern of individual behavior that realizes the institution of punishment in each case. If this is so, then the attempt to identify social institutions in terms of patterns of culturgen use—where culturgens are things that *individuals* adopt or abandon—will be defeated.

If we stretch the notion of culturgen, we can easily force a distinction between the Tabuit and the Moralmaj. Among the Tabuit there is a form of behavior—a punishment ceremony—that is absent among the Moralmaj. Count the ceremony as a culturgen, and the problem dissolves. However, the cost of this move is to cut the concept of a culturgen loose from its connection to the individual. Unless the reductionist work is done and we are shown how to identify the group property in terms of a pattern of behavior among the group members, the specification of cultural states in terms of culturgens cannot be achieved without a crucial ambiguity. To "solve" the problem of characterizing the cultural state of the Tabuit by invoking a new culturgen—the punishment ceremony—alters the notion of culturgen so that the apparatus of gene-culture translation is no longer applicable.

The problem is deep and general. Those who responded to Wilson's earlier version of pop sociobiology by claiming that the influence of culture had been forgotten were troubled primarily by the thought that particular social institutions play an important causal role in the development of individual behavior (see Sahlins 1976; Bock 1980). If their criticisms are to be adequately addressed—and it seems that the move to the theory of gene-culture coevolution was intended to address them—then Lumsden and Wilson will need a theory that accounts for the presence of social institutions, public ways of regulating the behavior of individuals. The challenge will be to specify the group property of having a social institution in terms of the frequency of use of culturgens, conceived as things that individuals can adopt. Unless that challenge is met, there will be no reason to think that Lumsden and Wilson have solved the problem they set for themselves. Even if we grant that their theory enables us to understand the patterns of behavior in societies—*when those patterns are conceived as frequency distributions of culturgen use*—a further step is needed to show that it can tackle the genuinely interesting phenomena, the institutions of different cultures. That step consists in showing how social institutions may be conceived in terms of frequency distributions of culturgen use. (I should note that these are the interesting phenomena in the context of defending pop sociobiology against the objections I have set forth. Students of animal behavior will rightly insist that there are all kinds of interesting questions—questions, for instance, about the evolution of patterns of mate choice and, perhaps, about the evolution of animals' attitudes toward the mate choices of others.)

Lumsden and Wilson are explicit in claiming that culturgens are things that individuals can adopt or refrain from adopting. Not only is this thesis presupposed in their mathematical analysis (as we shall see later), but it also figures in their popular account of their enterprise:

> [The mind] searches for new solutions and occasionally invents additional culturgens to be added to the repertory. Out of a vast number of such decisions across many categories of thought and behavior, culture grows and alters its form through time. The flux of culture comprises an unceasing torrent of changes in individual decision making. (1983a, 127–130)

Yet there is some evidence that they do not appreciate the burdens that the thesis imposes on them. Consider the following passage:

> . . . culture is expected to be richest in those categories of behavior where the rules most favor it. We would expect culture to pile up as nodes around the conventions most affected by biased epigenetic rules, such as incest avoidance, courtship, and discrimination between in-groups and out-groups. The most ritualized forms of culture will not tend to replace the epigenetic rules, as many social scientists have thought, but to reinforce them. (1981, 21)

Insofar as Lumsden and Wilson have an argument for the result they announce here, it depends on conflating the kinds of conclusions that their theory can generate (or, more exactly, the ones it stands a chance of generating)—namely, conclusions about the frequency distributions of forms of individual behavior within cultures—with the kinds of conclusions they were challenged to account for—to wit, statements about the presence of social institutions.

Lumsden and Wilson would like to dispose of a simple objection. The critic asks why cultures find it necessary to introduce taboos against incest if incest is something that we naturally avoid anyway (see Lewontin, Rose, and Kamin 1984, 137n). The discussion of human incest avoidance begins by stressing the fact that "nearly all" societies forbid marriages between siblings, as if this were something that Lumsden and Wilson intended to explain. It is not what they actually explain. Their model of gene-culture translation yields the expected distribution of societies with frequencies of incest ranging from 0 to 1, on the basis of individual propensities for incest avoidance. Until they are able to identify the state of having an incest taboo in terms of culturgens that individuals have a propensity to adopt, Lumsden and Wilson cannot even begin to put their machinery to work. Moreover, they have not even accounted for the presence of the widespread response to incest in others, the kind of response that we imagined for the Moralmaj. The apparently exciting conclusion that culture "piles up as nodes around the conventions most affected by biased epigenetic rules" amounts to something much less novel, something that is inept at defusing the simple objection: if people have strong propensities to avoid a particular form of behavior or to choose a particular artifact, then the frequency distributions in various cultures will reflect those propensities; there will be few cultures in which the behavior is common and many in which the artifact is widespread. Whether societies can be expected to introduce sanctions against the behavior or prescriptions for using the artifact is an entirely separate and open question.

Sophisticated reductionists might formulate the claim that it is possible to identify the presence of social institutions with frequencies of culturgen use while simultaneously honoring the idea that culturgens are things that individuals choose to adopt, and they might even attempt to deliver the goods on the identification. Lumsden and Wilson do not achieve this level of sophistication. They write as if incest avoidance, disapproval of incest, and the presence of an incest taboo could all be covered by a single "anti-incest" culturgen, so that to explain the widespread avoidance of incest would be to solve some really important problem about culture. If this is the mistake that moves the critics to expostulate about conceptual slovenliness and reductionist approaches to culture, then, at bottom, the critics are right.

There is another aspect of the atomistic conception of a cultural state that has emerged briefly in our discussion so far and deserves further scrutiny. Lumsden and Wilson simplify their analysis by supposing that culturgens can be adopted in isolation. However, their official definition and the paradigms that we have considered make it apparent that no culturgen is an island. Because the universe of culturgens is structured, culturgens are not chosen separately. To take an elementary example, to choose chocolate ice cream is to choose ice cream. A more interesting case concerns the different culturgens that we reviewed in discussing incest: how does the propensity to avoid incest relate to the propensity to criticize others who engage in incest, and how do both of these relate to the favoring of a public ceremony of punishment?

Consider how the complex relations among culturgens affect our decision making. Your propensity for adopting a particular culturgen c is not simply a function of your genetically determined propensity to favor c and the number of people around you who are adopting c. It is also dependent on your propensities for adopting a host of culturgens that your adoption of c might enable you to attain and of your propensities for adopting another vast collection of culturgens that the choice of c might prevent you from using. Moreover, your propensity for choosing *c* may depend on your past history of culturgen choice. The person who has learned how to make one kind of fishhook may be unmoved by the fact that all those around are making something very different.

Hence, if we are to pose the problem of choice of culturgens in its full generality, then we should consider a person as having an initial set of propensities to adopt culturgens that responds *as a whole* to the frequencies of culturgen adoption in the surrounding population. Intuitively, if two culturgens are connected so that the probability of using one, given present choice of the other, is small, then the propensity for one may act as a brake on the bandwagon in favor of the other. Moreover, in the context of decision making it is not the actual probability but the *perceived* probability that counts. Our choices are crucially affected by the relations we believe to hold between the things we want.

The moral of these points is that the Lumsden-Wilson analysis is not only inadequate as a means of explicating culture. The failure to show how to understand social institutions in terms of patterns of culturgen use is matched by a failure to take into account the interactions among choices of culturgens in the decision making of individuals. Only when little of long-term importance turns on the present decision to adopt a culturgen is it likely that the interdependence of our propensities can be ignored.

Lumsden and Wilson occasionally allude to this point, as if it were a small inconvenience that could be ignored for the purposes of developing a first model (1981, 55–56, 109). The costs of neglect need careful weighing. It is easy to imagine that a strong genetic predisposition to adopt a particular culturgen might coexist with a very low frequency of use of the culturgen because individuals in the society constantly satisfy other preferences in ways that preclude the use of the culturgen in question. Hence, where there are significant relations among culturgens, a derivation of societal patterns of use drawn from study of the genetic predispositions to adopt single culturgens and of the tendency to respond to the choices of others in the surrounding society may go wildly astray.

So if we are to test and confirm the theory of gene-culture coevolution, where should we look for results? One source is out. We cannot hope to predict the distribution of cultural institutions. Lumsden and Wilson are not even aware of the difficulties involved in doing that. Nor can we expect the theory to yield accurate predictions in any case in which individuals might perceive connections among distinct culturgens in ways that constrain their propensities to adopt the culturgens. The domain of the theory has shrunk considerably. It is difficult to see how even some resounding successes in accounting for societal patterns of adopting culturgens (in cases where the adoption was relatively isolated from other decision making) would answer the charge that Wilson's early pop sociobiology neglected both mind and culture. True, concentration on limited problems is often fruitful in the early days of research in a domain (witness the simple models of population genetics and the analogs recently articulated by Cavalli-Sforza and Feldman). However, further disappointments await those who see gene-culture coevolution (Lumsden-Wilson style) as the salvation of pop sociobiology.

Gene-Culture Theory Made Easy

To test a theory it is important to formulate it. Lucky expositors can state the theories they expound by quoting the authors. Lumsden and Wilson do not lend themselves to this kind of treatment. There is a common tendency, even among those who hail the theory as a major new intellectual departure (perhaps especially among those who so hail the theory—see Barash 1982 and van den Berghe 1982), to describe the mathematical formulations as opaque, even unintelligible. Perhaps this is one reason why Lumsden and Wilson have been able to dismiss many of their critics as failing to address the substance of their work. If the substance is guarded by densely packed equations, bristling with ferocious symbolism, reviewers may be forgiven for seeking topics elsewhere.

Nevertheless, as I hope to show, the theory can be presented relatively simply. It will prove helpful to divide it into four main parts. The first part consists of an elaboration of the notion of an epigenetic rule and the distinction of two kinds of epigenetic rules. In the second part Lumsden and Wilson assemble some psychological results, with the intention of defending claims about genetically controlled propensities for certain behavioral traits. The symbols begin to appear in the third part, where the authors derive the equations that they take to govern gene-culture translation. In the fourth part Lumsden and Wilson close "the coevolutionary circuit" by deriving the equations that are meant to govern changes in gene frequencies under certain assumptions about cultural states and the determination of fitness. I shall attempt to explain these four parts in order, restricting my attention to the concepts that are necessary for understanding the advertised tests of the theory and to the evidence that is offered for various subsidiary assumptions.

Part 1

Epigenetic rules are conceived as processes that map genotypes and environments onto phenotypes. Indeed, we can identify them as the mappings whose properties we have discussed in earlier chapters. There is some ambiguity in the formulation given by Lumsden and Wilson, in that epigenetic rules are sometimes identified as processes, sometimes as mappings that record the dependence of the outcome of a developmental process on genotype and environment; but this ambiguity does not make any significant difference to the theory.

The primary epigenetic rules are those that direct the formation of our basic perceptual system. These rules are "more automatic," in the sense that they tend to yield the same phenotypes even in the presence of highly varied environments (1981, 36). The chief example of a primary epigenetic rule is the process that leads to our disposition to perceive the world in terms of four basic colors (43–48), although Lumsden and Wilson also discuss the epigenetic rules associated with other sensory modalities.

The secondary epigenetic rules are processes that take us from genotypes and environmental stimuli—which, given the action of the primary epigenetic rules, can be conceived as including perceptions of the environment—to more complex behavioral dispositions. The secondary epigenetic rules are as varied as human behavior.

Part 2

Lumsden and Wilson proceed by reviewing some work in developmental psychology in an attempt to show that some epigenetic rules operate relatively independently of the environment. The standard method of argument is to appeal to psychological experiments in which the same phenotype emerges in different environments. (See, for example, the discussion of color classifications in various cultures, 45–46.) Apparently the authors aim to establish two conclusions: (1) that "the primary rules are the more genetically restricted and inflexible" (52) and (2) that there are at least some secondary epigenetic rules that are "relatively rigid" (96).

The first thesis is both less controversial and more firmly supported than the second. As Lumsden and Wilson move toward areas in which psychologists debate the relative impacts of nature and nurture, some of the old habits of early pop sociobiology intrude on the discussion. So, for example, they invoke the famous experiment showing the human ability to memorize rapidly seven unrelated syllables, in support of a broad speculation about the significance of the number seven.

> Miller wondered if it is only a "pernicious, Pythagorean coincidence" that there are seven wonders of the ancient world, seven seas, seven deadly sins, seven daughters of Atlas in the Pleiades, seven ages of man, seven levels of hell, seven days of the week, and so forth. We suspect not, although evidence from non-Western cultures is conspicuously lacking. (62)

Vistas of research open up here. What, after all, could be the psychological basis for our preference for fours (four seasons of the year, four elements, four horsemen of the Apocalypse), for threes (the Trinity, triumvirates, three daughters of Lear, three strikes and you're out), for . . . ?

Such lapses are frequent. In a discussion of young children's reactions to odors, Lumsden and Wilson dismiss the possibility that children's responses might reflect a tendency to say "I like it" to just about anything (41). They see a possible parallel between Harlow's famous experiments on rhesus monkeys, in which infants were *deprived* of their mothers, and some human situations in which children are *abused* by their parents (62). They draw conclusions about "mother-infant bonding" in humans from samples of less than thirty women (80–81). They assume, without argument, that the smiling behavior that emerges in young infants is continuous with the smiling that indicates friendliness in later life (77–78). In some cases the psychological studies receive the additional blessing of an adaptationist argument. In discussing phobias, Lumsden and Wilson write,

> It is a remarkable fact that the phenomena that evoke these reactions consistently (closed spaces, heights, thunderstorms, running water, snakes, and spiders) include some of the greatest dangers in man's ancient environment, while guns, knives, automobiles, electric sockets, and other far more dangerous perils of technologically advanced societies are rarely effective. It is reasonable to conclude that phobias are the extreme cases of irrational fear reactions that gave an extra margin needed to ensure survival during the genetic evolution of human epigenetic rules. (85)

Perhaps. But there are puzzles. Why are so many sufferers afraid of *open* spaces? Why crowds? How frequent was the danger posed by running water, so that it became adaptively advantageous to develop an irrational fear reaction to it? Why do some people have an irrational fear reaction to flying? Or to mice? Why are specific phobias directed at lions, leopards, and tigers uncommon? It is hard to resist the impression that the fond eye of the sociobiologist sees just those instances of phobias that would fit preconceptions about the shaping hand of evolution.

The achievements of the review of psychological results are as we might have expected. Lumsden and Wilson manage to show developmental inflexibility in the unexciting cases (such as color vision) where it is already relatively well appreciated. As they venture into controversial territory, they are forced to rely on old-style pop sociobiological arguments to buttress their conclusions about the rigidity of the epigenetic rules. There is one respect, however, in which the whole discussion is surprising. We might have anticipated that the investigation of psychology would have been directed at the construction of a picture of the mind, which could then be applied in understanding the details of the interaction between genes and culture. Psychology promises insights that can be used when one attempts to explain how human decisions are made, how those decisions are shaped by the social environment, and how they affect the social environments of others. Because they are bent on campaigning for the inflexibility of epigenetic rules, Lumsden and Wilson ignore any indications from psychology about the way in which the coevolutionary process should be analyzed. As we shall see, their equations are divorced from any articulated psychological theory.

Part 3

The first collection of mathematically formulated hypotheses is given in the study of gene-culture translation. Lumsden and Wilson imagine an egalitarian society with N members. Two culturgens "compete for use in the group" (110). The task is to relate the epigenetic rules of the individuals who make up the society to the pattern of use that emerges in the society as the result of individual decisions.

According to Lumsden and Wilson, the appropriate way to formulate the epigenetic rules is to suppose that they set "*probabilities of transition* rather than fixed usage patterns" (111). So they introduce probabilities u_{ij}, understood as the probability that a person who has been using culturgen c_i will choose c_j at a decision point. To understand the way the pattern of use in the society will change, it is also necessary to know the rates at which decision points arise. Lumsden and Wilson suppose that these rates may be different for the users of different culturgens, and they take the rate at which decision points occur for users of culturgen c_i to be r_i. (If Cadillacs last longer than Chevrolets, then the times at which Chevrolet owners are forced to make decisions about new cars occur more frequently than the corresponding times for Cadillac owners.)

The goal of the enterprise is to work out the probability that, at time t, exactly n_1 of the N members of the society are using c_1. (Because we are dealing with a situation in which just two culturgens are available, the state of the society is defined when we know how many of them are using one of the two.) Suppose that $0 \leq n_1 \leq N$ and that $P(n_1,t)$ is the probability that, at t, n_1 people use c_1. Then the aim is to express $P(n_1,t)$ in terms of the transition probabilities (u_{12}, u_{21}) and the rates at which decision points occur. Lumsden and Wilson show that it is possible to give a differential equation for $P(n_1,t)$ in terms of n_1 and the functions v_{12}, v_{21}, where

$$v_{ij}(n_1) = u_{ij}(n_1) \cdot r_i(n_1).$$

(See technical discussion L for details.)

It is not hard to obtain from this differential equation an explicit solution for $P(n_1)$ at the steady state. So we derive what Lumsden and Wilson call "the ethnographic curve," which displays the expected distribution of societies in which, for values of n between 0 and N, exactly n members use culturgen c_1. We seem to have solved the problem with which we began—to have found the relation between the transition probabilities for individuals (which, we recall, are supposed to represent the epigenetic rules) and the probability of finding a particular pattern of behavior in the society. What more could we ask for?

Lumsden and Wilson introduce two refinements. First, they want to be able to compare societies of different sizes, and they point out, quite correctly, that an analysis in terms of the number of people who use a culturgen will not achieve their goal. Instead, they introduce a scaling variable, ξ, defined as $n_2/N - n_1/N$. The initial problem is now re-posed: the new variable ξ takes values between -1 and $+1$, and we can ask for an account of how the probability that ξ will lie in a particular interval in this range depends on the transition probabilities and the decision rates. Lumsden and Wilson announce that the solution already derived for the steady-state situation does not produce "rapid insight into the dependence of the ethnographic curve on the epigenetic rules" (122). Their remedy—the second refinement—is to consider values of N so large that ξ can be treated as if it took on a continuous sequence of values. The old differential equation is now converted into a partial differential equation of "the forward diffusion, or Fokker-Planck, form." Lumsden and Wilson solve the equation in the standard way, thus showing how the probability that the value of ξ in a society will be found in a specified interval depends on the functions $u_{ij}(\xi)$ and $r_i(\xi)$. (As in the earlier case, the probability is expressed as a function of the v_{ij}, where $v_{ij} = r_i \cdot u_{ij}$.)

Technical Discussion L

Consider a small interval of time of length dt. If this is sufficiently small, then we can suppose that, at most, only one person in the society faces a decision point during the interval. The difference between $P(n_1,t+dt)$ and $P(n_1,t)$ will be the probability that there is a decision in the interval that changes the frequency of c_1 users to make it exactly n_1, less the probability that there is a decision in the interval that changes the frequency from n_1 to some other number. Given our assumption that the interval is so small that only one person encounters a decision point within it, we know that the frequency of c_1 users can only become n_1 during the interval if it was $n_1 - 1$ at the beginning and one c_2 user switched to c_1, or if it was $n_1 + 1$ at the beginning and one c_1 user switched to c_2. Let $n_2 = N - n_1$. Then we can derive the equation

$$\begin{aligned} &P(n_1,t+dt) - P(n_1,t) \\ &\quad = P(n_1-1,t) \cdot (n_2+1) \cdot r_2(n_2+1) \cdot dt \cdot u_{21}(n_1-1) \\ &\qquad + P(n_1+1,t) \cdot (n_1+1) \cdot r_1(n_1+1) \cdot dt \cdot u_{12}(n_1+1) \\ &\qquad - P(n_1,t)[n_1 \cdot r_1(n_1) \cdot dt \cdot u_{12}(n_1) + n_2 \cdot r_2(n_2) \cdot dt \cdot u_{21}(n_1)]. \end{aligned} \tag{1}$$

(Here the terms $nr_i dt$ express the probability that a c_i user faces a decision point during the interval.) Since r_i and u_{ij} always occur together, we can simplify the algebra by taking $v_{ij} = r_i \cdot u_{ij}$.

From (1) we can obtain a differential equation for $P(n,t)$:

$$\begin{aligned} d/dt[P(n_1,t)] = {} & (n_2+1)P(n_1-1,t)v_{21}(n_1-1) \\ & + (n_1+1)P(n_1+1,t)v_{12}(n_1+1) \\ & - [n_1 v_{12}(n_1) + n_2 v_{21}(n_1)]P(n_1,t). \end{aligned} \tag{2}$$

This is the main component of the "master equation" (4-21; see 1981, 120). The other components, which deal with the situation when $n_1 = 0$ or $n_1 = N$, are easily obtained by a similar analysis. The population reaches a steady state when $d/dt[P(n_1,t)] = 0$. It is not hard to obtain from (2) an explicit formula for $P(n_1)$ at the steady state. This formula represents the ethnographic curve.

Let $\xi = n_2/N - n_1/N$. Suppose that N is sufficiently large that ξ can be treated as a continuous variable. By using a well-understood technique in applied mathematics, (2) can be converted into the equation

$$\frac{\partial}{\partial t} P(\xi,t) = -\frac{\partial}{\partial x}[X(\xi)P(\xi,t)] + \frac{1}{2}\frac{\partial^2}{\partial \xi^2}[Q(\xi)P(\xi,t)], \qquad (3)$$

where

$$X(\xi) = (1-\xi)v_{12}(\xi) - (1+\xi)v_{21}(\xi),$$

$$Q(\xi) = \frac{2}{N}(1-\xi)v_{12}(\xi) + \frac{2}{N}(1+\xi)v_{21}(\xi),$$

$$-1 < \xi < 1.$$

(3) is the continuous equivalent of (2), and Lumsden and Wilson provide a steady-state solution:

$$P(\xi) = C/Q(\xi) \exp\left[2\int_{-1}^{\xi} X(y)/Q(y)\, dy\right] \qquad (4)$$
$$\text{for } -1 < \xi < 1.$$

(For a heuristic argument connecting (3) with (2), and for the solution (4), see Feller 1968, 354–359.) Here $P(\xi)$ represents the probability that the value of ξ in a society will be found in a small interval around ξ; more exactly, the probability that the value of ξ lies between w and z is $\int_w^z P(\xi)d\xi$.

Given the solution, Lumsden and Wilson can proceed to explore the structure of ethnographic curves. That is, they can look at the way in which the function $P(\xi)$ depends on the functions X and Q and thus, ultimately, on the functions v_{12} and v_{21}. The v_{ij} can be written as $v_{ij,o} \cdot f_{ij}(\xi)$ where the $v_{ij,o}$ are constants, representing the "raw" or "innate" biases (133) and the f_{ij} are updating functions. For the "exponential trend-watcher case,"

$$f_{12}(\xi) = e^{a_1\xi},$$
$$f_{21}(\xi) = e^{-a_1\xi}.$$

In their later discussions Lumsden and Wilson undertake to show how the functions that represent the frequency of transitions, the v_{ij}, can be written in terms of "raw" or "innate" biases and "updating" functions. As the name suggests, "raw biases" reflect innate propensities to switch to using a particular culturgen. The updating functions represent the ways in which the frequencies of transition respond to surrounding patterns of culturgen use. The results are as one might have expected. Lumsden and Wilson are especially interested in the "exponential trend-watcher case," where the updating functions are exponential. Intuitively, as more and more people switch to one culturgen, the propensity of the rest to switch to that culturgen goes up ever more dramatically. In "exponential trend-watcher" cases, rather small differences in the "raw biases" to switch culturgens in one direction as opposed to the other can lead to a high frequency of societies dominated by a single culturgen (see 140–141).

Lumsden and Wilson announce a triumphant conclusion: "a barely detectable amount of selectivity in an epigenetic rule operating during the behavior of individuals can strongly alter social patterns" (144; italicized in original). For the reader who is bemused by the talk of culturgens, ethnographic curves, Fokker-Planck equations, and all the rest, let me translate. If people are strongly inclined to follow what others around them do, small initial differences in switching propensity can be dramatically magnified at the level of the group. If we all have a slightly greater propensity for switching from chocolate to vanilla than for switching from vanilla to chocolate, and if our switching propensity responds to the choices of those around us, then it is hardly surprising to learn that our society is likely to go to one of the extremes, with the vanilla end being more probable than the chocolate end. Did we need the steady-state solution to the Fokker-Planck equation to tell us that?

It would be wrong, however, to think of Lumsden and Wilson as simply drawing up some equations that allow them to play with possibilities. What they would really like to do is to put this machinery to work in actual cases. The theory of gene-culture translation will be worth having if we can put forward hypotheses about epigenetic rules and updating functions that enable us to derive ethnographic curves that can be compared with the data drawn from an empirical investigation of various societies. Because it is possible to determine the forms of ethnographic curves with respect to different culturgens—at least those that are present in a sufficiently large number of societies—we can hope to confirm particular hypotheses about the epigenetic rules and the updating functions by showing how computations employing these hypotheses and fitting the pattern set forth by Lumsden and Wilson enable us to derive the empirical findings. This is the strategy that we shall discover in Lumsden and Wilson's favorite examples. We shall see if it works in the next section.

Part 4

The last part of the theory consists in an attempt to show how the propensities for culturgen choice and the relative fitnesses of the use of different culturgens together lead to changes in gene frequencies within populations. The study begins with a review of the standard equations of population genetics. Lumsden and Wilson state the equation for the change in gene frequency for two alleles at a single locus under selection:

$$p_{t+1} = \frac{p_t W_{AA} + p_t q_t W_{Aa}}{p_t W_{AA} + 2p_t q_t W_{Aa} + q_t W_{aa}}. \quad (5)$$

(For equations (1) through (4), see technical discussion L.) Here p and q are the frequencies of the A and a alleles respectively, so that $p + q = 1$ (for all t).

Lumsden and Wilson would like to focus on the ways in which culturgen use is implicated in fitness. To this end they introduce two vectors. The *absolute fitness value vector* for genotype ij is defined as

$$\mathbf{W}_{ij} = [W_{ij}(c_1), W_{ij}(c_2)],$$

where the components $W_{ij}(c)$ are just the fitnesses of a person with the genotype ij who uses the culturgen c. The second vector is the *usage bias vector*, defined as follows:

$$\mathbf{L}_{ij} = [u(c_1/ij), u(c_2/ij)],$$

where $u(c/ij)$ is just the probability that after enculturation a person with genotype ij will use culturgen c.

How should we represent the classical notion of fitness in terms of these vectors? You guessed it. For a person with genotype ij the fitness is

(probability of using c_1 × fitness of using c_1)
+ (probability of using c_2 × fitness of using c_2).

This is easily represented in terms of vector multiplication:

$$W_{ij} = \mathbf{W}_{ij} \cdot \mathbf{L}_{ij}$$

or, more prosaically,

$$W_{ij} = W_{ij}(c_1)u(c_1/ij) + W_{ij}(c_2)u(c_2/ij).$$

(Even more prosaically: the average fitness of a genotype is the average fitness in each of the environments it encounters, weighted by the frequency of the environments.) Armed with this identity, it is now possible to take any of the familiar equations of population genetics in which the terms W_{ij} occur and rewrite them by substituting the vector product for W_{ij}. So, for example, Lumsden and Wilson replace equation (5) by

$$p_t + 1 = \frac{p_t W_{AA} \cdot L_{AA} + p_t q_t W_{Aa} \cdot L_{Aa}}{p_t W_{AA} \cdot L_{AA} + 2p_t q_t W_{Aa} \cdot L_{Aa} + q_t W_{aa} \cdot L_{aa}}. \quad (6)$$

Lumsden and Wilson happily occupy themselves for an entire chapter playing this simple substitution game and conclude by congratulating themselves on having developed a "concrete model of the coupling between genetic and cultural evolution" (230). Fortunately, more interesting things are to come.

Substantive results that do not simply inherit the bland, ecumenical character of standard population genetics can be derived only by making assumptions about the forms that the two vectors are likely to take. Lumsden and Wilson embark on the more difficult task of introducing plausible assumptions and articulating the mathematics by offering an account of an idealized life cycle for a human population.

> . . . the offspring produced in a large, randomly mating population are socialized by both their peers and the older generation. Alternative culturgens are learned and evaluated through exploration, play, and observation. Later they are employed during a prereproductive period in the gathering of a resource, which can be broadly defined according to circumstances to mean either food and other limiting resources or territorial ownership and other modes of control by which resources can be gathered in an unimpeded manner. Individuals choose between two culturgens under the influence of epigenetic rules. Genetic variation is permitted in the degree of bias in the epigenetic rules, the amount of sensitivity to peer and parental usage, and the function by which resources are converted to genetic fitness. (265; italicized in original)

They go on to fortify the reader for passage through the thicket of equations to come by promising that there will be some surprising results at the end.

Let me begin by noting that part of the life cycle is not represented in the formal apparatus that Lumsden and Wilson provide. Although they note that the alternative culturgens are "learned and evaluated," there is no representation of the idiosyncrasies of individual experience and individual learning in the eventual model. (Here, where we might have anticipated insights from psychology, the appeal to psychology is notably absent.) Instead, we start with the process of geneculture translation, conceived now as somewhat more complex. The first complication results from the fact that "the cultures of two generations impact the epigenetic rules" (274): people respond both to their parents and to their peers. The second is a consequence of the fact that the state of the culture is no longer adequately represented by the sheer number of users of one of the two available culturgens (say c_1). We have to keep track of the pattern within each of the three subgroups consisting of members with the same genotype. We need to know the number of *AA* individuals who are using c_1, the number of *Aa* individuals, and the number of *aa* individuals. (Why? Because the fitness of the use of c_1 may be different for each of the three groups, and if we want to attend to the dynamics of selection we have to attend to all three frequencies.)

A little reflection reveals that the equation for the steady-state probability distribution will now be more complicated. Lumsden and Wilson point out that the equation (275, equation no. 6-6) can, in principle, be solved exactly. However, "the resulting formulation is not concise and generally requires the aid of the computer. In this initial study we need an approach that gives concise, readily grasped, analytic solutions . . . even if some accuracy is lost" (277). They propose concentrating on examples in which the societies studied are "not too small" and for which the steady-state probability distribution is unimodal and sharply peaked. Then the values of the variable n_1 (or ξ in the more refined version) will differ only slightly from the cultural mode, and we can assume that the individual preferences are determined as they would be if the pattern assumed the mode. In other words, we abandon the idea of considering the individual propensities to switch culturgens as sensitive to the pattern of use in the societies around them. We suppose that virtually all the societies approximate the same pattern of use, and consider individuals in any one society as if they were responding to some "average cultural state."

Armed with this assumption, Lumsden and Wilson are able to derive a "*coevolutionary equation* for the culture pattern of c_1." This is to serve the purpose of relating the mean frequency of c_1 use to the gene frequencies and the functions v_{ij} that represent the frequency of switching. (For details, see technical discussion M.) On the way to their equation Lumsden and Wilson make a further simplification. They argue that individuals make independent decisions, so that the distribution of c_1 use within each genotype is a binomial distribution. The overall distribution is taken to be the product of the three distributions within the genotype classes.

The coevolutionary equation is as follows:

$$\bar{\nu} = \frac{p_t^2 v_{21}^{AA}(\bar{\nu})}{v_{21}^{AA}(\bar{\nu}) + v_{12}^{AA}(\bar{\nu})} + \frac{2p_t q_t v_{21}^{Aa}(\bar{\nu})}{v_{12}^{Aa}(\bar{\nu}) + v_{21}^{Aa}(\bar{\nu})} + \frac{q_t^2 v_{21}^{aa}(\bar{\nu})}{v_{12}^{aa}(\bar{\nu}) + v_{21}^{aa}(\bar{\nu})}. \quad (7)$$

Let me translate. The mean frequency of c_1 use among the *AA* is simply the probability that an *AA* person will choose c_1. This probabil-

Technical Discussion M

The general problem of gene-culture translation is to relate the probability distribution $\mathcal{G}(n_1^{AA}, n_1^{Aa}, n_1^{aa})$, which takes account of the frequency distribution of culturgen use within the three genotype classes, to the individual preferences. As noted in the text, Lumsden and Wilson do not try to give a solution to this general problem. Instead, they suppose that the individual preferences are those at the (unique) mode of the probability distribution.

The story develops by introducing a new set of probabilities:

$\mathcal{J}_k^{ij}(t/\bar{\nu})$ = the probability that an organism of genotype ij is using culturgen c_k at time t given that the "mean cultural order" is $\bar{\nu}$.

Lumsden and Wilson now introduce new differential equations governing these probabilities, and solve them to give the steady-state solution

$$\mathcal{J}_1^{ij} = v_{21}^{ij}(\bar{\nu}) / v^{ij}(\bar{\nu}), \quad \mathcal{J}_2^{ij} = v_{12}^{ij}(\bar{\nu}) / v^{ij}(\bar{\nu}),$$

where

$$v^{ij}(\bar{\nu}) = v_{12}^{ij}(\bar{\nu}) + v_{21}^{ij}(\bar{\nu}).$$

This is not very surprising. At steady state the relative frequencies of the c_1 users to the c_2 users are in proportion to the rate at which people switch to c_1 and inversely to the rate at which people switch away from c_1. If 5 people take up chocolate for every 4 that switch to vanilla, then, at steady state, we expect 5/9 of the population to be consuming chocolate. The same goes within individual genotypes.

Lumsden and Wilson now argue that the probability distribution $\mathcal{G}(n_1^{AA}, n_1^{Aa}, n_1^{aa})$ is the product of three independent distributions $\mathcal{G}(n_1^{ij})$, and that each of these is a binomial distribution. Intuitively, we suppose that the decisions of individuals are independent. Hence the probability of a certain number of *AA* people choosing c_1, a certain number of *Aa* people choosing c_1, and a certain number of *aa* people choosing c_1 is just the product of the probabilities that that number of *AA* will choose c_1, that number of *Aa* will choose c_1, and that number of *aa* will choose c_1. Moreover, within each genotype the decisions of individuals are independent. Hence, if the probability that an *Aa* person will choose vanilla is p, and if there are N Aa people, the probability that exactly n of them will choose vanilla is

$$\binom{N}{n} p^n (1-p)^{N-n}.$$

This extremely elementary result about probabilities of independent events is almost lost in Lumsden and Wilson's complicated symbolism.

Once we know the probability distribution $\mathcal{G}(n_1^{AA}, n_1^{Aa}, n_1^{aa})$, it is a straightforward matter to calculate the mean frequency of c_1 users, $\bar{\nu}$, in the population. In this way Lumsden and Wilson are able to obtain a "*coevolutionary equation* for the culture pattern of c_1":

$$\bar{\nu} = \frac{p_t^2 v_{21}^{AA}(\bar{\nu})}{v_{12}^{AA}(\bar{\nu}) + v_{21}^{AA}(\bar{\nu})} + \frac{2p_t q_t v_{21}^{Aa}(\bar{\nu})}{v_{12}^{Aa}(\bar{\nu}) + v_{21}^{Aa}(\bar{\nu})} + \frac{q_t^2 v_{21}^{aa}(\bar{\nu})}{v_{12}^{aa}(\bar{\nu}) + v_{21}^{aa}(\bar{\nu})}. \tag{7}$$

Again, the complicated symbolism almost obscures a simple line of argument. Given the assumption that $\mathcal{G}(n_1^{AA}, n_1^{Aa}, n_1^{aa})$ is the product of three binomial distributions, the expected frequency of c_1 use within each genotype class is simply the probability that a person of that genotype will choose c_1, $\mathcal{J}_1^{ij}$. As we have seen, this probability is

$$v_{21}^{ij}(\bar{\nu})/[v_{12}^{ij}(\bar{\nu}) + v_{21}^{ij}(\bar{\nu})].$$

The grand mean, $\bar{\nu}$, is just the average of the expected values within each genotype, weighted by the frequencies of the genotypes. Thus (7) follows at once.

The next step is to compute the rewards for people of different genotypes. Let J_k be the rate of harvesting if a person uses the culturgen c_k. Suppose that the harvesting period is of length T. Since the net rate at which a person of genotype ij harvests is a weighted sum of the rates obtained by using c_1 and c_2, where the weighting factors are proportional to the times spent using c_1 and c_2, the naive version of the reward equation is

$$R^{ij}(\mathrm{T}) = \mathcal{J}_1^{ij} \cdot J_1 T + \mathcal{J}_2^{ij} \cdot J_2 T. \tag{8}$$

As noted in the text, Lumsden and Wilson think that there are costs due to the building, maintenance, and use of cognitive equipment. So they replace (8) with

$$R^{ij}(T) = (J_1 - L^{ij})\mathcal{I}_1^{ij}\, T + (J_2 - L^{ij})\mathcal{I}_2^{ij}\, T + C^{ij}T/\tau. \qquad (9)$$

Given their explanations, the last term is puzzling. In the first place, the sign seems to be wrong. The transition costs ought to be reflected in a loss, not in a profit. (As we shall see, Lumsden and Wilson take C^{ij} to be positive.) Second, τ is the time between successive decision points (assumed to be the same for c_1 users and for c_2 users). However, it is quite unclear that there ought to be *transition* costs except when the person *switches* culturgens. It follows that the value of the transition costs ought to be

$$(C^{ij}T/\tau) \cdot [2v_{21}v_{12}/(v_{21} + v_{12})].$$

(The probability that a decision point is a switching point is the probability that a person uses c_1 and switches to c_2 or uses c_2 and switches to c_1. This is

$$\mathcal{I}_1 \cdot v_{12} + \mathcal{I}_2 \cdot v_{21} = 2v_{21}v_{12}/(v_{21} + v_{12}).)$$

In our examination of the theory in action (technical discussions N and O and accompanying text), we shall consider not only the equations that Lumsden and Wilson present but also the effect of the modifications that I have suggested.

ity is the probability that an AA person will choose c_1, *given that the cultural order is* $\bar{\nu}$. (Recall the simplifying assumption.) It is easily calculated to be

$$v_{21}^{AA}(\bar{\nu})/[v_{12}^{AA}(\bar{\nu}) + v_{21}^{AA}(\bar{\nu})].$$

(See technical discussion M.) Similar results hold for the means among Aa and aa people. The grand mean in the population, $\bar{\nu}$, is a weighted mean of the means within the genotype classes, so we are to multiply the mean within each genotype by the relative frequency of that genotype. Recall that if the frequency of A is p and the frequency of a is q, then the genotype frequencies are p^2 (AA), $2pq$ (Aa), and q^2 (aa).

The next step is to introduce natural selection into the picture. To this end, Lumsden and Wilson propose to determine fitnesses by comparisons of fertility (there is no prereproductive mortality in their model) and to relate the fertility of an individual to the harvesting of resources during the prereproductive period. Assuming that the prereproductive period is of a fixed length, we are to assign rates of harvesting that vary with the culturgen that is in use. Intuitively, the rate at which an *AA* person stores calories (that can be used to make gametes) when that person consumes chocolate may be different from the rate when the flavor chosen is vanilla. But, given that an *AA* person and an *Aa* person both use chocolate, or any other culturgen, their rates of harvesting are the same. Different rates of harvesting, and thus differential fertility, for different genotypes depend only on the different patterns of culturgen use.

You might think that it would be easy to write down the equation that determines "the reward structure," the equation showing how the amount of resource harvested depends on genotype. Apparently, the net rate at which a person with a particular genotype harvests is a weighted sum of the rates obtained by using c_1 and c_2, where the weighting factors are proportional to the times that people with that genotype spend using c_1 and c_2. You might think this is obvious. Lumsden and Wilson disagree. Their version of the "reward equation" builds in costs that naive expectations do not recognize.

> . . . the cognitive processing required to evaluate usage at each decision point requires brain tissue, time, and energy. The computation apparatus must first of all be built from neurons; its maintenance requirements in energy (or resource) units may be designated the *bearing costs* L^{ij}. Whenever the apparatus is employed at a decision point to carry the individual from usage state $k=1,2$ to usage state $m=1,2$, a cost in energy (or resource) units is exacted that may be termed the *transition cost* C^{ij}. (281)

We shall discover later how important these "costs" are to Lumsden and Wilson's claims about gene-culture coevolution.

The reward equation is related to the standard machinery of population genetics by supposing that there is a function F that assigns a number of gametes produced depending on the reward harvested during the prereproductive period. (Intuitively: the more inclined you are to use the superior culturgen, the more resource units you harvest; the more resource units you harvest, the more gametes you produce.) Suppose that the number of gametes produced by a person with genotype ij is $2F^{ij}$, where F^{ij} is a function of R^{ij}, the reward obtained by a person with genotype ij. Then we can use the ordinary equations of population genetics to compute the new frequency of the A allele:

$$p_{t+1} = (F^{AA}p_t^2 + F^{Aa}p_tq_t)/F, \tag{10}$$

where

$$F = F^{AA}p_t^2 + 2F^{Aa}p_tq_t + F^{aa}q_t^2.$$

As Lumsden and Wilson point out, the F^{ij} depend on the reward structure. The reward structure depends on the probabilities of switching between culturgens. The probabilities of switching are functions both of the epigenetic rules and the "patterns of cultural usage" (284). Hence, Lumsden and Wilson claim that they have exhibited a connection between the selection of alleles, reflected in gene frequency changes, and cultural patterns.

Before we become too enthusiastic, it is well to remember a simplifying assumption. The entire discussion is predicated on the idea that the societies who use the culturgens with which we are concerned are all sufficiently similar that we can treat individuals as responding to some average culture. If we want a sober formulation of what Lumsden and Wilson have done, then we can say that, at most, they have revealed a link between the effects of an average culture and changes in gene frequencies, in cases where cultures are all relatively homogeneous in their usage.

The general coevolutionary equations simply consist in formulating (7) and (10) with dependencies on the right generations. To put them to work, we should begin with (7). The function of (7) is to give us the mean frequency of use of c_1. Once we know this, we are supposed to figure out the probabilities of switching between culturgens for people with different genotypes. These are plugged into the reward equation to yield conclusions about the harvesting returns for people with different genotypes. Then we apply the fertility map to compute the values of F^{ij}. Finally, from (10) we derive the changes in gene frequencies.

However, before we can do very much with this equipment, so painfully fashioned for us, we need some assumptions about the forms of the epigenetic rules (more exactly, about the ways in which the transition probabilities, supposedly related to the epigenetic rules, respond to the surrounding pattern of culturgen use), we need assumptions about the alleged costs that enter into the reward equation, and we need assumptions about the form of the fertility map. As we make assumptions of these kinds, it will be possible to construct a test of the theory. Ideally, we might try to confirm hypotheses about updating functions, costs, and fertility, use them to elaborate the coevolutionary equations, and come up with a confirmable prediction about changes in gene frequencies. In this way, like the standard equations of population genetics, Lumsden and Wilson's coevolutionary equations would prove their worth. As we shall see later, they do not attempt anything quite so ambitious.

Our review of the theory has brought us to the point where we can evaluate it. There are two ways in which it can be judged. The simpler treatment of gene-culture translation can be tested against ethnographic data. The coevolutionary equations can be supplemented with extra hypotheses and put to work. Having staggered up Olympus, let us see what we can see.

Translational Trivialities

When sociobiologists trot out their favorite examples, incest leads the parade. Lumsden and Wilson believe that the data about incest avoidance can be put to a new use. They advance hypotheses about updating functions and about the epigenetic rule underlying incest avoidance, construct the epigenetic curve, and find a correspondence with the data. Here are the hypotheses:

- The updating functions are constants, namely 1. Propensities to switch to incest or to stop engaging in incest are unaffected by the surrounding societal pattern.
- The "raw" propensity to switch away from incest is very close to 1; the "raw" propensity to switch to incest is very close to 0.

Given these hypotheses, and letting c_1 be the culturgen of not engaging in sibling incest, c_2 be the culturgen of engaging in sibling incest, the value of u_{21} can be set close to 1 and the value of u_{12} at close to 0; using the exact solution to equation (2) (see technical discussion L), Lumsden and Wilson can compute the ethnographic curve. They find that "anecdotal accounts of a wide range of societies . . . suggest that the true curve is closest [*sic*] to that generated by $u_{21} = 0.99$ than to the others displayed" (155). Before we exclaim at the wonder of the fit, let us approach the problem from a slightly different angle.

Suppose that by the age of six or seven, children have acquired, by whatever means, a strong inclination not to engage in incest with siblings. Suppose that the inclination is so strong that, at any given time, if we confront a society and pick a member out of that society, the probability that the person we have picked is currently engaged in an incestuous relationship is p, where p is close to 0. Now suppose that all the societies with which we are concerned are of size N. We ask the following questions: What is the probability that a society within the group is incest free? What is the probability that exactly n of the members of a society picked at random are currently engaging in incest?

We answer these questions by supposing that all the members of each society make their decisions independently. (There is a slight problem here, in that incest is not something one can do by oneself. However, like Lumsden and Wilson, I shall temporarily ignore this inconvenient fact about culturgen choice when the culturgen is incest.) Applying a little probability, we find that the probability of an incest-free society is $(1-p)^N$; the probability that exactly n people in the society are engaged in incest is

$$\binom{N}{n} p^n (1-p)^{N-n} = \frac{N!}{n!\,(N-n)!} p^n (1-p)^{N-n}.$$

These are just the results obtained by Lumsden and Wilson. There is no need to introduce epigenetic rules, updating functions, the complicated equations, or any of the rest of the machinery. We simply assume that people make their decisions independently and that the probability that a person engages in incest is independent of the pattern of incest behavior in the surrounding society, in the sense that the propensity to incest, once formed, does not adjust to the incestuous behavior of others. The latter assumption is surely reasonable, given the typically private character of incestuous intercourse. (However, the propensity may adjust to the *attitudes* of surrounding people, their disposition to tolerate or condemn incest, and so forth.)

Moreover, it is possible to refine our calculations to take into account the obvious point that it takes two to have incest. Imagine a society consisting of N individuals, divided into groups of siblings, each of size $2m$, consisting of m males and m females. Suppose, as before, that each person chooses independently of each other person and that for each person the probability of consenting to incest (at a particular time) is p. If we pick a family group at random, the probability that it is incest free (assuming that incest occurs only by mutual consent) is $(1-p^2)^{m^2}$. For there are m^2 potential incestuous relationships; incest occurs in anyone with probability p^2; hence each is incest free with probability $(1-p^2)$, and all are incest free with probability $(1-p^2)^{m^2}$. If the total number of individuals in the society is N, so that there are $N/2m$ groups of siblings, then the probability that the society is incest free is $(1 - p^2)^{\frac{1}{2}mN}$.

Let us now compare the new analysis with Lumsden and Wilson's "anecdotal data." According to Lumsden and Wilson, the data for societies of about 25 people fit the ethnographic curve predicted by taking $p = 0.01$ in *their* equation. So the observed probability that a society is incest free is $(0.99)^{25} = 0.778$. What value of p would we obtain if we compared this observed probability with our refined model? Plainly, we have to make an assumption about the value of m. If we take $m = 2$, then we know that $(1-p^2)^{25} = 0.778$, so that $1 - p^2$

= 0.99 and p = 0.1. If m = 1, then p is about 0.141. Taking into account the family structure of a society suggests that the propensity to consent to incest is far higher than Lumsden and Wilson would allow.

This result is hardly surprising. After all, if sibling incest occurs by consent (and I hasten to point out that empirical research will be needed to assess how frequently coercion is involved), then, given that families are relatively small, the effective probability that a person engages in incest will be smaller (maybe considerably smaller) than that person's propensity to consent to incest.

The principal point, however, is that the theory of gene-culture coevolution is completely idle in this example. Forget, for the moment, that relatively elementary probabilistic computations can lead us to a more realistic conception of the situation than that which Lumsden and Wilson propose. An even simpler analysis yields their result directly, without any hypotheses at all about the role of the genes. However good the ethnographic data were, we could account for them by attributing to the people involved a propensity to engage in incest, and leave entirely open the issue of whether that propensity is the result of enculturation or whether it signals some rigid genetic action. If that issue cannot be settled by relating the data from Taiwan and from the kibbutz, then it cannot be settled by dragging in the theory of gene-culture coevolution. If it can be resolved by citing the evidence about Taiwanese marriages and sexual relations in the kibbutz, then no increase in confirmation is brought by applying the Lumsden-Wilson analysis. In either case, specific claims about epigenetic rules and updating functions receive no confirmation from this example.

Worse still, the cumbersome machinery of the theory blocks us from seeing the simplicity of the calculation and so finding a way to improve our analysis. Once we simplify the issue, recognizing that what is really accomplished is the connection of individual propensities to social patterns, we can think about fashioning a more realistic account of the relationship. There is no need to blind ourselves by talking about genes and epigenetic rules, because these things do not really enter into the story. Gene-culture translation is a complete misnomer. Lumsden and Wilson have found a complicated way to struggle to an inexact solution to a rather easy mathematical problem.

Let us move on to Lumsden and Wilson's second example, Chagnon's account of the fissioning of villages among the Yanomamo (1976). Chagnon explains how Yanomamo villages expand until they reach a critical size. At this point the frequency and intensity of internal tensions and hostilities increase rapidly. Eventually the village splits, roughly in half, with some members leaving and others staying. Typically, groups of close relatives leave together, and groups of close relatives remain behind.

Lumsden and Wilson propose to "account for significant features of the village fissioning process" (167) by invoking two culturgens. There is a "stay" culturgen and a "depart" culturgen. Given the actual findings, in which the new pioneers leave in a group, we cannot interpret acceptance of the latter culturgen as actually departing. For if the acquisition of the "depart" culturgen is to be accelerated by the adoption of that culturgen by others, then the resultant process would be one in which a few people leave initially and are followed by more and more. Hence the Lumsden-Wilson analysis can only be applied to the example if we suppose that adopting the "depart" culturgen involves announcing one's willingness to depart. We can think of the situation as follows. As the population size comes close to 200 (which Lumsden and Wilson take to be the critical size), the probability of advocating departure is increased. As people around are advocating departure, the probability of advocating departure goes up. When a critical number of people advocating departure is reached, those who advocate departure pack up and leave.

Lumsden and Wilson dutifully formulate some culturgen assimilation functions, compute the values of X and Q for equation (4) (see technical discussion L), and offer a number of ethnographic curves. Unfortunately, they have no detailed ethnographic data with which to compare their curves. As a result, there is no basis for identifying the results as anything more than a mathematical exercise. If the theory is supposed to account for "significant features" of the process of village fissioning, it is reasonable to ask, What features?

Their ethnographic curve (1981, 169, figure B) tells us that in a small village just about everyone will advocate staying and in a large village just about everybody will urge departure. Setting on one side the obvious point that this looks like a better explanation of mass exodus than of village fissioning, it is relatively easy to see how Lumsden and Wilson obtain the results. If everybody becomes more disposed to advocate departure when the village size comes close to 200, and if people increase their propensity to advocate departure when those around them are doing so, then it is not hard to see that such qualitative conclusions will be generated. Since we have no precise data, it is impossible to tell if anything is gained by elaborating the decision-making process along the lines favored by Lumsden and Wilson. I shall argue in a moment that it is relatively clear that certain obvious and important aspects of the situation have been lost.

As in the example of incest avoidance, the problem is not properly seen as one of *gene*-culture translation. The question is, Given that the probability that a person will advocate departing increases both with the size of the village and with the number of people who are urging departure, what is the probability that, in a society of size N, exactly k people will advocate departure? Once again we can begin by assuming that, by whatever means, adults have arrived at a disposition to advocate departure that is described by a function $p(k,N)$—the probability that an adult will advocate departing, given that k members of the society (whose total number of members is N) are doing so. To satisfy our qualitative requirements, $p(k,N)$ should be very small when N is very small, it should increase with k for any value of N, and it should be appreciable for all values of k when N is large. What Lumsden and Wilson actually show is that there is a function that fits these conditions. There is no reason to invoke epigenetic rules of *any* form to generate their conclusions.

But, in any case, the Lumsden-Wilson analysis of the situation is highly implausible, once we take a close look at Chagnon's account of the fissioning process. As villages begin to break up, groups of relatives band together; and when the fission eventually takes place, it is often led by a family that subsequently achieves positions of power and wealth in the new village. Given Chagnon's description of the process, it seems clear that any simple dependence of probabilities on the size of the village and the number of advocates of departure misses the factors that dominate the decisions.

> Fissions are the consequence of internal fighting among adult males, ultimately over sexual infidelity and the possession of females. It is the adult males who decide on fissions and who align the group in such a way that when they depart to create a new village, they will create a village that has a composition congenial to their own political and social interests. (1982, 305–306)

Chagnon's analysis makes it clear that the probability for an individual of joining a move to depart is not any simple function of the number of people who advocate departure. For example, those in power in the old village are likely to want to keep the village together, so long as they can find ways to avoid continued internal hostilities that prevent the village from functioning effectively as a unit. Some others who are not in power in the old village may have increased probabilities of staying when they see their political enemies planning to depart. Families that are relatively large, but not in power in the old village, can expect to gain control of a new settlement. Hence they can be expected to urge others to follow them. We should expect the probability distribution to be highly skewed and to depend critically on the relations between those who are advocating leaving and those who are currently advocating staying. Contrary to the expectations of Lumsden and Wilson's ethnographic curves, it is extremely unlikely that the culturgen "depart" should ever be adopted by *all* the members of the village. What we would expect to find is a division of the village into two roughly equal groups, with the members of one advocating departure and the members of the other proposing that they and their friends remain, while encouraging the malcontents to leave.

Furthermore, as Chagnon notes, "most villages fission when they reach a population of about 125 to 150 individuals, but a few manage to grow larger—upwards of 400 individuals" (1982, 305). Lumsden and Wilson base their analysis on the idea that the critical size is about 200 individuals. It would be relatively easy for them to adjust their parameters to accommodate the lowered critical value; what is far less clear, though, is how their model can account for the variance.

I conclude that there is nothing in the Yanomamo village fissioning example to reveal the value of gene-culture theory. In the first place, there are no detailed data with which the ethnographic curves can be compared. Second, there is no need to invoke hypotheses about genes. As in the case of incest, the theory does no more than establish a connection between the dispositions of individuals and the pattern of behavior in the society. (It is interesting, at this point, to recall the bold advertisements about "beginning with the genes and the mechanisms that the genes actually prescribe" [Lumsden and Wilson 1981, 349]). Finally, there is every reason to think that the particular way in which Lumsden and Wilson choose to analyze the social expression of individual propensities—namely, in terms of response to the *frequency* of those around who advocate departure—will prove inaccurate, quite possibly wildly inaccurate. In sum, we have an implausible solution for a problem about the social expression of human preferences, in a case where there are no detailed results that could be used to distinguish the solution from the most elementary qualitative analysis of the situation.

One last example beckons us on. Lumsden and Wilson propose to understand changes in women's formal dress, using the data of Richardson and Kroeber (1940), "who took measurements from European and American paintings and fashion magazines from 1605 on" (Lumsden and Wilson 1981, 170).Waive any concerns about whether this is a reliable way of discovering the main features of formal dress in earlier centuries, or whether it is fruitful to think of a single category of "formal dress," restricted in earlier centuries to the gentry and becoming more widespread in our own times. Let us accept the findings as a revelation of what women have worn. Lumsden and Wilson reproduce the Richardson-Kroeber data (see figure 1), and they describe the results as follows:

> The principal features were claimed to oscillate back and forth through periods lasting about 100 years. During each century, for example, waist height went from high, near the bustline, to short, hugging the hips, and then back up to high again. Similar excursions occurred in dress length and decolletage. There appears to be an ideal although largely unappreciated pattern within these fluctuations toward which fashion gravitates: a long full skirt, the waist as slender as possible and in its true anatomical position, and much of the shoulders, arms, and upper portions of the bosom exposed. (170)

Lumsden and Wilson concentrate on two style culturgens: c_1 is "high waist" and c_2 is "low waist." They suppose that the switching probabilities v_{12} and v_{21} are functions of ξ (the scaling variable: recall ξ is $n_2/N - n_1/N$) and t. However, these probabilities change sufficiently

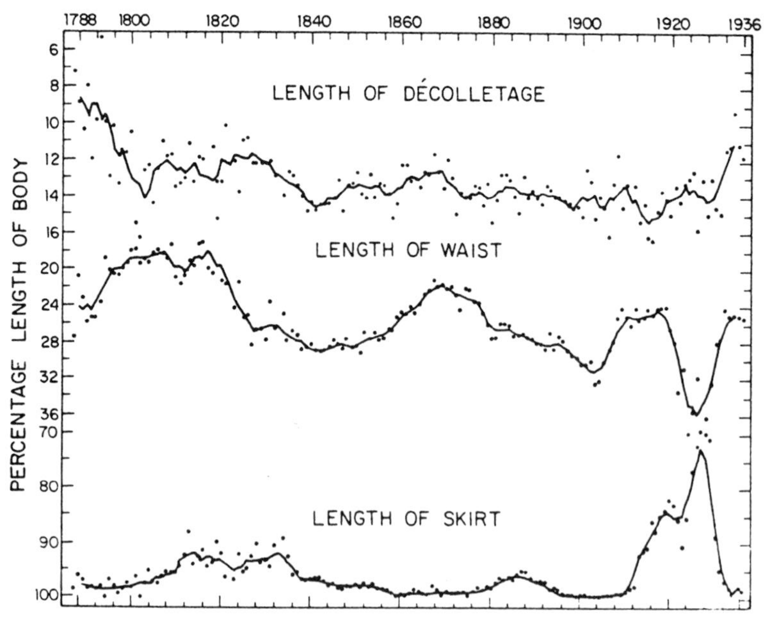

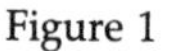

Figure 1
Variation in three measures of style in female formal dress between 1788 and 1936. (From Lumsden and Wilson 1983a, who adapt a figure of Richardson and Kroeber 1940)

slowly with time, so that the "group is at or close to the steady state P(ξ) that would apply to the current $v_{ij}(\xi,t)$" (173). High-speed communication is supposed to make this possible. (In 1800?) Plainly, much depends on the way the propensities to choose a high waist or a low waist change with time. It is not easy to appreciate the periodicity from the data that Lumsden and Wilson reproduce, since the period is 100 years and the data cover a little less than 150 years (1788–1936; the choice of this period reflects the understandable reluctance of Richardson and Kroeber to put great weight on the findings about earlier decades, for some of which they have no data).

There are other wrinkles. One of the minima in the "cycles" in skirt length, achieved in 1886, is greater than the maximum attained in a previous cycle. (In 1886 skirt length was 96 percent of body length; in 1823 skirt length was 95 percent of body length; the former is a "minimum," the latter a "maximum"; see Richardson and Kroeber 1940, 131.) We might also find it hard to recover all the features of Lumsden and Wilson's ideal pattern from the evidence: the idea of a *slender* waist may represent the influence of *Gone with the Wind* rather than the actual measurements. However, Richardson and Kroeber advance a hypothesis about changes in fashion that does propose the existence of such an ideal. They suggest that there is "a basic or ideal pattern, for Europe of the last two or three centuries," but that social disturbances (wars, revolutions, and so forth) "dislocate or invert" this pattern (1940, 149–150). Of course, this hypothesis, with its explicit disavowal of individual "psychological factors" (150), is quite at odds with Lumsden and Wilson's proposal. Hence Lumsden and Wilson can hardly appeal to Richardson and Kroeber's study to support their suggestions about ideals of female fashion.

Let us not quibble. Let us ignore all the peculiar annual fluctuations, the sudden rise of the short skirt in the 1920s, the relative constancy of decolletage length throughout a long period. We accept not only the data but also Lumsden and Wilson's interpretation of it. What insight will their analysis bring? Lumsden and Wilson introduce their hypothesis in a way that is surely unparalleled in the annals of the scientific study of fashion:

> Although the adiabatic approximation can be used on $v_{ij}(\xi,t)$ of general form, to accord with the basic Richardson-Kroeber model we make the assimilation functions cyclic processes $v_{ij} = v_{ij}(t) \in \mathbf{R}$ with period 100 years. A generalization of the model to incorporate complex ξ-dependence, overtones, and randomness is feasible (see for example Wang and Uehlenbeck, 1945), but it will require careful attention to the statistical properties of the time-

series data as well as to the specific mechanisms of prestige competition. (173)

We now come to the crunch. The ethnographic curve turns out to have "the time-dependent structure"

$$P(n_1,t) = \binom{N}{n} \rho(t)^{n_1} [1-\rho(t)]^{N-n_1},$$

where $\rho(t)$ is $v_{21}(t)/[v_{12}(t)+v_{21}(t)]$. Although they do not give the details, Lumsden and Wilson could easily go on to account for the periodicity of the Richardson-Kroeber data in terms of their model. Suppose that we assign c_1 a value $+d$ (the distance of "high waist" above the natural waist) and c_2 the value $-d$. Then the mean value of waist height at time t is

$$d[v_{21}(t)-v_{12}(t)]/[v_{12}(t)+v_{21}(t)],$$

since the expected value of n_1/N at time t is just $\rho(t)$, and the expected value of n_2/N at t is $1 - \rho(t)$. Because the periodic functions v_{12} and v_{21} can be chosen in a suitable way, it is possible to achieve a (rough) fit to the periodicity displayed (at least, displayed to Lumsden and Wilson) in the Richardson-Kroeber data.

Before we collapse in a frenzy of self-congratulation, it is well to take a close look at Lumsden and Wilson's "steady-state solution." Isn't it our old friend, the binomial distribution? Of course. So it should be. For we have, once again, given up the idea that the decisions of others exert an influence (in fashion, no less!). The solution is generated in a moment by realizing that the women are being treated as independent decision makers and that the expression $\rho(t)$ is just the probability that a woman will choose "high waist" at time t. Because they are compelled to take *transition* probabilities as primitive, Lumsden and Wilson have to express this probability in terms of the v_{ij}. As we saw above, in connection with their "simplifying approximation" to the problem of gene-culture coevolution (see the identification of $\mathcal{I}^{ij}$ in technical discussion M), the expression is easy. However, if we wanted to make life really easy for ourselves, we could achieve Lumsden and Wilson's results as follows. Let $p(t)$ be the probability that, at time t, a woman will choose c_1. Then, in a society of size N, assumed constant over time, the expected distribution for choice of c_1 at time t is given by

$$P(n_1,t) = \binom{N}{n} p(t)^{n_1}[1-p(t)]^{N-n_1}.$$

Mean waist height at time t is

$$p(t)d - d[1-p(t)] = d[2p(t)-1].$$

Choosing p to be a periodic function of t, we can obtain a periodic variation in mean waist height, if that is what we are out to show.

So behind Lumsden and Wilson's references to adiabatic approximations lurks an elementary idea. Notice that the model gives absolutely no insight into the reasons behind the periodicity. It is simply imposed from without. There is no attempt to relate the mathematics to the dynamics of innovation on the part of couturiers or enterprising women, about which Lumsden and Wilson offer some sketchy remarks. When the model is stripped of its obfuscatory trimmings, it is seen as extremely naive.

The first point to recognize is that, as usual, we are not discussing a process of gene-culture translation but a relation between individual preferences and a social pattern of behavior. Second, the choice of a periodic function for the probability bears no relation to any psychological mechanisms. But there are new failures in the example.

Lumsden and Wilson make the implausible assumption that women choose independently. They suppose that the mean cultural waist height is determined by the proportion of women who choose one of two extremes. Most surprisingly of all, they apply the model of gene-culture translation to data on *mean* values of waist length, skirt length, and so forth. If that model can do anything, then it can account for patterns of *variation within* a society. Intuitively, the model is an account of how individual propensities are reflected in the frequencies of use of different culturgens found in a society. Hence, to obtain a success for it, one must consider variations in choice of dress and show how decisions by some women affect those of others.

To appreciate the full intellectual poverty of the analysis offered by Lumsden and Wilson, let me offer a rival model that sheds exactly as much (or as little) light on the Richardson-Kroeber data. Suppose that there is a woman—call her "Queen B"—who leads fashion. Each year Queen B determines waist height. Everybody else imitates Queen B. So the mean waist height in the society is exactly that determined by Queen B. Oddly enough, variation in waist height turns out to be periodic, with a period of 100 years. (Let me hasten to point out that when Queen B loses her eminence her office is taken over by another.) Please do not complain that the hypothesis of periodicity is unmotivated. It is no more unmotivated than Lumsden and Wilson's conclusions about the periodicity in the switching probabilities for every woman. Please do not suggest that it is implausible to think that everybody will do exactly as Queen B does. Is that any more implausible than to think of a society in which women opt either for high waists or for low waists, and for nothing in between? Anyway, if you want variance, we can easily build that in. Simply suppose that the members of the society arrange themselves in a binomial distribution about Queen B's mean.

Obviously, any serious discussion of the relation between individual choices and changes in fashion will have to take into account a host of factors. Styles are presumably connected with ideals of beauty and, as these change, can be expected to change with them. In a similar way fashion has been constrained by the prevailing ideas about freedom of sexual expression, and, as these too alter, there is an effect on what is viewed as acceptable style in dress (even in formal dress). Moreover, the dynamics of fashion leadership and of the ways in which women accommodate their dress both to their own conceptions of themselves and to their conceptions of others' views of them are tasks for psychological investigation. Lumsden and Wilson dress up a trivial solution to a poorly posed problem and pretend that they have done something useful with their theory of gene-culture translation. They have not.

The three examples I have reviewed are the only cases in which the formal machinery assembled by Lumsden and Wilson is put to work on actual data about human behavior. In all three of the cases the formal tools yield nothing that could not have been obtained by applying the ordinary theory of probability to some hypotheses about human preferences and computing the pattern of behavior that is likely to result. In the first and last examples, the unnecessary mathematical apparatus actually stands in the way of offering better analyses of the situation. In the second example, Lumsden and Wilson manage only to provide a conclusion that is indistinguishable from results that we can reach by qualitative argument, given the data available. Moreover, the analysis they offer rests on assumptions about the society under study that we know to be radically incorrect. Similarly, in the last example, not only is Lumsden and Wilson's analysis ad hoc and directed at a problem for which their theory is not designed, but their simplifying assumptions are also so obviously defective that one can only wonder why they bothered to handle the example at all. Finally, we should recognize clearly that no link is ever forged between hypothetical genes for specific forms of human behavior and the resulting behavioral dispositions, that the analysis of decision rests on claims about the responses in probability of culturgen choice to frequencies of culturgen use in the surrounding society—claims that are uninformed by any insights from contemporary psychology—and that the end result is not the study of social institutions but the description of patterns of behavior.

Touring the Coevolutionary Circuit

We come to the last area in which Lumsden and Wilson's account might bring us illumination. If their theory is to be worth having, it must be able to provide some insight into the coevolution of genes and culture. We must be able to take the general coevolutionary equations, elaborate them with hypotheses, and so trace the modification of cultural patterns and human gene frequencies. Here we face disappointment, however. Lumsden and Wilson are not in any position to display a human population evolving in a way that would conform to their equations. They cannot offer even the kinds of empirical findings that were presented in the case of translation. The best they can do is advance some claims about the bearing costs, the transition costs, and the fertility map, and argue on this basis that certain types of coevolutionary scenarios are to be expected. Because their analysis leads them to a few surprising results, it is important to understand the assumptions that are fed into it.

Lumsden and Wilson announce five main conclusions:

(A) Pure tabula rasa is an unlikely state. That is, selection will eventually lead to a bias in favor of switching to a culturgen whose use promotes fitness.
(B) Sensitivity to usage patterns increases the rate of genetic assimilation. In a population whose members are sensitive to the patterns of culturgen use of their fellows, an epigenetic rule biased toward a fitness-promoting culturgen will spread more quickly.
(C) Culture slows the rate of genetic evolution.
(D) Changes during the coevolutionary process can nevertheless be rapid.
(E) Gene-culture coevolution can promote genetic diversity.

(For Lumsden and Wilson's formulations and explanations, see 1981, 290–297.)

In a penetrating review, John Maynard Smith and N. Warren have analyzed these conclusions and the arguments for them (1982). My own treatment will parallel theirs in many respects, although I hope that it will show even more clearly the bankruptcy of the Lumsden-Wilson enterprise.

Let us begin with (A). There is no need for any complicated coevolutionary theory to derive this result (a fact that is signaled by the presence of an argument for it in Lumsden and Wilson's introductory chapter). Suppose that choosing c_1 instead of c_2 promotes fitness. Then a person who has a higher probability of choosing c_1 will have higher fitness (other things being equal). Hence, if *AA* individuals are indifferent with respect to c_1 and c_2, while *Aa* and *aa* individuals have a propensity to choose c_1 more often than c_2, then (other things being equal) the allele *a* will be favored by selection. This is the argument for (A) that Lumsden and Wilson dress up in their complicated terminology. We have no need of the theory to obtain the conclusion, and the theory gives us no more insight into the conclusion than we had already. It simply restates a simple argument in a complicated way.

As Maynard Smith and Warren point out, principle (B) is initially counterintuitive. If we consider a strictly deterministic model, in which *AA* individuals always choose c_1, *Aa* individuals are indifferent between the two culturgens, and *aa* individuals always choose c_2, then, if c_2 usage increases fitness relative to c_1 usage, selection will favor the *a* allele. Indeed, the course of selection under a situation in which there are no cultural effects on culturgen choice will be faster than that in cases in which people adjust by imitating the majority. (The one way in which culture could accelerate genetic change in this case is if the *Aa* individuals acquired some propensity to use c_2 more frequently as c_1 use increased in the population.)

Lumsden and Wilson arrive at their principle by formalizing the intuitive idea that people update their "raw" biases by following trends, so that they respond to the pattern of use in the society around them. This elaboration should only increase our perplexity. For if people watch trends, then those with "raw" biases toward using inferior culturgens will apparently be protected from their own inclinations by the superior choices of their fellows. As the successful culturgen becomes more widely used, those who are genetically predisposed to use the inferior one will watch the trend and will imitate the trendy. In this way their shortcomings will be hidden from the keen eye of selection, and they will continue to pass on their genes. Trend watching should slow down the rate of genetic change.

At this point the reward equation arrives to save the day. Lumsden and Wilson offer an accounting of the costs that figure in their reward equation. The fundamental idea is that there is a "cost-free tabula rasa state." In this state people (or other animals) have no "raw" bias toward either culturgen, no response to peer use, and no response to parental use. The bearing costs and the transition costs are supposed to increase as a person (or animal) departs further from the tabula rasa state. This accounting procedure seems to embody two odd ideas. First, any kind of biasing—even "raw" biases—imposes costs. Second, as Douglas Futuyma pointed out to me, even animals incapable of learning have brains, and it is far from clear why there should be any direct connection between the *use* of brains and costs to the animal. (For more details on the accounting procedure and its difficulties, see technical discussion N.)

Lumsden and Wilson conclude their filling in of the relevant parameters by supposing that c_1 is the more efficient of the two culturgens and that it is five times as efficient as c_2. They also offer a specification of the fertility function (see technical discussion N).

Consider now Lumsden and Wilson's own statement of their principle (B):

> Suppose that two culturgens are adopted by a species for the first time. One culturgen provides higher fitness than the other, but at first there is no clear preference between them because no biasing rule exists in the cognition of the species members. In time new genotypes appear by mutation or immigration and direct cognition in favor of the more efficient culturgen; the constituent alleles now compete with the older, tabula rasa alleles. In species where individuals are already sensitive to the usage patterns of other members of their society, the rate of replacement is increased. In other words, genetic assimilation of the favored culturgen proceeds more quickly. (290)

Let *aa* be the tabula rasa genotype, and suppose that *AA* and *Aa* have "raw" biases, $v_{21,o} = 0.6$, $v_{12,o} = 0.4$. Lumsden and Wilson's basic claim amounts to the following: Consider the ratio of F^{AA} to F^{aa} under two sets of conditions, first where there is no response to patterns of behavior in the surrounding culture, and second where people watch trends; the ratio will be greater in the latter case; this means that where people respond to the practices of their fellows, the relative fitness of the *AA* genotype is increased and the course of evolution under selection goes more quickly.

It is easy to lose sight of what is going on. Lumsden and Wilson use the term "tabula rasa" in two different ways. The genotype *aa* is a tabula rasa genotype only in the sense of being indifferent between the two culturgens. In the case where people watch trends, all people depart from the tabula rasa *state* and incur the costs associated with the development of equipment for assessing the surrounding behavior. The opportunities for confusion are evident when we understand that it is possible to formulate the intended conclusion as follows: Tabula rasa *genotypes* will be replaced more quickly when all the members of the population depart from the tabula rasa *state* by imitating the behavior of those around them.

Does the principle follow from the Lumsden-Wilson theory? It depends on the reward equation that is used. When we apply the

Technical Discussion N

Lumsden and Wilson suppose that the crucial functions v_{ij}^{AA} can be written as

$$v_{ij}^{AA} = v_{ij,o}^{AA} \exp \{-\alpha_{ij}^{AA}[\beta_{ij}^{AA}\xi_t + (1 - \beta_{ij}^{AA})\xi_t^p]\}.$$

Here $v_{ij,o}^{AA}$ is a constant, representing the "raw" bias in favor of switching from c_i to c_j for a person with genotype AA; ξ_t represents the frequency of c_1 usage in the peer generation (recall: $\xi = n_2/N - n_1/N$); ξ_t^p represents the frequency of c_1 usage in the parental generation. The α_{ij}^{AA} and β_{ij}^{AA} are constants; α reflects the tendency to follow trends and β embodies the disposition to follow peers rather than elders.

According to Lumsden and Wilson, the "cost-free tabula rasa state" is one in which $v_{12,o} = v_{21,o} = 0.5$, $\alpha = 0$, and $\beta = 1$. The "costs" involved in assessing and responding to the behavior of others are

$$L = e^d - 1\,;\, C = e^{0.1d} - 1,$$

where d is the "distance" of the genotype from the tabula rasa state. Thus

$$d^{AA} = [(v_{12,o}^{AA} - 0.5)^2 + (v_{21,o}^{AA} - 0.5)^2 + (\alpha_{12}^{AA})^2 + (\alpha_{21}^{AA})^2 + (1-\beta_{12}^{AA})^2 + (1 - \beta_{21}^{AA})^2]^{1/2}.$$

Unfortunately, this definition will not do what Lumsden and Wilson intend. If we consider a case in which $v_{12,o}^{AA} = 0.4$, $v_{21,o}^{AA} = 0.6$, $v_{12,o}^{aa} = v_{21,o}^{aa} = 0.5$, all the α's and β's are 0 and 1 respectively, and the harvesting rates are as in Lumsden and Wilson's preferred example (see below), the reward to AA is substantially less than the reward to aa (4 as opposed to 6). The imposition of costs on "raw" biases tells against the idea that departures from tabula rasa will always be favored by selection. Using the definitions actually given, we discover that (B) is false because the costs may be so high that tabula rasa genotypes are not even replaced!

So we have to amend the account. In the spirit of Lumsden and Wilson's attempt to motivate the costs, we can assume that the abilities to respond to parents and peers are crucial. Let us suppose that

$$L = e^d - 1\,;\, C = e^{0.1d} - 1.$$
$$d^{AA} = [(\alpha_{12}^{AA})^2 + (\alpha_{21}^{AA})^2 + (1-\beta_{12}^{AA})^2 + (1 - \beta_{21}^{AA})^2]^{1/2}.$$

(I shall waive obvious concerns about whether this sort of accounting is appropriate. It is a straightforward amendment of the Lumsden-Wilson proposal, advanced only in the interests of understanding their reasoning.)

When neither AA nor aa individuals are sensitive to surrounding usage, all the α's and β's are 0 and 1, respectively.

$$\begin{aligned} &\text{Let } v_{12,o}^{AA} = 0.4,\ v_{21,o}^{AA} = 0.6,\ v_{12,o}^{aa} = v_{21,o}^{aa} = 0.5, \\ &J_1 = 1, \quad J_2 = 0.2, \\ &T = 10, \quad \tau = 0.1. \end{aligned}$$

(c_1 is five times as efficient as c_2; the prereproductive period is ten units long; decision points arise ten times per unit. These are Lumsden and Wilson's specifications of the parameters.) Since $d^{AA} = d^{aa} = 0$, there are no costs and all versions of the reward equation give the same result:

$$R^{ij}(T) = \mathcal{G}_1^{ij} J_1 T + \mathcal{G}_2^{ij} J_2 T.$$

Substituting, we have

$$\begin{aligned} R^{AA} &= 6.8, \\ R^{aa} &= 6. \end{aligned}$$

Rewards are converted into fitness as follows:

$$F^{ij} = F_{\max}\,[1 - \exp\,(-b^{ij}\,R^{ij})],$$

where b^{ij} is 0.1 for all genotypes. The critical ratio F^{AA}/F^{aa} in the case of no sensitivity to surrounding usage is therefore $(1-e^{-0.68})/(1-e^{-0.6})$, which is about 1.09.

One obvious feature of the computation is that if R^{AA} and R^{aa} are both small, then $\exp(-b^{ij}\,R^{ij})$ will be approximately $1 - 0.1\,R^{ij}$. Hence the critical ratio will be approximated by R^{AA}/R^{aa}. As we shall see, when people adjust their behavior to the surrounding culture, Lumsden and Wilson's preferred reward equations enable them to decrease the values of R^{AA} and R^{aa} so that this approximation becomes useful.

In the situation where there is sensitivity to surrounding usage, all the parameters α,β are set at 0.2 and 1 respectively. (Note that the dependence on usage in the parental generation never figures in the accounting.) Under these circumstances the value of d for both genotypes is $\sqrt{0.08}$, which is about 0.28. The bearing cost L is $e^{0.28}-1$, which is about 0.32; the transition costs C are $e^{0.028}-1$, about 0.028. We are to insert these values in the reward equation. We can consider the three versions distinguished in technical discussion M:

$$R^{ij} = \mathcal{P}_1^{ij} J_1 T + \mathcal{P}_2^{ij} J_2 T \text{ (naive);} \quad (8)$$

$$R^{ij} = \mathcal{P}_1^{ij} (J_1 - L^{ij}) T + \mathcal{P}_2^{ij} (J_2 - L^{ij}) T - C^{ij} T/\tau$$
$$\text{(Lumsden-Wilson; sign corrected);} \quad (9)$$

$$R^{ij} = \mathcal{P}_1^{ij} (J_1 - L^{ij}) T + \mathcal{P}_2^{ij} (J_2 - L^{ij}) T$$
$$- C^{ij} \cdot T \cdot 2v_{12}^{ij} v_{21}^{ij} / \tau(v_{12}^{ij} + v_{21}^{ij})$$
$$\text{(Lumsden and Wilson; refined).} \quad (9A)$$

It is now possible to appreciate how the rewards are so dramatically diminished. Given that c_2 is a relatively inefficient culturgen, J_2 is less than L (0.2 as opposed to 0.32), and the second term, in (9) and (9A), is negative. C is small, but the ratio T/τ (which is the number of decision points) is large enough to compensate. When we plug in all the values, we obtain

$$R^{ij} = \mathcal{P}_1^{ij} (6.8) - \mathcal{P}_2^{ij} (1.2) - 2.8; \quad (9)$$

$$R^{ij} = \mathcal{P}_1^{ij} (6.8) - \mathcal{P}_2^{ij} (1.2) - 2.8 \, [2v_{12}^{ij} v_{21}^{ij} / (v_{12}^{ij} + v_{21}^{ij})]. \quad (9A)$$

In fact, the equations are almost too perfect to be believed. Recall that $\mathcal{P}_1^{aa}$ is the probability that an *aa* individual will choose c_1. Since the population is composed of people who are indifferent between c_1 and c_2 and people who have a fairly small bias toward c_1, $\mathcal{P}_1^{aa}$ is always likely to be a little greater than 0.5. Notice that if $\mathcal{P}_1^{aa}$ is *exactly* 0.5, the value of R^{aa} is *exactly* 0. This means that as A alleles enter a population of *aa* individuals, the value of R^{aa} is extremely small and the ratio F^{AA}/F^{aa} is very large. (In a population consisting entirely of *aa* individuals, R^{aa} is 0; consequently F^{aa} is 0; presumably this is the end of the story, since nobody produces any gametes; I am grateful to Naomi Scheman for helping me to see this point.)

The numerical example in the text illustrates the point. If A is present at low frequency ($p = 0.1$), then the values of the probabilities are approximated by

$$\mathcal{P}_1^{AA} = 0.606, \ \mathcal{P}_2^{AA} = 0.394$$
$$\mathcal{P}_1^{aa} = 0.505, \ \mathcal{P}_2^{aa} = 0.495,$$
$$v_{12}^{AA} v_{21}^{AA} / (v_{12}^{AA} + v_{21}^{AA}) = 0.48,$$
$$v_{12}^{aa} v_{21}^{aa} / (v_{12}^{aa} + v_{21}^{aa}) = 0.5.$$

Substituting these values, we obtain the results about the F^{AA}/F^{aa} ratio presented in the text.

specifications of the parameters to a situation in which A is present at low frequencies ($p = 0.1$), then we obtain the following values for the rewards, depending on our preferred equation (see technical discussion N):

	Naive equation	Lumsden-Wilson (sign-corrected)	Lumsden-Wilson (refined)
R^{AA}	6.848	0.85	2.30
R^{aa}	6.040	0.04	1.50
F^{AA}/F^{aa}	1.090	20.40	1.53

The variation in the values of the crucial ratio is striking. Even without much analysis it is plain that the imposition of costs can radically alter the expected rewards; by doing so, it can make the ratio F^{AA}/F^{aa} in situations where people respond to surrounding behavior as high as we like. (The trick is to make the costs so large that the value of R^{aa} is barely positive.) If we use the naive equation, however, the value of the ratio is almost exactly the same as when there is no response to surrounding behavior. (In both instances the value is 1.09.) A more detailed analysis would show that when the naive equation is employed, we achieve the intuitive result that the ratio is reduced when the A allele is prevalent and when people respond to surrounding behavior. Hence, unless we have some reason to abandon the naive reward equation and to count the costs that Lumsden and Wilson invoke, there is no reason to maintain principle (B). We can rest content with our pretheoretical view that attention to surrounding behavior would slow the course of selection.

We can now give a more perspicuous statement of the principle, outline the argument offered for it, and diagnose what has gone wrong. First let us reformulate the intended conclusion:

> (B′) Let G_1 and G_2 be two groups of human beings. Suppose that the people in G_1 do not have the cognitive equipment required for adjusting their behavior to the behavior of those around them; suppose that those in G_2 do have this equipment. Genotypes that promote indifference between culturgens will be replaced by genotypes biased toward superior culturgens in both groups, and the replacement will be faster in G_2 than in G_1.

Let us write F_C^{ij} and F_N^{ij} for the values of the fertility function for a genotype ij in cases where there is cultural assimilation (adjustment of behavior in response to surrounding behavior) and in cases where there is not, respectively. We can view Lumsden and Wilson as deriving (B′) from

$$(\text{B}'') \quad F_C^{AA} \,/\, F_C^{aa} > F_N^{AA} \,/\, F_N^{aa}.$$

In effect, they can establish this inequality, because, given their particular reward equation, (B″) takes the form

$$(r_C^{AA} - k^{AA}) \,/\, (r_C^{aa} - k^{aa}) > r_N^{AA} \,/\, r_N^{aa},$$

where k^{AA} is approximately k^{aa} (the costs to both genotypes are about equal) and r_C^{ij} is supposed to be about the same as r_N^{ij} (the rewards in both situations are about equal). So the entire computation turns on a trivial arithmetical inequality, namely,

$$(r_1 - k) \,/\, (r_2 - k) > r_1/r_2.$$

This inequality always holds if $k > 0$ and $r_1 > r_2$. This is the deep theoretical result underlying Lumsden and Wilson's counterintuitive result (B′).

Maynard Smith and Warren note the point, and chide Lumsden and Wilson for burying a rather trivial arithmetical result under pages of complicated equations. While my own struggles at fathoming Lumsden and Wilson's argument lead me to sympathize with Maynard Smith and Warren, I think it is worth noting that Lumsden and Wilson have an obvious line of response. After all, their theory of gene-culture coevolution has linked an arithmetical truth to a rather surprising conclusion about evolution. Before we can dismiss the theory, we have to assess the hypotheses that were used to make the connection.

The rabbit emerges from the hat because of the way the reward equation is structured and the way values are assigned. At first sight the idea that there are costs involved in developing the cognitive equipment required for the assessment of surrounding behavior appears quite harmless. Resources that might have been used elsewhere have to be channeled into building a brain. When we see the sizes that Lumsden and Wilson assign to the costs, however, and when we see how dramatically the costs cut down the rewards, then it is natural to worry. After all, if the costs are this large, how would it ever have proved profitable for us to develop our cognitive equipment? As we shall see, there are two ways to make this worry precise.

Lumsden and Wilson motivate their accounting in the following passage:

> By trial and error it was found that the values . . . give realistic behavior over the range of the epigenetic rules we wish to explore. They result in bearing and transition costs per unit time on the order of 1 to 10 percent of the reward rate. These loads approximate the size of the drain that would be put on a total

energy budget by an organism with a ratio of brain weight to body wieght in the primate range. For every generation $\tau = 0.1$ and $T = 10$ time units. (289)

This is *all* the motivation they provide. Notice that the value of τ is chosen so that the person makes 100 decisions about which culturgen to use during the prereproductive period—a choice that ensures that the values of the rewards will decrease, thereby increasing the crucial ratio—and that no justification whatever is given for this choice. However, let us ignore this feature of the accounting, and let us grant to Lumsden and Wilson the point about brain/body ratios, even though it is far from evident just how these affect the bearing and transition costs. The most glaring feature of the situation is that the *entire* costs of building a brain are being levied with respect to a choice involving a *pair* of culturgens. (Once again, we should also remember that animals that do not make any choices still have brains.)

Small wonder that the rewards show a dramatic decrease when we subtract from them the full costs of developing our cognitive equipment and of keeping it in service! The costs have been inflated, and the inflation gives us the arithmetic inequality, which in turn leads to the desired conclusion. The first way of making the worry precise is to imagine that the comparison is between groups G_1 and G_2 with the following properties: G_1 members can assess the behavior of others and adjust their actions in all contexts in which G_2 members can do this, except those contexts that involve choice between c_1 and c_2. Intuitively, G_1 members have a little less cognitive equipment than G_2 members, but they are otherwise the same. If this is so, then we do have a basis for saying that the gross returns to both groups will be approximately the same. The differences between their harvesting of resources will be relatively small. However, we expect that the differences in the costs incurred by G_1 members and by G_2 members will be small as well. For people in G_2 will have only the additional bearing and transition costs associated with a small increment in cognitive equipment. Under these circumstances the basic inequality underlying (B″) is

$$(r_{1C} - k)/(r_{2C} - k) > r_{1N}/r_{2N},$$

where r_{iC} is approximately the same as r_{iN} and k is very small. Evidently, the ratios flanking the inequality will be very close in value, and if the difference between r_{iC} and r_{iN} is bigger than k, then the inequality need not be satisfied. (More exactly, if $r_{iC} = r_{iN} + m$, where $m > k$, for $i = 1,2$, then the equality will not be satisfied.) What this means intuitively is that if the extra costs involved in developing and maintaining the equipment for assessing the behavior of others in connection with the choice between c_1 and c_2 are smaller than the increase in returns from developing and maintaining the equipment, then Lumsden and Wilson will not achieve the result they want. But if the costs outweigh the increase in return, then it appears that the extra cognitive ability will not be favored by selection in either group. Hence it appears that the Lumsden-Wilson analysis is thoroughly mistaken.

The second way of making the worry precise is to suppose that the people in G_1 are completely bereft of cognitive equipment. They do not have the apparatus needed to assess the behavior of those around them—and to do all the other things that we are able to achieve with our brains. Under these circumstances we can hardly suppose that the difference between their harvesting of resources in the prereproductive period and the harvesting of the G_2 members turns simply on the choice of c_1 rather than c_2. The effects of the other culturgens that G_2 members are able to employ have to be taken into account. Now the inequality underlying (B″) is

$$(r_{1C} - k) / (r_{2C} - k) > r_{1N} / r_{2N},$$

where r_{iC} is expected to be very different (very much larger) than r_{iN}. Once again, if the difference between r_{iC} and r_{iN} is not sufficiently large to outweigh the cost k, then we shall expect that selection will act against the development of cognitive abilities. On the other hand, if this difference is big enough to outweigh the cost, then the inequality will fail (certainly there is no reason to think that it will hold, and if $r_{iC} - r_{iN} - k$ takes on the same value for both values of i, it will fail). Either way, principle (B′) will be defeated.

The irony of the situation is that when the reward equation is carefully analyzed, we find no good reason for departing from the naive equation. All Lumsden and Wilson's theoretical machinery goes for naught. That machinery obscures the workings of a very simple trick, and a sober look at the hypotheses that stand behind the trick shows that they are completely misguided. Lumsden and Wilson wax eloquent about the "autocatalytic quality of gene-culture coevolution in the human species, which led to an extraordinarily rapid evolutionary increase in brain size" (290), and they seem to think that their principle (B) sheds some light on the autocatalytic mechanism. However, all this is so much whistling in the dark, dark induced by proliferation of equations and careful number juggling. Principle (B), the only principle that distinguishes Lumsden and Wilson's theory from the simple story of gene-culture coevolution with which we began, turns out to be based on numerous errors.

As I have already noted, Principle (B) looks as though it contradicts Principle (C). However, a full statement of (C), together with our analysis of (B) as (B′), dissolves the mystery. In (B) we are concerned with comparisons between two groups with differing cognitive abilities for assessing the behavior of others and responding to the pattern of behavior. The idea behind (C) is that if we take two groups with the *same* cognitive abilities, then, in groups that contain individuals who have a tendency to follow fashion, alleles biased toward inferior culturgens will be replaced at a slower rate than in groups that do not contain such people. More exactly:

> (C′) If people in G_1 adjust their propensities to switch to c_1, so that they switch to c_1 more often if c_1 is widely used, and if *aa* individuals have a smaller "raw" propensity to switch to c_1 than do *AA* individuals, then, other things being equal, *A* will replace *a* in G_1. If G_2 is like G_1 except that people in G_2 do not follow any trend, then *A* will replace *a* more quickly in G_2 than in G_1.

In this form the principle amounts to the claim that imitation enables cognitively sophisticated hominids to slow down the course of selection. However, there is no need for an elaborate theory to obtain it. We can derive the thesis by noting the argument rehearsed earlier: as *A* genes spread, people who are *aa* acquire increased protection from selection if they have a tendency to imitate those around them. Hence the trendy manage to preserve inferior genotypes.

We have a straightforward and intuitive result, derivable by simply using a general picture of gene-culture coevolution, such as the one offered earlier in this chapter. However, Lumsden and Wilson obscure the commonplace character of the result by linking it to something different: "*Any* propensity to acquire the more successful culturgen, in other words any value of u_{21} above zero, will slow the rate of replacement of the prescribing genes below what would be the case if $u_{21} = 0$" (295). This statement is also true, and is readily derivable by an intuitive argument. Suppose that *AA* individuals choose c_1 with some lower probability and that *bb* individuals choose c_1 with probability 0. In a population in which *A* and *a* are present, *A* will replace *a* under selection; in a population in which *A* and *b* are present, *A* will replace *b* under selection, and the process will go faster here, because the relative fitness of *bb* people is lower than that of *aa* people (even a little use of c_1 helps). This result is very obvious, given elementary population genetics. However, it only has significance for gene-*culture* coevolution if the difference in the propensity of *aa* individuals to use c_1 results from the response to surrounding behavior (or cultural institutions). Suppose that the propensity of *aa* individuals is "raw" and that it is unaltered by response to surrounding behavior. Then the replacement of *a* by *A* would still be slower than the replacement of *b* by *A*, but it would be folly to proclaim this as another case in which *culture* slows the rate of genetic evolution.

Of course, the ordinary conclusion that culture slows the rate of genetic evolution is not one that Lumsden and Wilson like very much. So they continue by announcing a "thousand-year rule." Populations able to exploit "highly efficient new culturgens" (295) are able to evolve so that genes predisposing people to use the superior culturgens can become prevalent in about 50 generations (approximately 1,000 years). Lumsden and Wilson take pains to stress the fact that "the estimate is by order of magnitude" (295). Thus they arrive at a second relatively provocative result, (D). Again, it is worth exploring to see how the principle is derived.

Lumsden and Wilson support their claim by drawing some graphs that simulate evolutionary trajectories under various conditions. Let us focus on the case that appears to yield replacement most swiftly and is therefore the best from their point of view. In this example *AA* and *aa* individuals have "raw" biases as follows:

	$u_{12,o}$	$u_{21,o}$
AA	0.4	0.6
Aa	0.4	0.6
aa	0.6	0.4

Now Lumsden and Wilson have chosen the crucial parameter that determines the way in which transition rates are "updated" so that these rates always remain relatively close to the "raw" rates of transition (see technical discussion O). When we compute the values and substitute them in the naive reward equation, the relative fitness of *AA* to *aa* is of the order of 1.2 throughout the course of the evolution of the population. Hence it is no surprise that when we consult the standard equation of population genetics governing the elimination of a recessive allele under selection (see Roughgarden 1979, 33), we find that it takes about 70 to 100 generations to achieve a situation in which the frequency of *A* goes from less than 0.1 to more than 0.9. However, if we consider a somewhat less extreme example than that which Lumsden and Wilson provide, supposing that the superior culturgen is less than 5 times as efficient as the inferior culturgen, then the situation is quite different. Suppose that J_2 (the harvesting rate for the inferior culturgen) is 0.8 instead of 0.2. Then the fitness ratio falls to about 1.02. Under these circumstances the time required for elimination of the *a* allele is greatly extended—it takes more than 1,000 generations (that is, more than 20,000 years).

Technical Discussion O

Lumsden and Wilson suppose that

$$v_{ij}^{AA} = v_{ij,0}^{AA} \exp\{-\alpha_{ij}^{AA}[\beta_{ij}^{AA}\xi_t + (1-\beta_{ij}^{AA})\xi_t^P]\}.$$

(See technical discussion N.) Recall that the α's and β's are 0.2 and 1 respectively. Hence, $v_{ij}^{AA} = v_{ij,0}^{AA} \exp(-0.2\xi)$, where ξ lies between -1 and $+1$.

It follows that

$$v_{21}^{AA} \leq 6e^{0.2} \leq 7.33, \qquad v_{21}^{AA} \geq 6e^{-0.2} \geq 4.9,$$
$$v_{12}^{AA} \geq 4e^{-0.2} \geq 3.25, \qquad v_{12}^{AA} \leq 4e^{0.2} \leq 4.9.$$

Given the symmetry of the initial biases, it is easy to see that the bounds on the transition rates for *aa* people are similar.

To understand the fitness ratio through the process in which *a* is replaced by *A*, we should recognize that when *A* is rare, the relevant probabilities are

$$\mathcal{J}_1^{AA} = 0.5, \qquad \mathcal{J}_1^{aa} = 0.31.$$

Using the naive reward equation, we obtain

$$R^{AA} = 6,\ R^{aa} = 4.5,\ F^{AA}/F^{aa} = 1.25.$$

When *A* is common, the probabilities are

$$\mathcal{J}_1^{AA} = 0.69, \qquad \mathcal{J}_1^{aa} = 0.5.$$

Now,

$$R^{AA} = 7.5,\ R^{aa} = 6,\ F^{AA}/F^{aa} = 1.18.$$

Of course, if the efficiency of c_2 is closer to the efficiency of c_1, the difference between the values of R^{AA} and R^{aa} is decreased. Suppose that J_2 is altered from 0.2 to 0.8. Then, when *A* is rare, the values are $R^{AA} = 9$, $R^{aa} = 8.6$, $F^{AA}/F^{aa} = 1.02$.

When *A* is common, we have

$$R^{AA} = 9.38,\ R^{aa} = 9,\ F^{AA}/F^{aa} = 1.02.$$

However, Lumsden and Wilson obtain their result (D) not just by supposing that there is a great difference in the efficiency of the culturgens. If we look at the bounds on the transition rates (see technical discussion O), we can quickly convince ourselves that the probabilities of using a particular culturgen never stray very far from the initial values. Even when c_1 is rare, *AA* people use it more than half the time; when c_1 is common, *aa* people barely manage to use it half the time. Hence, even though the people whose behavior Lumsden and Wilson are attempting to understand respond to the pattern of behavior around them, they do not modify their actions very much. As a result, it is hardly surprising that the rate of evolution should not be slowed very much.

We have found that the thousand-year rule depends on two crucial assumptions. People are not supposed greatly to modify their propensities for using culturgens in the light of what they see around them, and the differences in success of different culturgens have to be rather large. If we combine the two assumptions and begin to think about a realistic situation in which they might be satisfied—for example, a situation in which two tools are available to a society for use in cultivation of crops—then we can find an easy way to understand what Lumsden and Wilson have accomplished. The thousand-year rule is a theorem of the theory of gene-culture coevolution, when it is applied to the evolution of hypothetical people of extraordinary stupidity. It is assumed that these people are able to change their probabilities of using a tool, so that they opt for a tool that is five times more efficient than its competitor, provided that this tool is becoming popular in their group. But they do not modify their behavior very much; those with a genetic propensity for the inferior tool continue to use it quite frequently and thus fall prey to selection.

While we may wonder whether principle (D) applies to any realistic situations in the evolutionary history of our species, our understanding of the way in which (D) was derived also enables us to see that the elaborate machinery of the theory was quite unnecessary. Consider any situation in which use of c_1 promotes far greater fitness than use of c_2. Suppose that the extent of imitation does not go so far as to make any significant alteration in the difference between the probability that *AA* people use c_1 and the probability that *aa* people use c_1. Then, while cultural imitation will slow the rate at which *a* alleles are replaced, *aa* individuals will always be sufficiently distinguishable from *AA* and *Aa* individuals to be visible to selection. So, by constraining the extent of imitation sufficiently and by making the differences in fitness sufficiently high, we ought to be able to make the difference between the rate of evolution with cultural dampening and the rate of evolution without cultural dampening as small as we like. It is simply a matter of juggling the numbers. *Any* way of developing our general, pretheoretical picture ought to yield the consequence that, with some carefully designed manipulation, we can arrange for replacement in 1,000 years.

The final result, (E), states that gene-culture coevolution can promote genetic diversity. (Lumsden and Wilson might ponder the fact that, among mammals, humans have relatively low genetic diversity.) To obtain this conclusion, Lumsden and Wilson have to resort to a modification of their fertility function. They suppose that individuals who harvest too much are subject to "fitness suppression." That is, people around them begin to object and interfere. In order to give an analysis of this process, Lumsden and Wilson propose that the fitness function takes the following form:

$$F^{ij} = F^{ij}\,[1-\exp(-b^{ij}R^{ij})] \qquad \text{for } R^{ij} \leq R_{max};$$
$$F^{ij} = F^{ij}\,\{\exp[-D(R^{ij} - R_{max})]\} \quad \text{for } R^{ij} > R_{max}.$$

By supposing that the value of R_{max} is right, Lumsden and Wilson can now arrange for frequency-dependent selection and the establishment of polymorphisms. The intuitive idea is that if the expected return for *AA* individuals (who tend to use the more efficient culturgen) lies beyond R_{max}, while the expected return for *aa* people lies below R_{max}, then, in a society dominated by *AA* individuals (and heavy use of c_1), *aa* people will be favored. As in the case of (D), the result depends on juggling the numbers. Not only must the value of R_{max} be chosen correctly, but there must not be too great a propensity for imitation.

In any case, it is extremely difficult to give a realistic interpretation of what Lumsden and Wilson propose. If we assume that the pressure against those who harvest too much results from the existence of a limit on resources, then the natural expectation is not that an inferior culturgen and a genotypic propensity for it will survive, but that people will adopt the superior culturgen and devote less of their time to harvesting. It would surely be ludicrous to suppose that selection will favor the retention of inferior tools because the use of superior implements is too efficient for the public to bear. If resources are limited, one would expect everybody to use the most efficient tools available and to compete for what they can get. Even if resources are not limited and we can envisage the peer pressure against greedy acquisition that Lumsden and Wilson hypothesize, it is hard to see why selection would not favor the use of efficient tools and the enjoyment of well-earned leisure.

So we reach the end of our evaluation. We are offered five conclusions. Two, (A) and (C), are easily obtained by simple arguments from the general, pretheoretical analysis of gene-culture coevolution. One, (B), is based on mistakes. Another, (D), is the result of some careful number juggling; its general possibility is apparent in our pretheoretical model. Finally, (E) is also the result of artifice, and in this case the contrivance is so strained as to defy any realistic interpretation of the model. Lumsden and Wilson offer us a specific way of developing a simple picture of gene-culture coevolution with which almost everyone will agree. We are entitled to ask what results and insights this particular proposal will bring. The answer is, None.

Moreover, the two principles, (B) and (D), that are initially surprising are just the kinds of claims that enable pop sociobiology to neglect culture and its history. Both are designed to show that culture makes little difference to the course of selection, that superior genotypes remain visible to selection, that rates of replacement continue to be rapid. When we recall the criticisms that provoked Lumsden and Wilson to offer the theory of gene-culture coevolution, we can begin to see the ends for which the numbers were juggled and the peculiar assumptions about costs of cognitive equipment were made. The numerical contortions are not accidental. The trickery has the effect of playing down the objections. Armed with (B) and (D), Lumsden and Wilson could reply that they have shown that incorporating the effects of culture makes little difference to the results of previous pop sociobiological analyses. It seems that culture has been introduced into pop sociobiology only that it may henceforth be ignored.

Before we conclude our tour of the coevolutionary circuit, it is important to make one point explicit. Lumsden and Wilson announce that they have integrated work in psychology with evolutionary biology and so achieved a genuinely new coevolutionary theory. Consider the following remark from their popular account:

> . . . we hoped to do better than previous scientists, who lacked any means of joining the evolution of genes and culture in a way that also takes into account the facts of psychology. Our aim was to find solid steps that lead from the world of psychology to the world of cultural anthropology. (1983a, 122)

Even some of their more astute readers (for example, Flanagan 1984) have taken Lumsden and Wilson at their word and have supposed that there has been genuine integration. But the review of psychological findings serves only one purpose—to underwrite the old saw about the fixity of epigenetic rules. The grand equations of the coevolutionary circuit owe nothing to any psychological insights into human decision making. Where they depart from the standard equations of population genetics, they do so either by offering curious ideas about fitness costs for cognitive capacities—ideas that turn out to be riddled with errors—or by adopting very neatly contrived claims about limited powers of assimilation and extreme differences in culturgen efficiency. Only those who are blinded by the symbols will be lulled into thinking that grand new insights have been obtained by a rare synthesis. Not only are there no insights. There is no synthesis either.

Abusing Mathematics

Lumsden and Wilson set out to answer the objection that pop sociobiology has ignored culture and its history. We have seen that their impoverished conception of culture does not touch the real concerns of the critics. The entire line of reply is also permeated by an extremely primitive understanding of the scientific enterprise. It is as if the authors had declared, "We can answer the detractors. If they want to stress the role of the human mind, we will give them minds. If they want to emphasize the impact of culture, we will give them culture. If they decry sociobiology as science, we will present it in a form that will silence all objections. Mathematics is the mark of good science, and we will give them mathematics. Indeed, we will give them more mathematics than they could ever ask for. We will give them more mathematics than they could ever use. We will give them more mathematics than they could ever understand."

Genes, Mind, and Culture is an extreme example of a certain type of work. Complex mathematics is employed to cover up very simple—often simplistic—ideas. In many instances beside the ones on which I have focused, Lumsden and Wilson occupy themselves by applying symbolism of mind-numbing obscurity to the solution of unimportant problems. For example, one appendix investigates the possibility of a computer program for modeling the initiation of young males among the Warao: the initiation involves smoking hallucinogens and reporting dream experiences. In another appendix the authors, obsessed with the desire to demolish the notion that human beings are tabulae rasae, compute the waiting time for departure from the state of indifference.

What is irritating, and occasionally amusing, about these uses of mathematics is that they serve to disguise the poverty of the thought. Twentieth-century biology has benefited from the introduction of mathematical ideas and techniques. There is a great tradition, inaugurated by Fisher, Haldane, and Wright and continued in our day by Hamilton, Lewontin, Maynard Smith, Kimura, Crow, Cavalli-Sforza,

Feldman, and many others. In the work of people in this tradition we see the possibility of advancing biological understanding through precise formulations, designed to highlight crucial features of the processes and situations that are under study. The mathematics corrects naive expectations, leads us to new problems, and makes us aware of assumptions that we had not previously recognized.

Lewontin has characterized *Genes, Mind, and Culture* as containing "no genes, no mind, no culture" (1983b). The charge sounds flippant, but we have found it to be apt. There are no genes: for the examples studied in detail, the examples of "gene"-culture translation, are independent of the genetic bases of our preferences, and the authors have no other new information to offer about the genetics of human behavior. There is no mind: for, as we have seen, psychological insights about human decision making play no role in the theory. There is no culture: for Lumsden and Wilson do not even *see* the problem of identifying social institutions in their preferred idiom, much less solve it. So indeed there are no genes, no mind, no culture. But there are lots of equations.

Chapter 11

The Last Infirmity

Taking Up the Biologist's Burden

Ever since Aristotle, many great biologists have felt the lure of philosophy. After struggling with problems about organic nature, they are tempted to close their volumes with a brief survey of the human implications, to reward themselves and their readers by providing some speculation about human freedom or human destiny. Pop sociobiologists follow the trend. The final part of our inquiry will consider their efforts at reforming our general ideas—and ideals—of human nature.

At first glance it might seem that claims to have answered the large questions about ethics, free will, and human worth—questions that reflective people have posed almost throughout recorded history—would constitute the part of the sociobiological program most in need of examination. What can be more important, or more profound, than these vast questions? I think, however, that pop sociobiological proposals about the nature of value are far less likely to be influential than the more particular claims about aggression, gender differences, sexual preference, and the rest of the headline-grabbing suggestions. When pop sociobiologists speak on these latter topics they can don the mantle of the expert. Ironically, the very ease with which they come to pronounce on philosophical issues that go beyond their professional expertise tells against their having much influence on our understanding of those issues. Biologists may believe that they have a license to advance views about human freedom and morality without considering what philosophers and other humanists have written about these subjects. They are not alone, and the readers who are most interested in philosophical questions may feel that they can grant the same license to themselves.

In this chapter I shall look at the pop sociobiological treatment of three philosophical questions: Are people capable of genuine altruism? Are people able to act freely? Are there objective moral principles? These questions have been addressed in technical works by contemporary thinkers, as well as in the masterpieces of some of the greatest minds in the history of civilization. They are difficult questions. They have not submitted to the same definitive style of answering that we find in some areas of the natural sciences. Philosophy's apparent lack of progress can provoke impatience, especially among those who have contributed to the advance of some area of science. Perhaps what philosophy needs is an injection of scientific knowledge and rigorous method. So the pop sociobiologists go to work. Give them a wet Sunday afternoon and they will unriddle humanity.

It is easy to respond to the colonizing efforts, difficult to respond to the predicament that prompts them. A little analysis of the sociobiological "solutions," comparing them with the history of research on the problem, shows that a Sunday afternoon's reflection delivers just about what might have been expected. Yet if the matter is left there, it will still appear that there is a confused cluster of significant questions in need of invasion by rigorous thinkers. Nor is it possible to dispel the impression by offering firm answers to philosophical questions. To see how the present crop of breezy scientific suggestions does not touch the root of the issues and to show how to resist the blandishments of the next group of would-be colonizers, we need to show how philosophers have made progress on the questions that pop sociobiologists yearn to answer, how more recent discussions avoid the errors of earlier solutions, how the options that now compete for our attention are more sensitive to difficulties and more refined than those that were previously available. When it is understood that those who are ignorant of the history of philosophy, including the recent history of philosophy, are doomed (at best) to repeat it, then the notion that the biology of behavior (or the next aspiring biological subdiscipline) provides a quick fix will be greeted with the skepticism it deserves.

Looking Out for Number One

Most people believe that human beings are capable of behaving altruistically. We seem to do things to serve the interests of others, and, while there may be disagreement about the extent of altruistic behavior, it would seem unduly cynical to deny that any such behavior exists. Nevertheless, psychological egoism, the thesis that all human behavior is selfish when it is properly understood, has attracted a number of adherents in the history of Western thought. A prominent example is Thomas Hobbes, who claimed that when our motives are fully revealed we are exposed as calculating creatures, adjusting our behavior to what we perceive as our own interests.

Any discussion of human altruism must begin by noting an ambiguity in the phrase "doing something to serve the interests of another." For it is obvious that we sometimes perform actions that have the effect of advancing the welfare of others. Even the most malevolent and self-serving actions may go awry, bringing benefits to those we wish to harm at costs to ourselves. Shylock's insistence on the formula of the bond leads to his own undoing. Angelo's sexual blackmail of Isabella culminates in his unwitting redemption of his promise to Mariana.

Angelo and Shylock do not act altruistically, even though what they do has the effect of advancing the welfare of others at cost to themselves. Genuine altruism depends crucially on facts of intention. Doing the right deed for the wrong reason is not enough.

There is an ordinary notion of altruism, embedded in our practice of assessing the actions of ourselves and others, for which the following (partial) analysis recommends itself: a person acts altruistically when he or she acts with the intention of advancing the welfare of another person and in recognition of some disadvantage to himself or herself. I hasten to point out that the analysis will need to be refined if it is to avoid some obvious difficulties. Those who form their beliefs hastily and plan a course of aid to others on the basis of opinions that are unreasonable in the light of the available evidence are dubious examples of genuine altruism. Altruistic plans do not have to be correct. Altruistic people can be ignorant of relevant facts. But altruistic plans ought to be rational.

Even without refinement our preliminary analysis enables us to note some obvious points. Sometimes altruistic actions succeed. They bring benefits to those who were the intended recipients. Sometimes altruistic actions fail. A person may strive to help others in the rational expectation of personal cost, but the action may fail to bring help and the expected costs may even be replaced by unexpected benefits. Imagine a conscientious person, far from rich, who decides to give a large sum of money to support famine relief. In the common fashion of charitable schemes, this person also organizes a raffle, and he himself happens to win the grand prize. Meanwhile, the contribution is used to buy food, which accidentally spoils during transport. In this scenario the original donor reaps an unexpected benefit and the generous action ultimately fails to achieve its purpose. Irrespective of what the person subsequently does with the winnings, the original gift should count as an altruistic action. (Interestingly, if the person failed to give any of the prize money away, we might be inclined to invoke the commonplace that wealth blunts the sensibilities.)

Similarly, as the cases of Angelo and Shylock make clear, actions that are not altruistic can achieve good effects. Unfortunately, we know all too well that selfish actions can sometimes succeed in their purposes. Hence, if we introduce the term "effectively altruistic" to cover cases in which the action of a person has the effect of promoting the welfare of another at a cost to self, then we can summarize the discussion so far by noting that there seem to be four possibilities: altruistic actions may be effectively altruistic or they may not; nonaltruistic (selfish) actions may be effectively altruistic or they may not. (For a similar assessment, see Mattern 1978, 464–465.)

The Hobbesian rejoinder is that things are not as they seem. It *appears* to us that people act with the intention of helping others, but this is only because we have an inadequate understanding of human motives. Psychological egoists delight in proposing that there are springs of human conduct that we typically hide from ourselves and that show the role of self-interest in even the most apparently altruistic actions. Sometimes the egoist's case rests on the idea that those who act freely act from their own desires (an idea that is not as simple as it first appears—see Dworkin 1970, and the next section). We single out altruistic actions as valuable because we suppose them to be performed freely. If the actions are performed freely, then they proceed from the desires of the agent. If they proceed from the agent's desires, then they must surely be selfish and not altruistic after all.

Such arguments are sleight of hand. They depend on confusing two obvious ways in which people can be related to their desires. All of my desires are mine in the trivial sense that I am the person who has them. However, unless I deceive myself, not all of my desires are directed at securing my own welfare. On some occasions I seem to put the interests of others above my own wishes. Conspicuous cases of altruism are those in which we recognize in the agent a desire to subordinate personal wishes to the interests of others. Mother Theresa acts from her own desires. What excites our admiration, even awe, is the extent to which she desires the well-being of others above material comforts for herself. (For a clear discussion of related attempts to argue for egoism, see Feinberg 1981.)

The egoistic challenge must go beyond the glib line that free people really do what they want to do, so that apparent altruists are merely satisfying their own desires. What has to be shown is either that the apparent intention to promote the well-being of another over the interests of self is just a screen that masks a deeper, self-centered motive, or that when the intention is properly understood, our estimate of its importance will be found to be mistaken. Here there are genuine possibilities of scientific discovery. Empirical investigations might reveal to us that our motives, the proximate causes of our actions, are not as we take them to be.

Shorter ways will not do. Students of the biology of behavior have adopted the term "altruism" to cover cases in which an animal acts so as to promote the fitness of another animal at some costs (in units of fitness) to itself. Even if this notion were clearly defined, and even if it were shown that there are no instances of "animal altruism," arguments would still be needed to draw conclusions about human altruism. (For a sensitive discussion of "animal altruism" that appreciates the problems of giving a definition, see Bertram 1982.) The biological concept is, at best, a concept of effective altruism. Without linking it explicitly to conclusions about human motives, we can gain no insight into the possibility of altruism.

Pop sociobiologists plainly think that they can advance our understanding of human altruism. Dawkins begins his book *The Selfish Gene* by claiming that people are "born selfish" (1976, 3) and, more generally, that we would expect anything evolved by selection to be selfish (4). Wilson suggests, with some qualifications, that sociobiology discloses the selfish roots of apparently altruistic actions: "The evolutionary theory of human altruism is greatly complicated by the ultimately self-serving quality of most forms of that altruism. No sustained form of human altruism is explicitly and totally self-annihilating" (1978, 154). Barash is more forthright: "Real, honest-to-God altruism simply doesn't occur in nature" (1979, 135); "Evolutionary biology is quite clear that 'What's in it for me?' is an ancient refrain for all life, and there is no reason to exclude *Homo sapiens*" (167).

There are two ways of interpreting such passages. The first is to read pop sociobiologists as debunking the ordinary view of human altruism by buttressing the Hobbesian challenge. On this construal the sociobiological claim is that the apparent intention to promote the well-being of another person is a screen that hides some deeper motive. People really calculate the consequences of their actions in terms of their own inclusive fitness and act so that their inclusive fitness is maximized. The alternative interpretation acknowledges that human beings sometimes do intend the well-being of others and that such intentions sometimes move people to action; it suggests, however, that the evolutionary explanation for the presence of such intentions and for their efficacy in moving us to action lies in considerations of the maximization of inclusive fitness. When we understand the evolutionary roots of the mechanisms that lead us to help others, then we may no longer be so impressed with the practice of evaluating people as altruistic or selfish.

The two styles of interpretation correspond to the ways of articulating Alexander's program (discussed in chapter 9). We shall examine the merits of each version separately.

Consider first the more radical claim that human actions are really driven by mechanisms of which we are normally unaware and that serve the interests of fitness maximization. We have already seen that appeals to the idea that people actually maximize their inclusive fitness by acting in ways that initially appear to detract from their reproductive success provide no basis for the hypothesis that human actions are driven by unconscious calculations of inclusive fitness (or by proximate mechanisms that simulate such unconscious calculations). Analogous appeals fare no better in the context of attempts to debunk our ordinary views of human altruism.

Alexander takes "the outstanding example" of altruistic behavior toward relative strangers to be "the event we call 'falling in love' " (1979, 123). Emphasizing the sudden and dramatic way in which people fall "across the breach from social strangeness to social intimacy," Alexander suggests that mate selection is evidence for the "existence and importance of the kinds of gene effects . . . postulated to underlie social learning" (123). A propensity to fall in love quickly with a stranger is supposed to maximize inclusive fitness because it combines the virtues of "outbreeding and long-term commitment" (123).

There is much that is sketchy in Alexander's story—in its supposition that there is a single pattern of human "falling in love" that literature records, in its suggestion that if it were done 'twere best done quickly, in its neglect of the fact that, in the context of human affairs, love, sex, and reproduction have no automatic connections with one another. But what is crucial is the failure to respond to the real problem. Even if we were to grant the highly speculative ideas about the mechanisms of mate choice, the genuine puzzle of "extreme altruism" is that lovers are willing to sacrifice themselves for the beloved—even though there may be no possibility of leaving direct descendants. At most, Alexander may show that there are secret springs that prompt us to choose our mates in a particular fashion (although that is highly dubious). He provides no reason to think that our understanding of individual self-sacrifice in the interests of love is to be replaced by recognition of secret fitness-maximizing calculations. What is most dramatic about the cases of Haemon and Sydney Carton is not the speed with which they fall in love, but what they are prepared to do once they have fallen. (Perhaps the fitness calculator misfires in such cases. But there is no reason to believe in the calculator, and no reason to invoke the misfiring.) Literature, art, and music, to which Alexander appeals to support his ideas about the dramatic way in which people fall in love, provide far more evidence of the human capacity for self-sacrifice in the cause of love and friendship.

Wilson attempts to confront a broader range of cases. He suggests that we should distinguish between "hard-core altruism" and "soft-core altruism." Roughly, hard-core altruism is found in cases where people (or nonhuman animals) are moved in ways that are "relatively unaffected by social reward or punishment beyond childhood" (1978, 155). Hard-core altruism is "likely" to have evolved through "kin selection," and the behavior is thus expected to be directed at close relatives. Soft-core altruism, on the other hand,

> is ultimately selfish. The "altruist" expects reciprocation from society for himself or his closest relatives. His good behavior is calculating, often in a wholly conscious way, and his maneuvers are orchestrated by the excruciatingly intricate sanctions and demands of society. (155–156)

Within this dichotomy Wilson hopes to capture all cases in which people appear to act altruistically. Those who sacrifice themselves to help strangers will be understood as "soft-core" altruists, operating from conscious or unconscious calculations of future benefits. Those who sacrifice themselves for their kin will be viewed as engaging in actions that are little more than reflexes, responses lacking in the autonomy we take to be a prerequisite for altruistic behavior.

Neither category appears to fit the most prominent examples of human altruism. When we recall cases of altruistic actions toward kin, we do not primarily think of the instinctive responses of parents who pluck their children from danger almost before they appreciate the threat to themselves. We focus instead on the political prisoners who submit to torture in order to shield their kin, on Cordelia accompanying her father to prison, on Antigone's resolve to bury her brother. These are not cases that we are inclined to dismiss as involving responses "relatively unmodified beyond childhood." Instead, they seem to reveal courageous self-sacrifice after deep reflection. Moreover, there are numerous examples that seem to show less dramatic personal sacrifices: parents sacrificing their own comfort for the sake of educating their children, children caring for their elderly and sick parents, and so forth. What reasons are there to think that our ordinary assessment of such actions is mistaken?

If Wilson intends to make the radical claim that we are wrong about the proximate mechanisms of altruistic action (and we shall consider the alternative construal afterward), then there are two kinds of arguments that he might give. The more general line of argument is to contend that evolution favors forms of behavior that maximize inclusive fitness, so that we should take people to be acting in ways that actually enhance their own inclusive fitness, even when they do not appear to be doing so. The more specific line of argument is to try to explain exactly how the great altruists really fit into the two categories.

The reply to the more general argument is that there is no reason to accept the picture of the evolutionary process as one in which animals and humans are equipped with fine-tuned mechanisms for identifying the actions that would maximize their inclusive fitness and for behaving accordingly. Pop sociobiology has no reason to prefer that picture to the rival view on which evolution equips us with certain cognitive capacities and basic propensities, which combine in the social environments that we experience to produce the beliefs, desires, and intentions that we normally attribute to ourselves. Thus there is no evidence to lead us away from the natural idea that, given the traits with which evolution has equipped us, we are able to set ourselves personal goals and to perform actions that detract from our inclusive fitness. It is possible to take the evolution of *Homo sapiens* seriously and yet to deny that natural selection has fashioned dispositions to behavior that lead us always (or almost always) to maximize our inclusive fitness.

So we are left with the more specific argument, the attempt to show that even the most striking instances of apparent altruism can be understood as stemming from secret, selfish desires. Prompted by a question from Malcolm Muggeridge, Wilson considers the work of Mother Theresa. He admits that Mother Theresa lives a life of "total poverty and grinding hard work" (1978, 165). What explanation can be given for her apparent willingness to give herself for others?

> In sobering reflection, let us recall the words of Mark's Jesus: "Go forth to every part of the world, and proclaim the Good News to the whole creation. Those who believe it and receive baptism will find salvation; those who do not will be condemned." There lies the fountainhead of religious altruism. Virtually identical formulations, equally pure in tone and perfect with respect to ingroup

> altruism, have been urged by the seers of every major religion, not omitting Marxism-Leninism. All have contended for supremacy over others. Mother Theresa is an extraordinary person but it should not be forgotten that she is secure in the service of Christ and the knowledge of her Church's immortality. (165)

The tactic is familiar. Subtract the references to the twentieth-century context, and the passage might have been written by Hobbes.

Notice that there is no suggestion of a new biological insight (nor of the results of any other empirical investigation). Wilson does not propose that Mother Theresa is "really" maximizing her inclusive fitness. He simply offers a Hobbesian speculation about motives. Indeed, he does not quite offer even that. We are to remember that Mother Theresa has been promised salvation by her church. So maybe (nudge, nudge) it is her desire for personal salvation that moves her to tend the poor of Calcutta. Mother Theresa is really a "soft-core" altruist after all.

On the radical interpretation, pop sociobiological claims about human altruism embody the idea that biology teaches us new insights about the proximate mechanisms that underlie apparently altruistic behavior. But we have seen in earlier discussions that pop sociobiology does nothing to subvert our self-image. When we examine the pop sociobiological treatment of human altruism, it is found to dissolve into gratuitous Hobbesian speculations that have no basis in biology or any other science.

Let us turn to the conservative construal. We now suppose that the intentions that we normally take to underlie altruistic actions are genuinely those that move the agents to act as they do. Those who sacrifice personal comfort, health, even their lives, in the service of relative strangers are viewed as acting from concern for the well-being of those to whom they bring aid. Nevertheless, it may still be suggested that the actions are "ultimately selfish." Pop sociobiologists may argue that the concern for the well-being of others arises from mechanisms that were originally shaped by natural selection in the interests of maximizing inclusive fitness. Instead of supposing that the apparent motives hide the secret springs of human action, they may grant that the apparent motives are the real causes of altruistic behavior but deny our normal assessment of their worth. Once the motives have been seen in the light of our evolutionary history, there is no reason to find them especially worthy of commendation.

The right reply to this version of the attempt to debunk altruism is that when it is articulated in detail it is seen to be innocuous. Consider a concrete example. A childless couple spends many years in adopting and caring for children with a variety of birth defects. They do whatever they can to promote the happiness of the children they adopt, at considerable sacrifice to their own health and comfort. The pop sociobiologist arrives on the scene to give a diagnosis. There is no denying the fact that the parents are moved by compassion and love for the children. But we are invited to consider their motives in the light of evolutionary biology. The concern for children that they manifest is a propensity that other people show in caring for (biological) offspring. That propensity was fixed in our species (and in many others) because it enhanced fitness. Therefore the propensity and the motives of the childless couple are "ultimately selfish."

Does this "ultimate selfishness" cause us to revise our appreciation of the moral worth of the couple's actions? Surely not. In the first place, the links to the maximization of inclusive fitness are too tenuous and indirect. Even if we were to suppose that all the dispositions and capacities that figure in the parental behavior derive ultimately from basic dispositions and capacities that have been shaped by selection (the propensity to care for helpless young, the ability to recognize the needs of others, and so forth), that would not detract from the unselfishness of the actions. Why should the remote action of evolutionary forces be relevant here? Second, it is far from clear that in cases where fitness-maximizing dispositions function directly to enhance inclusive fitness the actions should be classified as selfish. Parents who make large personal sacrifices on behalf of their children may enhance their chances of spreading their genes into future generations. But what exactly is the relevance of this fact? If we thought of the parents as consciously calculating that a large personal sacrifice would transmit more copies of their genes into future generations and deciding on this basis to make the sacrifice, then our reaction would not be that they are being relentlessly egoistic. People who act in this way strike us as crazy rather than selfish (Midgley 1978, 129). When pop sociobiologists announce that apparently altruistic behavior is ultimately selfish because it proceeds from propensities that typically maximize inclusive fitness (for example, concern for children), it is appropriate to point out that the personal sacrifices of the altruists are no less real and that the ultimate selfishness is irrelevant to the assessment of moral worth.

Nevertheless, there may be a deeper issue. As we noted at the beginning of the discussion, the concept of altruism that underlies our moral evaluations emphasizes the importance of human intentions to give aid to others. Perhaps what really motivates the pop sociobiological pronouncements is the idea that such intentions are merely the means that evolution has employed in shaping humans to maximize their inclusive fitness. Natural selection has favored animals who assist their kin and who are prepared to cooperate with others when it maximizes the inclusive fitnesses of the participants to do so. Our impulses to sacrifice ourselves for our children, to give up our own comfort in the service of an ideal, and the like are simply the peculiar ways in which the job of fitness maximization has been accomplished in our own species. There is no reason to value this particular mechanism for promoting aid to kin and cooperation among unrelated individuals above the different mechanisms that, we assume, achieve similar ends in different groups of animals.

Any such reasoning achieves its force by raising doubts about human freedom and human autonomy. The debunking argument must forestall the idea that humans, unlike animals, freely form plans and intentions, and that this dissimilarity in proximate mechanisms grounds the distinction between genuine altruism and the effective altruism that occurs throughout the animal kingdom. We must be led to think of people as puppets, whose basic design is fixed by evolution under selection. Sometimes the mechanisms produced by the evolutionary process lead us directly to the maximization of our inclusive fitness. Sometimes their effect is modified by the constraining actions of others and the manipulations of society. In neither case is there anything especially valuable in the presence of intentions to help others and in actions that proceed from those intentions.

The divide-and-conquer strategy is illustrated in Barash's discussion of the kamikazes (1979, 167–168). Barash is willing to try any line that leads to the conclusion that the kamikazes were actually selfish in sacrificing themselves, ostensibly for their country. He speculates that "their cost-benefit equation would have favored altruism"—allegedly because the families of the pilots gained enhanced status and because the pilots may have been given "sexual privileges" (I am not sure how seriously this is intended). But if the kamikazes are not to be understood as blindly following their fitness-maximizing proclivities, there is always the possibility of social coercion. Barash reminds us that the kamikazes were shot if they refused to fly.

Wilson hints at a similar theme in his remark that the actions of soft-core altruists are "orchestrated by the excruciatingly intricate sanctions and demands of society" (1978, 156). Perhaps the counterpoise to the idea that Mother Theresa has been bribed by the Church is the notion that people everywhere are bludgeoned into "altruistic" behavior by the surrounding society. Our evaluation of actions as altruistic depends on the assumption that the desires that prompt action are genuinely those of the agent. If the person who makes sacrifices for others does so under duress, then, while we may sympathize with the predicament, we are not so impressed. Similarly, if the apparent altruist is revealed as one who has been conditioned over many years to accept a particular scale of values, then what previously seemed awe-inspiring can come to appear faintly pathetic.

This is a curious line for a pop sociobiologist to take, for it suggests that our nature is sufficiently plastic to allow society to foist off on us values that lead us to contravene our fitness. There are hints of contradictions in the pop sociobiological picture of human nature. When the context suits, we are warned that the dictates of fitness maximization are so strong that we oppose them at the risk of forfeiting our "essential humanness." When there are difficulties in understanding apparently altruistic human behavior, there are vague references to the power of society to "orchestrate" our maneuvers. Perhaps, after all, social pressure is able to overcome the fitness-maximizing impulses. Wilson writes that "human beings are absurdly easy to indoctrinate—they *seek* it" (1975a, 562). Yet he does not pause to consider how the tendency to indoctrinability (allegedly produced by evolution) fits with the confident claims about limits on our social behavior. Has natural selection favored a propensity to be influenced by those around us that can override the dispositions that would otherwise lead us to maximize our fitness?

Wilson and Barash do not say. Their picture of human nature is not thought out in any serious detail. There is a broad suggestion that human beings are ultimately selfish, but we have discovered no argument for any interesting version of that conclusion. When all the confusions and gratuitous speculations are cleared away, we are left with the dim suspicion that, between our evolutionary heritage and the demands of our culture, there is no room for human autonomy. This is not a result produced by a new and powerful science, but the perennial first reaction upon thinking about the problem of free will. We may move on from the pop sociobiological treatment of altruism and consider the underlying question of human freedom.

Half-Truths about Human Freedom

There is a traditional line of dispute in the clash between determinism and human freedom. If a person acts freely, then, it is claimed, he or she could have done otherwise. If the action was determined by antecedent states or events, no alternative action was possible. Hence, to the extent that our actions are determined by prior conditions, we do not act freely. According to the thesis of psychological determinism, all human actions are determined by antecedent conditions. So, if psychological determinism is true, human freedom is illusory.

The hoary puzzle has lines of solution that are almost as venerable. David Hume pioneered the way. Free action does not require any unrestricted ability to do otherwise, he suggested. All that is needed is that we should have been able to do otherwise if we had wanted to. Free actions are those that stem from the desires of the agent. Psychological determinism is perfectly compatible with the existence of human freedom, for freedom does not consist in the absence of determination but in the way in which the action is determined.

Hume's proposal can easily be motivated by rehearsing a familiar philosophical fiction, designed to show that indeterminism is no help. Imagine a random robot—call it Oscar—whose limbs move in accordance with a sequence of random events. Inside Oscar's head is a collection of pigeonholes; each pigeonhole contains a radioactive nucleus; the nucleus sits on a platform, attached to Oscar's limbs with sensitive springs and levers; as the nucleus decays, the platform moves and one of Oscar's limbs jerks. The sequence of Oscar's actions is not determined. Yet Oscar is not free.

The example suggests that Hume's point is well taken. A prerequisite for free action seems to be that the actions proceed from the agent. They cannot be mere burps, tics, or random jerks. Instead, we must think of them as at least causally dependent on the agent's attitudes, goals, intentions, beliefs, and desires. Perhaps it will even be possible to think of them as *determined* by the agent's intentions, beliefs, and desires.

Hume hoists us over an initial obstacle, providing us with an approach that can be elaborated in response to more sophisticated worries. Wilson alludes to the trickier issues in giving his own formulation of the free will problem.

> The great paradox of determinism and free will, which has held the attention of the wisest of philosophers and psychologists for generations, can be phrased in biological terms as follows: if our genes are inherited and our environment is a train of physical events set in motion before we were born, how can there be a truly independent agent within the brain? The agent itself is created by the interaction of the genes and the environment. It would appear that our freedom is only a self-delusion. (1978, 71)

There are two common sources of anxiety that are brought together here. The first is the problem of *historical remoteness*. If my current actions are determined, no matter how, by events or states that occurred long before I was born, then how can those actions be genuinely free—or even genuinely mine? Apparently, the past has fixed the course of action for me, and I simply play the role that was determined for me. The second problem is the problem of *external imposition.* If my dispositions to behavior are the products of genes that were passed on to me by my parents and of the environment in which I developed, then in what sense do these dispositions lead me to free actions? I may sometimes act in accordance with the desires and intentions that I have acquired, but there is a residual worry that the desires and intentions are not genuinely mine. They were foisted on me by the genetic propensities with which I was born, genetic propensities that were shaped by selection in response to some remote ancestral environment, and by the environment in which I grew up. Am I to blame if the results are dispositions that lead me to socially undesirable behavior?

The general difficulty for elaborating the Humean account is to resolve the worry that the psychological attitudes underlying those actions that we identify as free are really external to the agent. There is a related problem—which Wilson does not mention—deriving from the fact that people sometimes act from their own desires and intentions without acting freely. Compulsive pyromaniacs and people addicted to dangerous drugs may do what they want to do. The actions that satisfy their compulsive desires are not free. Hence, if Hume's solution to the problem of freedom is to be made to work, it must be possible to distinguish those cases in which the agents act from desires and intentions that are genuinely their own and to block the skeptical challenge that our attitudes are always thrust upon us from without.

To appreciate the force of the problems requires a clearer formulation of the notion of determination. Let us say that a state of the world S_1 occurring at an earlier time determines that a state S_2 will occur at a later time, if there is no possible way (given the laws of nature) for the world to continue after the earlier time such that S_2 would fail to occur at the later time. The problem of historical remoteness poses a threat because we imagine it to be quite possible that the state of the world some very long time ago (say 100, 1,000, or 10,000 years) determined that we shall now act in just the ways that we actually do. If we offer the Humean counter that our present actions are determined by our own beliefs, desires, and intentions, then the skeptic can challenge the relevance of the point. Even though we play a proximate role in the action, it appears that what we do is ultimately determined by states of the world that occurred long before we were born. This, it will be suggested, is enough to explode the idea of freedom.

When the problem is properly posed, we see that indeterminism is no help. Suppose it is not true that earlier states of the world determine later states; instead, the earlier states only fix a probability distribution of later states. Given that a state S_1 occurs at the earlier time and given the laws of nature—*all* the laws of nature—there is a probability p_1 that S_{21} will occur at the later time, a probability p_2 that S_{22} will occur at the later time, and so forth. S_{21} actually occurs. Should the fact that it was not determined reassure us that human freedom is here revealed? I think not. By hypothesis, either the intentions of the agent were not determined by S_1 or those intentions did not determine the coming about of S_{21}. If the latter is true, then, given that *all* the laws of nature have been taken into account, there is a random element in the genesis of S_{21}. If the former is true, then there is a random element in the genesis of the agent's intentions. In either case it is hard to see how the substitution of random processes for determined states holds the key to freedom. Our ability to act freely is just as much a mystery if the present is related to the remote past by probabilistic laws instead of by strictly deterministic laws.

Let us return to the threat posed by psychological determinism. The skeptical challenge about the remote past is readily developed. If we point out that the actions we take to be free proceed from the desires and intentions of the agents, then the skeptic will respond that those desires and intentions were themselves determined by historically remote states. We can reply that this does not detract from the fact that the desires and intentions grew out of the agents' psychological development. They reflect individual encounters with and reactions to the environments in which those agents grew up. The skeptic's next gambit is obvious. It should be equally evident that to defend freedom it will be necessary to introduce some new idea.

We cannot deny that our present desires and intentions, the processes through which we acquire them, even the basic propensities that enable us to undergo those processes, could all have been determined by historically remote events. To explain why this does not matter, we must discover in the character of the processes through which our desires are formed and through which they become effective in governing our actions some quality that separates free action from the doings of automata and of those who are coerced.

Hume's insight is that it is the character of the determination, not the absence of determination, that counts in freedom. To suppose that the desires, or some part of their etiology, must escape the deterministic (or probabilistic) order of nature is to hand the skeptic the keys to the citadel. Instead, we must ask for the distinctive features of those cases in which we are confident that people act freely. We discover that agents sometimes act from desires with which they themselves identify—desires that they wish to be effective in leading them to action. We find that an agent's intentions are sometimes sensitive to reason and evidence. We find that they sometimes conform to a system of values that the agent rationally accepts. Perhaps in one of these features, or in a combination of them, we can isolate the crucial element for human freedom.

At this point it is helpful to take up the challenge from the biology of behavior. Both genetic and cultural determinism advance the idea that our desires and intentions are imposed upon us and that we are, in consequence, not free. For the cultural determinist, people are so plastic that their goals are entirely the product of the environment in which they are reared. The desires of agents cannot be viewed as genuinely theirs, for they might have been led in quite contrary directions had they only been given different experiences at the crucial stages. Genetic determinists, by contrast, emphasize the rigidity of an agent's inclinations. There is nothing environmental variation can do to affect a genetically determined trait. Should it not therefore be considered a type of reflex, something that develops in indifference to the experiences people have as they grow up and to their own reflections on those experiences? On either account, the prospects for freedom, conceived as grounded in desires that are genuinely those of the agent, begin to look dim.

Neither cultural determinism nor genetic determinism is likely to be the truth about any interesting behavioral trait. Yet it is legitimate to wonder whether the difficulties of the extremes do not combine in the more plausible intermediate positions. If our behavioral dispositions are determined by the genes we inherit and the environments in which we develop, does this allow any greater scope for human freedom? (As we saw in the last section, if there is a point about altruism raised by pop sociobiologists like Wilson and Barash, it consists in raising this question.)

Setting aside the myths of the blank mind and the iron hand of the genes brightens the prospects for freedom. The picture of a formless creature, stamped with whatever aspirations society chooses to impose, and the rival picture of a being who acts by a disjointed series of reflexes both give way to something quite different. Everything depends on the ways in which our dispositions to behavior vary with the environment. It is possible that our genes and environments fix our dispositions to behavior in ways that preclude the possibility of free action. If I was born with a propensity to develop a craving for a particular kind of thing (perhaps harmful, perhaps harmless), given one type of developmental history, or to develop a repugnance for the same thing, given any alternative type of developmental history, and if my adult dispositions are quite impervious to any evidence about the value of the thing, then my actions in choosing or refusing the thing are not free. But it is equally possible that my adult dispositions should be the outcome of a process in which reasoned evaluation plays an important causal role. If my choices are produced through my reflections on what would be best, if they are modifiable in the light of new information about the consequences of my planned actions, then, although those choices are ultimately fixed by my genotype and the sequence of environments I encounter, they may nonetheless be free.

Behavioral dispositions are the product of a process involving our genes and our environments. The mere fact that both genes and environments play a role does not guarantee the possibility of freedom. But if the process involves the development of capacities to respond to evidence and to evaluate alternative courses of action, then we may avoid the debunking conclusions. Our actions are not simple reflexes. Nor are they the expressions of desires imposed on us by the surrounding culture. They are determined by the gene-environment interaction. If that interaction involves our appreciation of what is valuable and our modification of our desires in accordance with the appreciation, then the behavioral dispositions that result can lead us to free action.

The perspective I have been advocating can be motivated by thinking of cases in which people act from intentions that are formed in the *wrong* ways, so that they fall short of freedom. Imagine someone who is in the grip of a compulsive desire—say the desire for a dangerous drug or the urge to set fire to buildings. It matters very little whether the compulsion results from the action of the environment alone, from a genetic propensity, or from an interaction between genes and environment. What is crucial is the rigidity with which the desire governs the person's actions. There are a number of important differences between this person and those whom we take to be free agents. Unlike a free agent, our imaginary subject will typically be at odds with the desire. Most of the time he or she will prefer that the desire not be effective in leading to action. This preference is cast aside, however, when the compulsion has the agent in its grip. Moreover, the person's attitudes are not modifiable in the light of evidence. Whatever others do to show the undesirability of the drug or to exhibit the evil consequences of the fire setting, the urges persist. The person is unresponsive both to considerations of publicly recognized value and to points about the promotion of his or her own well-being.

It is easy to refine the account of what distinguishes the compulsive by recognizing that the kinds of sensitivity that the compulsive person lacks are matters of degree. Those who are compelled by their desires are only able to prevent the desires from issuing in action when the preference that they should not be effective is very strong (perhaps when it is artificially fortified by concentrating on the harmfulness of the consequences). They are responsive to evidence only when the evidence is presented in the most graphic fashion. In these ways we could develop the notion of degrees of sensitivity that are relatively high in those agents we take to be free and relatively low in those we perceive as compelled.

From a certain minimal perspective, the compulsive look the same as free agents. For any genotype we assume that there is a mapping that assigns a unique behavioral phenotype to each developmental environment. (For each environment there is a corresponding phenotype, but the phenotypes corresponding to different environments may be different.) The assumption applies equally to the compulsive and to the free agents. Differences are revealed only when we attend to the details of the correlation of behavioral phenotypes with genotypes and environments. The behavioral dispositions of the compulsive are generated in ways that evade the authority of the power to form rational evaluations, and this is reflected in the insensitivity of the behavioral phenotype to changes in the environment that amass evidence against the desirability of the things to which the compulsive are drawn. By contrast, because the developmental route to the dispositions of the free lies through the employment of their cognitive capacities, those whom we count as free agents will have behavioral phenotypes that respond to such environmental variations.

To articulate the general idea that human freedom consists in the presence of dispositions to behavior that are somehow responsive to the environment is no easy matter. Three main ways of proceeding have been popular in the recent philosophical literature (see Frankfurt 1970; Dworkin 1970; Watson 1975; Wolf 1980; Nozick 1981). One is to focus on the fact that those who act under compulsion act from desires with which they do not genuinely identify. We can introduce the idea of a *second-order* desire, conceived as a preference about which of our ordinary (first-order) desires should be effective in action. Pyromaniacs would prefer that their desires to set fires be overridden: they would like to be able to "stop themselves." Yet when temptation strikes, the preference is powerless, and its powerlessness signals their lack of freedom. Others are more fortunate. They act from desires that are in harmony with their second-order desires. A generous person chooses to give money for famine relief. The choice accords with a second-order desire that desires for the welfare of others be effective in action. Conflict in the former case and concord in the latter give us a basis for distinguishing the compulsion of the one from the free action of the other.

Alternatively, we can attribute to agents an overall plan for themselves and their lives, which constitutes their conception of value and of their own happiness. Those who are genuinely free are able to adjust their actions in accordance with this plan. They can prevent those desires that threaten its implementation from being effective in action. A disciplined person, for whom a morning run is simply a means to continued health and the ability to attain long-term goals, can overcome the temptation to laze in bed. Similarly, those who are free can modify their desires and their intentions as they receive evidence about the consequences for their overall plan. As they learn that desires previously viewed as harmless might damage their chances of obtaining what they value, they are able to curb those desires. On this perspective the person is viewed as a parliament of inclinations and aspirations, and freedom consists in the possibility of adjusting voting power in the light of reason.

Finally, the notion of a free person as one who is able to achieve coherence in a subjective conception of what is good, valuable, and desirable can be replaced by the idea that freedom consists in having the ability to respond to what is objectively good. According to this approach, those who are free have the capacity for recognizing what is objectively worth pursuing and for adjusting their actions accordingly. The benevolent person who supports the cause of famine relief acts freely because the action responds to the objectively good. The pyromaniac fails to act freely because the actions are performed even in the presence of acknowledgment that what is done runs counter to what is good.

Despite the sophistication with which these proposals have been elaborated, none of them is completely immune from problems. There is an obvious worry about the first two conceptions. Freedom cannot simply be understood in terms of harmony between the inclinations of the moment and second-order desires or overall conceptions of value, for the latter might themselves be infected with unfreedom. Can we not imagine that a person compulsively desires that certain (first-order) desires be effective in action, or has rigid propensities for holding certain things to be valuable? The final approach avoids this particular difficulty, but it faces problems of its own. The existence of human freedom is made to rest on the objectivity of value (for discussion of this issue, see the next section), and this may cause alarm in some quarters. Moreover, it appears that the approach will generate an unanticipated asymmetry. If freedom consists in responding to what is objectively good, then perhaps only those actions that promote the good will be free. Those who act badly do not act freely. (In an important article [1980], Susan Wolf explicitly develops and defends the asymmetry.)

None of these problems counts as checkmate. In each case proponents of the approaches I have sketched can seek to expose the objection as confused or to refine their accounts to accommodate it. I shall not pursue the philosophical dialectic further. Nor shall I respond to those skeptics who complain that the Humean tradition misses the real point, that the issue of historical remoteness remains alive and unresolved. (I think that the beginnings of a response can be found in scrutinizing the idea that tracing an action to historically remote causal factors always supersedes explanations in terms of proximate causes. Why should the "ultimate" perspective always be the appropriate one for assessing the autonomy or the value of an action?) My main aim is to understand the bearing of advances in the biology of behavior on the problem of human freedom. That aim can now be achieved.

If Humeans are wrong, then the details of determination do not matter. The crucial problems arise irrespective of the deliverances of biology. If, on the other hand, Humeans are correct, then the central question is how to understand that difference in mode of determination that decides whether or not a person acts freely. Here biology is potentially relevant. Philosophy faces the task of explaining the role of evaluation, of sensitivity to evidence or sensitivity to the objectively good, in the development of those desires that are expressed in free action. Given an exact account of what is required, biological and psychological investigations could show us whether or not we satisfy the conditions for freedom. In the present state of our understanding, however, we are in no position to stage any such confrontation. Philosophers have a general theory that free action involves the ability to make choices in the light of evaluation—that those who act freely are not slaves to the passions of the moment—and they have some particular ideas about how this might be articulated. We also have a biological picture that our behavioral dispositions, including our preferences and our capacities for adjusting them, result from an interaction between genotype and environment. Nothing in this picture threatens the thesis that people sometimes (perhaps quite frequently) are able to base their choices on an appreciation of what is valuable.

For the Humean there is an obvious verdict on the current state of the free will problem. Considerable philosophical work is needed to elaborate the general theoretical view of our freedom. At present there is no indication from biology that if conditions for freedom can be successfully elaborated, they will turn out to be unsatisfiable. However, we must await the construction of a detailed account of freedom and a far more extensive understanding of the biological basis of behavior before making a definitive judgment on the question.

Let us now go back several rounds and pick up Wilson's discussion. The entire treatment is predicated on an idiosyncratic definition of determinism:

> It is a defensible philosophical position that at least some events above the atomic level are predictable. To the extent that the future of objects can be foretold by an intelligence which itself has a material basis, they are determined—but only within the conceptual world of the observing intelligence. (1978, 71)

Wilson goes on to appeal to his own reading of the Heisenberg uncertainty principle and to the complexities of neurophysiology; he concludes that people are unpredictable by one another and perhaps unpredictable by any physically possible intelligence.

> The mind is too complicated a structure, and human social relations affect its decisions in too intricate and variable a manner, for the detailed histories of individual human beings to be predicted in advance by the individuals affected or by other human beings. You and I are consequently free and responsible persons in this fundamental sense. (1978, 77)

Although Wilson does mention the Humean idea that free action occurs when people make decisions of their own accord (71), this idea is never explored nor are its difficulties noted. Instead, we find a simple line of reasoning. Unpredictability suffices for freedom. We are unpredictable. Therefore we are free.

There is a popular conception that determinism should be understood in terms of predictability. The definition I have offered breaks with this conception, and I think that it is correct to do so. Surely the root idea of determination is that, given the past state of the universe and the laws that govern the universe, the present state is now determined, in the sense that no alternative is possible. Whether there is actual or possible prediction seems irrelevant.

Nevertheless, one might think that the idea of possible prediction is simply a harmless way of making determinism more vivid. Pierre Laplace imagined a being who knew the laws of nature and who knew the earlier state. On the assumption of determinism, that being would then be able to derive descriptions of subsequent states. The

Laplacean image raises more problems than it solves, however. If the being is allowed to interact with the universe it is predicting, then it is possible to generate paradoxes. If its knowledge is represented in a way we find comprehensible, there are troubles with Gödel's first incompleteness theorem. (Intuitively, if the predictor is conceived as deriving theorems within a formal mathematical theory, it may be unable to generate some of the true descriptions of future states.)

Whatever the merits of thinking about determinism by imagining the predictions of knowledgeable beings, Wilson's conception is much more peculiar. Wilson seems to hold that a person is free relative to a type of being if it would be impossible for beings of that type to predict the person's behavior. The relevant types of beings are taken to be people, and, perhaps, other physically possible systems. But why should we be consoled by the thought that we are unpredictable to others? Someone who is worried by the thought that human actions are determined (in my sense) by earlier states of the universe should not take much comfort in the fact that people are not smart enough to figure out the determination. Wilson's remarks are entirely irrelevant to the problem of historical remoteness. They are equally beside the point when we consider the possibility that human desires and intentions are externally imposed. The mere fact that we cannot predict one another's actions brings no reassurance to those who want to know the difference between the apparently free decision to support famine relief and the compulsive behavior of the pyromaniac.

Yet the most glaring error in Wilson's discussion lies in its equation of unpredictability (by cognitively limited creatures) with some "fundamental sense" of freedom and responsibility. As I have been at pains to emphasize, the problems of freedom do not dissolve when we assume that human behavior is indeterministic. Still less do they disappear when we suppose that human behavior is unpredictable by humans. There are surely numerous physical systems whose workings outrun our predictive powers, either because they are too complex for us to compute their future behavior or because they involve some irreducible randomness. Perhaps even the behavior of relatively simple systems (such as flipped coins) would be impossible for us to predict in detail. This does not mean that freedom is rampant throughout the universe.

Wilson's assault on the problem of human freedom simply bypasses the difficult questions. The difficulties that arise from determinism, worries about historically remote determination or externally imposed desires, cannot be assuaged by misformulating them in terms of unpredictability and then taking consolation in the limited predictive powers of human beings. Salvaging the notion of human freedom requires us either to defend the general Humean approach against skeptics who are impressed by the possibility that our current actions, our intentions, and their etiologies were all fixed by the state of the universe long ago or to show in detail how indeterminism allows for autonomous action. (As I have emphasized, the mere fact that our actions are not determined would not guarantee that they were free.) It also requires us to succeed in the more difficult task of explaining just how to characterize those desires that are expressed in free actions. Those who are clear headed about the real issues will not expect to complete these projects by turning to them only in their leisure time.

The Hypothalamic Imperative

We come to the most ambitious of pop sociobiological adventures in philosophy. The opening lines of Wilson's *Sociobiology* announce a program. Given that "self-knowledge is constrained and shaped by the emotional control centers in the hypothalamus and limbic system of the brain" and that these systems evolved by natural selection, evolutionary biology must undertake "to explain ethics and ethical philosophers, if not epistemology and epistemologists, at all depths" (1975a, 3). At the end of the same volume Wilson returns to the theme, inviting his readers to consider "the possibility that the time has come for ethics to be removed temporarily from the hands of the philosophers and biologicized" (562). For, according to Wilson, "In the first chapter of this book I argued that ethical philosophers intuit the deontological canons of morality by consulting the emotive centers of their own hypothalamic-limbic system" (563). Wilson's self-knowledge seems constrained by his own optimism. Despite the frequency of assertion, there is no vestige of *argument* for any such conclusion.

The idea recurs with monotonous regularity (see, for example, Wilson 1978, 5–7, 196; Lumsden and Wilson 1983a, 175). Like many intellectual hobbyhorses, however, Wilson's *idée fixe* is not entirely easy to grasp. In what exactly does the biologicization of ethics consist? We can imagine four possibilities:

(A) Evolutionary biology has the task of explaining how people come to acquire ethical concepts, to make ethical judgments about themselves and others, and to formulate systems of ethical principles.

(B) Evolutionary biology can teach us facts about human beings that, in conjunction with moral principles that we already accept, can be used to derive normative principles that we had not yet appreciated.

(C) Evolutionary biology can explain what ethics is all about and can settle traditional questions about the objectivity of ethics. In short, evolutionary theory is the key to meta-ethics.

(D) Evolutionary theory can lead us to revise our system of ethical principles, not simply by leading us to accept new derivative statements—as in (B)—but by teaching us new fundamental normative principles. In short, evolutionary biology is not just a source of facts but a source of norms.

Wilson appears to accept all four projects, (A) through (D). I shall argue that (A) and (B) are legitimate tasks—although this is hardly news—and that Wilson's attempts to articulate (C) and (D) are thoroughly confused.

Since our ethical behavior has a history, it is perfectly appropriate to ask for the details of that history. Presumably, if we could trace the history sufficiently far back into the past, we would discern the coevolution of genes and culture, the framing of social institutions, and the introduction of norms. It is quite possible, however, that evolutionary biology would play a very minor role in the story. All that selection may have done for us is to equip us with the capacity for various social arrangements and the capacity to understand and to formulate ethical rules. Recognizing that not every trait we care to focus on need have been the target of natural selection, we shall no longer be tempted to argue that any respectable history of our ethical behavior must identify some selective advantage for those beings who first adopted a system of ethical precepts. It is entirely possible that evolution fashioned the basic cognitive capacities—*alles übriges ist Menschenwerk.*

So (A) is sanctioned as long as we do not overinterpret it. There is a legitimate enterprise of trying to reconstruct the history of our ethical behavior—although it must inevitably be fraught with speculation. Evolutionary biology will play some role at some level in whatever history we achieve. But we must allow for the possibility that that role is very indirect. The ethical attributes of formulating and obeying normative principles may have very little to do with the operations of selection.

Nor should we believe that to reconstruct the history of ethics, say by showing how ethical principles originated in myths used to buttress social arrangements (for example, the myth of a deity who would punish those who violated the precepts), is to cast doubt on the objectivity or the correctness of the principles we espouse. Just as a detailed history of arithmetical concepts and counting practices might show us a succession of myths and errors, yet would not lead us to question the objectivity of the arithmetical statements we now accept, so too reconstructions of the historical development of ethical ideas and practices do not preclude the possibility that we have now achieved a justified system of moral precepts. Wilson is far too hasty in assuming that the evolutionary scenario he gives for the emergence of religious ideas—a scenario that stresses the adaptive advantages of religious beliefs and practices—undercuts the doctrine that religious statements are true. Even if Wilson's scenario were correct, the devout could reasonably reply that, like our arithmetical ideas and practices, our religious claims have become more accurate as we have learned more about the world.

Lumsden and Wilson offer a cryptic rejoinder: "But the philosophers and theologians have not yet shown us how the final ethical truths will be recognized as things apart from the idiosyncratic development of the human mind" (1983a, 182–183). This is effectively to introduce a different type of argument. Independent of the question of our ability to trace the history of ethical (or religious) behavior, the task required of the philosopher who believes in the objectivity of ethics (or the believer who claims that certain religious doctrines are true) is to provide a convincing account of how the relevant claims are justified. As we shall see when we look at project (C), there is an important challenge here. (We shall also discover that the analogy with arithmetic proves helpful in responding to it.) The existence of this *separate* challenge does not entitle us to infer that a natural history of ethical behavior (or of religious practices) will automatically show that our ethical beliefs (religious beliefs) are not objectively justifiable. Conclusions reached through the acceptance of myths or dubious analogies may later obtain rigorous justification. Kekulé is supposed to have thought of the structure of the benzene molecule by staring into a fire. Kepler's discovery of the laws that bear his name was achieved partly through his acceptance of dubious ideas about the souls of planets and the music of the heavens.

So, while (A) is a reasonable enterprise, it hardly amounts to a dramatic removal of ethics from the hands of the philosophers. (B) is similarly innocuous. Ethicists have long appreciated the idea that facts about human beings, or about other parts of nature, might lead us to elaborate our fundamental ethical principles in previously unanticipated ways. Card-carrying Utilitarians who defend the view that morally correct actions are those that promote the greatest happiness of the greatest number, and who suppose that those to be numbered are presently existing human beings and that happiness consists in certain states of physical and psychological well-being, will derive concrete ethical precepts by learning how the maximization of happiness can actually be achieved. Analogous points apply to rival systems of ethical principles.

Sometimes biologists restrict themselves to helping in this legitimate and uncontroversial enterprise. Alexander writes, "I will argue, and I hope show, that evolutionary analysis can tell us much about our history and the existing systems of laws and norms, and also about how to achieve any goals deemed desirable; but that it has essentially nothing to say about what goals are desirable, or the directions in which laws and norms should be modified in the future" (1979, 220). As I understand him, Alexander—quite correctly—restricts himself to (A) and (B). These are projects that genuinely fall within the scope of the natural sciences—although, as we have seen (chapter 9), there is no reason to think that Alexander brings them to successful completion. Wilson, however, has his mind set on higher things.

There is an apparent tension in the pop sociobiological pursuit of (C) and (D). The central theme of Wilson's ideas in meta-ethics is that ethical statements are not objective; they simply record the emotional reactions of people. However, evolutionary theory is to allow us to *improve* ethics. The subject can no longer "be left in the hands of the merely wise" (1978, 7). Knowledge of human nature is needed if we are to make "optimum choices." Wilson appeals to biology to formulate a new system of principles. But if ethics stems from our emotions, what is the source of the imperative to alter it? In what sense can ethics be improved? What makes the optimal choices optimal? It is hard to wear the hats of skeptic and reformer at the same time.

For the moment I shall ignore the tension and concentrate on the two separate parts of the "biologicization" of ethics. Wilson's meta-ethics is prompted by impatience at the inability of philosophers to reach agreement about the systematization of ethical principles. It receives additional support from his conception of the possible theories of objectivity in ethics. The conclusion to which he is led is formulated in the following passage:

> Like everyone else, philosophers measure their personal emotional responses to various alternatives as though consulting a hidden oracle. That oracle resides in the deep emotional centers of the brain, most probably within the limbic system, a complex array of neurons and hormone-secreting cells located just beneath the "thinking" portion of the cerebral cortex. Human emo-

> tional responses and the more general ethical practices based on them have been programmed to a substantial degree by natural selection over thousands of generations. (1978, 6)

Stripped of references to the neural machinery, the account Wilson adopts is a very simple one. The content of ethical statements is exhausted by reformulating them in terms of our emotional reactions. Those who assent to "Killing innocent children is morally wrong" are doing no more than report on a feeling of repugnance. Ethical statements turn out to be on a par with statements we make when moved by our gastronomic preferences. Just as there is no objective standard against which those who like lutefisk are to be judged, so too there is no objective appraisal of those who disagree with us about the propriety of killing innocent children. Their hypothalamic-limbic complexes incline them to different emotional responses. That is all.

Wilson explicitly acknowledges his commitment to different sets of "moral standards" for different populations and different groups within the same population (1975a, 564). But I doubt that he appreciates the full character of the position to which his subjectivist meta-ethics leads him. Perhaps Wilson believes that the emotional responses of the deviant can somehow be discounted. How? One suggestion might be that they are to be dismissed because they disagree with the attitudes of the majority. To embrace this proposal not only embodies the dubious idea that majority responses are somehow correct or justified. It also reintroduces the very notion of objectivity that Wilson finds so mysterious. A different suggestion, perhaps more in the spirit of pop sociobiology, might be that emotional responses that maximize fitness are to be preferred to those that detract from fitness. If this is to succeed, however, it must be possible to defend both the thesis that the attitudes of the deviant detract from their fitness (in what environment?) and the claim that fitness-maximizing reactions are objectively preferable to other emotional responses. The notion of ethical objectivity, ostentatiously expelled through the front door, sneaks back through the rear.

Wilson is driven to a position that is very difficult to sustain. Faced with the deviants who respond to the "limbic oracle" by willfully torturing children, our reaction is strictly analogous to that evoked by people who consume food we find disgusting. We are revolted. Our revulsion may even lead us to interfere. Yet if we are pressed to defend ourselves, we shall have to concede that there is no standpoint from which our actions can be judged as objectively more worthy than the actions of those whom we try to restrain. They follow their hypothalamic imperative, we follow ours.

Given that the conclusions are unpalatable, we do well to look at the line of reasoning that leads to them. Should we be discomfited by the fact that there are rival systems of ethical principles? Should this lead us to conclude that there is no objective standard against which the rival systems can be judged, so that it is "every man his own emoter"? I think not. There are numerous areas of human inquiry in which theoretical disputes persist unresolved, and in which we do not abandon the idea that there is an ultimate possibility of objective solution.

Just as the fact that there have been, and continue to be, large theoretical disputes in evolutionary biology does not tell against the existence of a consensus on all kinds of important claims about the history of life, so too the presence of rival philosophical theories of the foundations of ethics should not blind us to the substantial areas in which reflective people agree in their moral appraisals. Philosophers who practice normative ethics rightly attempt to decide the *unclear* cases, for these are where philosophical advice is most needed. On the other hand, those who attempt to systematize our ethical judgments face all the problems inherent in any effort at theoretical systematization. Disagreement about difficult cases and foundational concepts can persist in the presence of agreement about easier cases and even when there is unanimity about generalizations.

Wilson's second line of argument is more persuasive, and does succeed in raising a serious philosophical question. The challenge for those who advocate the objectivity of ethics is to explain in what this objectivity consists. Skeptics can reason as follows: If ethical maxims are to be objective, then they must be objectively true or objectively false. If they are objectively true or false, then they must be true or false in virtue of their correspondence with (or failure to correspond with) the moral order, a realm of abstract objects (values) that persists apart from the natural order. Not only is it highly doubtful that there is any such order, but, even if there were, it is utterly mysterious how we might ever come to recognize it. Apparently, we would be forced to posit some ethical intuition, by means of which we become aware of the fundamental moral facts. Not only would it be necessary to explain just how this intuition works, but we would also be pressed to understand the refinement of our ethical views in the course of history. If we are able to apprehend the moral order, how is it that our ancestors so conspicuously failed to do so? And how is it that disagreements in ethics continue to this day?

Thus the skeptic. Before we abandon ethical objectivity in dismay, however, we should remind ourselves that a parallel argument goes through for mathematics. If mathematical truths are to be objective, then they must be objectively true or false. If they are objectively true or false, then they must be true or false in virtue of their correspondence with (or failure to correspond with) a realm of abstract objects (sets, numbers, and so forth). Even if we concede the existence of these objects, it is mysterious how we might ever come to interact with them and to appreciate their properties. Apparently, we are forced to postulate some mathematical intuition by means of which we become aware of the fundamental mathematical facts. Not only are we required to explain how this intuition works, but we are also pressed to understand the refinement of our mathematical views in the course of history. If we are able to apprehend the mathematical heaven, how is it that our predecessors failed to do so (how did they land themselves in muddles about infinite sets and infinitesimals, complex numbers, and so forth)? And how is it that disagreements in certain parts of mathematics (for example, in the foundations of set theory) continue to this day?

Few will sacrifice the objectivity of mathematics without a struggle. Philosophers disagree about exactly how to reply to the skeptic. Few advocate the view that the objectivity of arithmetic (and many other parts of mathematics) is thereby impugned. There are too many alternatives to be explored. Extreme Platonists will accept the position that the skeptic reconstructs and will try to answer the questions directly. Others will protest that there are important errors at early stages of the skeptic's argument. Some may wish to assert the objectivity of mathematics without claiming that mathematical statements are objectively true or false. Others may develop an account of the truth or falsity of mathematics that does not assume the existence of abstract objects. Still others may allow abstract objects but try to dispense with mathematical intuition.

Parallel moves are available in the ethical case. One possibility is to give up the idea of objective truth or falsehood for ethical statements in favor of the notion that some statements are objectively *justified* while others are not. Thus we may try to work out the view that, strictly speaking, "Killing innocent children is morally wrong" is neither true nor false, but that we are objectively justified in accepting this statement. (As Norman Dahl pointed out to me, we count a judge's verdict as correct or incorrect and we admit objectivity in the law, even though we do not think of verdicts as being true or false.) A variant on the same theme is to suppose that what truth and falsehood amount to in the case of ethics is not some correspondence (failure to correspond) with some independent moral order; instead, we may propose that an ethical statement is true if it would be accepted by a rational being who proceeded in a particular way. A distinct position is to declare that ethical statements correspond (or fail to correspond) to the moral order, but to attempt to understand the moral order in natural terms. For example, one may suppose that moral goodness is to be equated with the maximization of human happiness, and view morally right acts as those that promote moral goodness (or those that accord with procedures that can be expected to promote moral goodness). Yet another option is to claim that there are indeed non-natural values but that these are accessible to us in a thoroughly familiar way—for example, through our perception of people and their actions. Finally, as I have already noted, the defender of ethical objectivity may accept all the baggage that the skeptic assembles and try to explain the phenomena that the skeptic takes to be incomprehensible.

Wilson's rush to emotivism depends on slashing the number of alternatives. Only two possible accounts of ethical objectivity figure in Wilson's many pages on the topic. One of these, the attempt to give a religious foundation for ethics, does not occur in my list of options. Wilson mentions religious systems of morality only to dismiss them; his reason is spurious: "If religion . . . can be systematically analyzed and explained as a product of the brain's evolution, its power as an external source of morality will be gone forever" (1978, 201). The argument turns on a crucial ambiguity. If religious concepts are nothing but products of our brains, then, of course, religion is just a story. If, however, the history of religious belief shows human beings gaining knowledge of entities that actually exist, then there are no grounds for Wilson's conclusion. As we saw in discussing (A), there is no quick argument for debunking religion (or mathematics) on the grounds that it has a checkered history.

There are far better reasons for neglecting religious foundations for ethics. They have been familiar to philosophy ever since Plato. Suppose that we take actions to be good because they are prescribed in divine commands. Then either the character and prescriptions of the divinity are themselves objectively good, or they are not. If they are, then the fundamental notion of goodness is not fixed by the commands of the deity but is independent of them. If they are not, then it is appropriate to ask why people should obey the prescriptions. Either way, it appears that divine commands are incapable of grounding the moral order. (I should note that some philosophers have attempted to rebut Plato's argument and to refurbish divine-command theories. See, for example, Quinn 1981.)

The other alternative that Wilson considers is the possibility that ethical principles should be known by some mysterious intuition. In his most explicit account Wilson defines *ethical intuitionism* as "the belief that the mind has a direct awareness of true right and wrong that it can formalize by logic and translate into rules of social action" (1975a, 562). Having encumbered the defenders of ethical objectivity with all the problematic notions assembled in the skeptical argument, he can then announce that the philosophers have really been deceiving themselves. They have thought that they were intuiting mind-independent values. They were really consulting their own hypothalamic-limbic systems. If only we could replace their muddled emotings with some scientific system of ethics!

> For now, though, the scientists can offer no guidance on whether we are really correct in making certain decisions, because no way is known to define what is *correct* without total reference to the moral feelings under scrutiny. Perhaps this is the ultimate burden of the free will bequeathed to us by our genes: in the final analysis, even when we know what we are likely to do and why, each of us must still choose. (Lumsden and Wilson 1983a, 183)

We shall see later that Lumsden and Wilson are sometimes prepared to be more optimistic about the future of scientific ethics. Here, however, they toy with the possibility that there may be no objective standard for moral choice. The argument that inclines them to this possibility is based on skepticism about the prospects for giving an account of ethical objectivity.

Wilson (and, I assume, Lumsden with him) hurtles down the road to emotivism by ignoring the two dominant positions in the last two centuries of reflecting about ethics. Utilitarians attempt to explain what goodness is (and derivatively what the moral rightness of an action is) through the idea that goodness consists in the maximization of human welfare. Immanuel Kant and his philosophical descendants conceive of the problem of moral objectivity as one of showing that fundamental moral principles would be those adopted by rational beings, placed in a certain ideal situation. Their proposal can be elaborated in two slightly different ways. They may suggest that there is a moral order that is independent of our beliefs and feelings and that this moral order is determined by the judgments of rational beings who follow a certain kind of procedure. (On this construal, ethical statements are objectively true or false, and what makes them true or false is coincidence—or failure to coincide—with the outcome of a certain procedure.) Or they may give up the notion of ethical truth while emphasizing the concept of ethical justification. Some principles are objectively justified, and they have this status because they would be reached by beings who followed the special procedure.

In recent years the first type of Kantian view has received a detailed and profound elaboration by John Rawls, who states the central idea as follows:

> Apart from the procedure of constructing the principles of justice, there are no moral facts. Whether certain facts are to be recognized as reasons of right and justice, or how much they are to count, can be ascertained only from within the constructive procedure, that is, from the undertakings of rational agents of construction when suitably represented as free and equal moral persons. (1980, 519)

Both the Utilitarian approach to the question of ethical objectivity and the "Kantian constructivism" defended by Rawls are prefigured in my list of options for replying to the skeptic. At best, Wilson's argument for his emotivism is the line of reasoning I attributed to the skeptic. Why then are the genuine options, the ones philosophers have labored to find and to explore, absent from Wilson's discussions?

The answer is evident when we read Wilson's explicit comments on Rawls's views. Because he does not formulate the skeptical argument carefully, and because he does not attend to the formulations and arguments of the ethicists whom he dismisses as "blind emoters," Wilson fails to see any possibility for ethical objectivism except the most extreme form of intuitionism. All philosophers who believe in the objectivity of ethics turn out to be extreme intuitionists. Indeed, in his initial critique of theories that rely on ethical intuition of "true right and wrong," Wilson's prime target is Rawls! (If Rawls is an intuitionist in *any* sense, he certainly does not adopt the simple picture of the untutored intellect apprehending abstract values, the picture that Wilson attributes to him.) There is no subtle new argument for debunking ethical objectivity. Wilson reaches his conclusion by ignoring the serious alternatives. He ignores them because, apparently, he does not understand them.

Nevertheless, there is a genuine challenge here. Philosophers do owe a cogent response to the skeptical attack on ethical objectivity, and the lack of a broadly accessible discussion of the problem has probably encouraged many scientists (and other nonphilosophers) to think of Wilson's crude incursions as welcome relief from the unrigorous musings of philosophy. I hope that the analysis already given undercuts any such reaction. I shall return to the issue of ethical objectivity at the end of the chapter, after we have looked at the final phase of Wilson's project.

In the search for new normative principles it is not clear whether pop sociobiology is to offer promise or performance. The most recent expression of (D) supposes only that greater self-knowledge will be the key to an improved system of ethics:

> Only by penetrating to the physical basis of moral thought and considering its evolutionary meaning will people have the power to control their own lives. They will then be in a better position to choose ethical precepts and the forms of social regulation needed to maintain the precepts. (Lumsden and Wilson 1983a, 183)

Having examined the emotivist meta-ethics defended by Wilson (and, apparently, by Lumsden), we are now able to appreciate just how curious this idea is. If there is no sense to moral correctness, then what exactly is meant by the claim that greater self-knowledge might place us in a better position to choose ethical precepts? The obvious way to answer the question is to say that greater awareness of our hypothalamic-limbic systems might put us in a better position to satisfy our desires over the course of our lives. So indeed it might. But given that ethical statements simply record our emotional responses, why is one set of responses—even an "informed" set—any better than another? Is emotivism giving way to the position that ethical precepts are those that, if followed, would maximize long-term happiness? And whose happiness is to be relevant?

Such questions are academic. Wilson has some general project of "biologicizing ethics" and there is no serious attempt to distinguish separate enterprises such as (A) through (D). Hence it is hardly surprising that when routine philosophical distinctions are introduced, contradictions quickly emerge. It is still possible, though, that in his uncharted wanderings across the ethical map Wilson might stumble upon an important insight.

So let us consider the character of the intended advance. Our biological knowledge is not simply to be used to obtain derivative moral principles from more abstract normative premises, as in (B). It is also supposed to affect our most fundamental values. In his earlier writings Wilson was prepared to sketch the improved morality that would emerge from biological analysis. After one of his advertisements for the power of neurophysiology and phylogenetic reconstruction to fashion "a biology of ethics, which will make possible the selection of a more deeply understood and enduring code of moral values," Wilson allows us to sample his wares:

> In the beginning the new ethicists will want to ponder the cardinal value of the survival of human genes in the form of a common pool over generations. Few persons realize the true consequences of the dissolving action of sexual reproduction and

> the corresponding unimportance of "lines" of descent. The DNA of an individual is made up of about equal contributions of all the ancestors in any given generation, and it will be divided about equally among all descendants at any future moment. . . . The individual is an evanescent combination of genes drawn from this pool, one whose hereditary material will soon be dissolved back into it. (1978, 196–197)

As I understand him, Wilson claims that there is a fundamental ethical principle, which we can formulate as follows:

> (W) Human beings should do whatever may be required to ensure the survival of a common gene pool for *Homo sapiens*.

He also maintains that this principle is ethically fundamental in the sense that it is not derived from any higher-level ethical statement but is entirely justified by certain facts about sexual reproduction. (Wilson is going for bigger game than the uncontroversial (B), which allows for the use of biological premises in conjunction with ethical premises to yield new ethical consequences.) Hence there is supposed to be a good argument to (W) from a premise about the facts of sex:

> (S) The DNA of any individual human being is derived from many people in earlier generations and—if the person reproduces—will be distributed among many people in later generations.

Two questions arise from these Wilsonian claims. Is the argument from (S) to (W) a good one? Is (W) an objectively correct moral principle?

The first question leads into one of the most famous passages in the history of ethics, Hume's identification of the "naturalistic fallacy." According to Hume, normative conclusions are not deducible from factual premises. The point is not difficult to appreciate. Normative statements provide guidelines for action; factual statements apparently do not. (See Singer 1981, 74ff., for a good explanation of the point in the context of Wilson's example.) It is easy to add ethical premises to (S) that will enable us to deduce (W), but this is to give up the ambitious enterprise (D). Nor can we suppose that there is some other type of argument, nondeductive but nonetheless good, that enables us to infer (W) from (S) alone. The common types of reasoning that occur in arguments that we approve seem to be of no avail. No familiar form of inductive or statistical argument will lead from (S) to (W). Nor can we claim that (W) is somehow acceptable because it provides an explanation for the truth of (S). The explanation of (S) is provided by genetics, not by ethical theory. There may be a momentary charm in the idea that our DNA dissolves back into a common gene pool because it is a cardinal moral principle that we ought to ensure the survival of that gene pool, but it is the charm born of appreciation of absurdity.

Wilson is aware that a famous fallacy threatens his enterprise. He consoles himself with the idea that criticism on this basis "has lost a great deal of its force in the last few years" (1980a, 431) and with the claim that "the naturalistic fallacy is much less a fallacy than previously supposed" (1980b, 68). He amplifies his remarks by proposing that a system of ethics that had nothing to do with genetic fitness, and therefore (?) ran counter to the commands of the limbic system, would produce "an ultimate dissatisfaction of the spirit and eventually social instability and massive losses in genetic fitness" (1980b, 69). Do these dire possibilities hint at a possible bridge from biological facts to moral principles?

The most obvious ways to reconstruct a Wilsonian argument from (S) to (W) introduce further ethical claims. Perhaps Wilson intends to claim that acting against (W)—allowing the common human gene pool to cease to be—would inevitably cause "dissatisfaction of the spirit" or "social instability" or "massive losses in genetic fitness" or possibly all three. To parlay these claims into an argument for (W) apparently presupposes that we should not do those things that lead to dissatisfaction of the spirit or to social instability or to massive losses in genetic fitness. Whichever option (or options) we choose, the argument is not promising. Wilson has no reason to assert any of the factual premises: there is no evidence that people are unable to face the idea that a common human gene pool should no longer exist. Moreover, in each case the ethical work would be done by some background ethical theory to the effect that dissatisfaction of the spirit, social instability, and loss of fitness are always to be avoided (presumably at any cost). Since this ethical theory does not appear on stage, we have no idea about how Wilson would try to support it. Furthermore, if the construal that I have offered is correct, then the enterprise of biologicizing ethics seems to reduce to the entirely uncontroversial project (B), pursued in a muddled and ineffective way because crucial ethical assumptions are not stated clearly or developed in detail.

Scattered hints do not encourage the confidence of interpreters. It is possible that Wilson has something entirely different in mind. One thing, however, is plain. Pop sociobiology makes no serious attempt to face up to the naturalistic fallacy—to pinpoint the conditions under which normative assertions can be garnered from biological premises and to show that the moral principles of the new scientific ethics really do stand in the proper relation to the biological findings. (If the naturalistic fallacy is not a fallacy, then there will be *some* good arguments from factual premises to normative conclusions. That does not mean that *every* argument from fact to value compels our assent.) All we have been offered is a stark juxtaposition of a biological commonplace, (S), with a statement allegedly encapsulating a "cardinal value," (W). The connection is left as an exercise for the reader. After studying the exercise, it is not surprising that critics are tempted to exclaim "Fallacy!"

So far we have considered only the possibility of justifying an ethical principle on the basis of a biological premise. Let us now ask if the ethical principle itself is correct. There are grounds for doubt. Under most circumstances the exhortation to preserve the human gene pool requires very little of us. It is quite compatible with our allowing millions of people to die—or even with our hastening their ends. (Mass murder would even be compatible with Wilson's second principle, which enjoins us to favor genetic diversity. We just have to be careful to ensure that those we kill carry duplicates of genes already present.) When life gets tough, however, the principle implies some controversial recommendations.

Imagine the stereotypic situation after a holocaust. There are five survivors: four women, all between menarche and menopause, and one man. Assume that they know themselves to be the only survivors and that they recognize that the future of the species depends on their decisions and actions. After lengthy reflections and discussions each of the four women decides that she is not prepared to bear a child. All are sickened by the recent history of their planet, and, though all are aware of the biology of sexual reproduction, they resolutely oppose the idea of transmitting their genes. The man has the power to force at least one of them to copulate with him, to impregnate her, and to compel her to bear children. What should he do?

(W) offers a clear answer. The man should do whatever is needed to ensure the survival of the gene pool for *Homo sapiens*. If that includes rape and other forms of violence, so be it.

Yet there are ethical principles that point in a different direction. The women are autonomous agents. Their rights to make their own decisions should not be violated. They should not be treated as mere means, instruments for the preservation of hominid genes. Hence the man should not force them to copulate with him against their will.

It is easy to construct any number of similar examples. Suppose, for example, that we come to know that unless the world population is cut to a tenth of its present size, *Homo sapiens* will become extinct twenty generations hence. Assume further that any delay will make solution of the population problem impossible. It will not do to sterilize the living; most of them have to be killed. Unfortunately, the large majority of the human population refuses to go along voluntarily with schemes for the termination of their lives. What are those in power to do?

When we think about examples like these, it is no help to remind ourselves that our DNA was derived from many people and will be dispersed among many people in whatever future generations there will be. At stake are the relative values of the right to existence of future generations and the right to self-determination of those now living. The biological facts of reproduction do not give us information about that relationship. I shall return to this point shortly. It signals a crucial failure of the type of "scientific ethics" that Wilson hopes to develop.

Wilson's other attempts to advance substantive moral principles fare no better. In fact, one of them—the third—does remarkably worse. Wilson believes that we shall come to regard "universal human rights" as a primary value. Waiving any worries we might have about exactly what rights are to be regarded as universal, let us consider the alleged biological basis for this principle. According to Wilson, the important fact is that we are mammals; thus our society is built "on the mammalian plan" (1978, 199). The natural question is "*What* mammalian plan?"—for, as should be clear from the discussions of earlier chapters, mammalian societies are highly diverse. Moreover, we can only marvel at the claim that the "mammalian imperative" (199) counsels against the perpetuation of inequities. The societies of elephant seals, langurs, lions, and hamadryas baboons offer no such comforting indications. The example thus seems to combine the use of the naturalistic fallacy in argument for a vague principle with some peculiarly selective biology.

I conclude that the ambitious projects—(C) and (D)—end in failure. I want to end my discussion of pop sociobiological ventures into ethics by emphasizing one important feature in competent discussions of ethical issues that is entirely absent from Wilson's treatment. Recognizing this feature will enable us to see clearly why the maxims of the new scientific ethics are so naive; it will also shed some light on the problem of objectivity in ethics.

Pop sociobiology is completely insensitive to the issue of how the competing interests of different individuals are to be treated. The insensitivity becomes apparent when we reflect on Wilson's "cardinal values" and imagine situations in which people would be forced to make large personal sacrifices to promote those "values." In my scenarios for exposing the dubious nature of Wilson's principle (W), I was able to exploit the absence of any perspective from which the rights and duties of particular people can be assessed.

One part of the problem of ethical objectivity turns on precisely this issue. I have my feelings about the way I want the world to be. You have yours. There will undoubtedly be conflicts. Is there any set of principles for resolving such conflicts that all people ought to acknowledge? Is there an impartial perspective, a point of view from which the values and goals of others should enter into the decisions of a moral agent? The notion of ethical objectivity may appear somewhat less mysterious when we think of the possibility that there may be a *right way* for a person to recognize the inclinations of others and to give those inclinations their due in any conflict with selfish desires. The task of arriving at ethically correct decisions no longer appears as one of conforming to ethereal entities (abstract values); rather, it involves recognizing the existence of a standard beyond personal wishes, a standard in which the wishes of others are given their place.

In an extremely oversimplified way, this approach identifies the problem that Rawls sees as central to ethical theory. Rawls suggests that

> the objectivity or subjectivity of moral knowledge turns, not on the question whether ideal value entities exist or whether moral judgments are caused by emotions or whether there is a variety of moral codes the world over, but simply on the question: does there exist a reasonable method for validating and invalidating given or proposed moral rules and those decisions made on the basis of them? (1951, 177)

Rawls's suggestion meets skeptical challenges to the notion of ethical objectivity at an early stage of the argument. The task of exhibiting the objectivity of ethics is that of showing that there is the possibility of giving reasons for or against moral rules or particular decisions, *reasons that are valid for all parties.*

Rawls attempts to carry out this task by suggesting that the principles that govern the resolution of differences in the desires and interests of different people are those that would be accepted by rational beings placed in a hypothetical situation that Rawls calls "the original position." I shall not elaborate the details; the basic idea is that the beings in the original position have some knowledge of human motivation but must reach their decisions from behind a "veil of ignorance." Most importantly, they do not know in advance their own particular positions in the society whose arrangements are determined by their decisions. Rawls's thesis is that the principles of justice are those that would be reached by rational agents proceeding on the basis of this mixture of knowledge and ignorance.

Wilson's assessment of this proposal exposes the depth of his misunderstanding of it and of the important questions to which Rawls is responding: "While few will disagree that justice as fairness is an ideal state for disembodied spirits, the conception is in no way explanatory or predictive with reference to human beings" (1975a, 562). The criticism turns on the idea that Rawls's account of justice as fairness must be inadequate because it is not imbued with the biological knowledge that Wilson claims as the foundation for his own theory. Yet even if we were to suppose that pop sociobiologists have fathomed all the hypothalamic imperatives, the problem Rawls addresses would remain untouched. There would still arise an important set of issues that refused Wilson's biologicization. Can we find a set of reasons that are valid for all parties in a clash of interests? If so, how do we specify such reasons? Has Rawls succeeded in giving a method for discovering the reasons? Plumb the hypothalamic-limbic system as we may, the answers to these questions will not be forthcoming.

Rawls's original and profound suggestions provide one way of responding to the challenge that moral "correctness" is an utterly mysterious notion. (In his recent work Rawls has explicitly addressed one issue that might continue to appear vexing: "Why should the conclusions reached by parties in an ideal situation, the original position, prove binding on actual people?" See Rawls 1980 for an attempt to answer this question. Similar inquiries are undertaken with great thoroughness in Darwall 1983.) I sketch some of Rawls's ideas only for the purpose of showing how pop sociobiological ventures relate to serious studies in moral theory. The same point can be made by appealing to very different approaches to the foundations of ethics (see, for example, the lucid discussion in Singer 1981, which is also designed to expose the limitations of pop sociobiological ethics). What is needed is not a recapitulation of what contemporary ethicists—and some of their predecessors—have said, but the construction of a perspective from which the character of their enterprise can be appreciated.

That perspective reveals what is missing from the "ethical theories" promised by Wilson and his followers. A central task for any system of ethics is the construction of the impartial perspective. With its emphasis on the dictates of neural systems that have allegedly been fashioned to maximize the inclusive fitness of the individuals who possess them, pop sociobiological "ethics" lacks any theory of the resolution of conflicts. To the extent that people can be viewed as maximizing their own inclusive fitness through cooperation with others, apparent conflicts of interest may be diagnosed as situations in which all the parties maximize their inclusive fitness by coordinating their behavior. Yet there are innumerable situations—among them some of the most troubling—in which the reproductive interests of individuals do clash. For these situations pop sociobiology has nothing to offer. There is no higher standpoint than the dictates of the hypothalamus. There is no impartial perspective. There is only the conflict.

Postscript

Sociobiology has two faces. One looks toward the social behavior of nonhuman animals. The eyes are carefully focused, the lips pursed judiciously. Utterances are made only with caution. The other face is almost hidden behind a megaphone. With great excitement, pronouncements about human nature blare forth.

I have attempted to identify the two faces: to make it clear that there have been great advances in our theoretical understanding of evolution by natural selection, that some of the techniques developed have been carefully applied to the study of nonhuman animal behavior, that there are interesting results about the social lives of insects, birds, and mammals; and yet to show how the building of grand conclusions about ourselves is premature and dangerous. We have seen again and again how the assertions about human nature begin with unrigorous analyses of fitness, how they deal loosely with data about animal and human behavior, how they employ problematic concepts, how they rely on dubious connections between optimality and selection, how they offer spurious arguments for the inflexibility of the phenotype.

Cataloging these errors is important because the effects of accepting the pop sociobiological view of human nature are grave. That view fosters the idea that class structures are socially inevitable, that aggressive impulses toward strangers are part of our evolutionary heritage, that there are ineradicable differences between the sexes that doom women's hopes for genuine equality. None of these ideas should be adopted lightly. As I argued at the beginning of my discussion, the true political problem with socially relevant science is that the grave consequences of error enforce the need for higher standards of evidence. In the case of pop sociobiology, commonly accepted standards are ignored. The mistakes merely threaten to stifle the aspirations of millions.

We have no guarantee that our goals of justice, equality, and freedom can be obtained. Scientific studies might one day reveal to us that our nature precludes the possibility of the kind of society to which we aspire. That day has not yet arrived. We should beware the recommendations of the pop sociobiology salesmen, with their self-congratulatory claims to have looked the facts in the face and to have found us wanting. They have done no such thing.

Because of the complexity of some social problems, it is tempting to turn for advice to those who make free offers. Our understanding of the social order has been achieved through a long dialectic: thinkers as diverse as Plato and Locke, Marx and Rawls, have all helped to shape contemporary thought about justice, freedom, and equality. We would like to know if it is possible to implement whatever political ideas we draw from this tradition, and it is here that pop sociobiology holds out the hope of quick answers. Only certain kinds of solutions are possible for us—or, at least, possible without severe costs. We are far better off without the facile answers. Mindful of Socrates' famous dictum, we should acknowledge what we do not yet know.

What are the prospects for a genuine human sociobiology, a serious science that would bring to the study of human behavior not only the trimmings of the recent insights of evolutionary biology but the rigor that distinguishes the best work in nonhuman sociobiology? Here we must again confess our ignorance. At present we cannot confidently predict how far we shall be able to go. I have endeavored to show how serious students of nonhuman behavior presently face the exciting task of exploring numerous alternative hypotheses. There is an abundance of precise models, with further refinements galore waiting in the wings, and a current paucity of data. In the case of *Homo sapiens* the first rigorous analyses still await us.

The problems involved in developing such analyses are obvious. People have unparalleled abilities for assessing both their own situations and the strategies that are being pursued by those around them. Hence we can hardly expect to represent our own behavior by restricting ourselves to the simple, unconditional strategies often singled out in studies of animal behavior. Nor do we yet understand how to represent the interactions between the behavior of individuals and the surrounding culture. Until we can elaborate the general picture of gene-culture coevolution, conclusions about the institutions that evolution can be expected to favor are simply guesswork.

A serious human sociobiology needs help from many fields. If it is to be achieved, it will have to draw on the work of evolutionary theorists, behavior geneticists, developmental biologists and psychologists, sociologists and historians, cognitive psychologists and anthropologists. Any resultant discipline would be a real synthesis. But it is hard to synthesize when some of the contributors to the union are yet unformed.

Pop sociobiology thrives on the image of bold science, voyaging through strange seas of thought. Yet true pioneering science succeeds not only by virtue of intellectual daring but also because of awareness that the tools needed for the new ventures are at hand. To the pop sociobiologists' favored image we may counterpose another. Macbeth, on the verge of regicide, speaks:

> . . . I have no spur
> To prick the sides of my intent, but only
> Vaulting ambition, which o'erleaps itself
> And falls on the other.

At this point in the drama, Macbeth's own worthy service has obtained high honors for him. But these are not enough: the possibility of becoming king beckons him on to forget his double duty as host and as subject, to set in motion the train of events that will culminate in disaster at Dunsinane. The chain of dreadful consequences could so easily have been avoided. After all, to be Thane of Cawdor is no mean thing.

References

Alexander, R. 1974. "The Evolution of Social Behavior." *Annual Review of Ecology and Systematics* 5, 325–383.

Alexander, R. 1979. *Darwinism and Human Affairs*. Seattle: University of Washington Press.

Alexander, R., and Noonan, K. 1979. "Concealment of Ovulation, Parental Care, and Human Social Evolution." In Chagnon and Irons 1979, 436–453.

Allen, E., et al. 1975. "Against 'Sociobiology.' " In Caplan 1978, 259–264.

Ardrey, R. 1966. *The Territorial Imperative*. New York: Atheneum.

Axelrod, R. 1981. "The Emergence of Cooperation among Egoists." *American Political Science Review* 75, 306–318.

Axelrod, R., and Hamilton, W. D. 1981. "The Evolution of Cooperation." *Science* 211, 1390–1396.

Bachmann, C., and Kummer, H. 1980. "Male Assessment of Female Choice in Hamadryas Baboons." *Behavioral Ecology and Sociobiology* 6, 315–321.

Barash, D. 1976. "The Male Response to Apparent Female Adultery in the Mountain Bluebird, *Scalia currucoides:* An Evolutionary Interpretation." *American Naturalist* 110, 1097–1101.

Barash, D. 1977. *Sociobiology and Human Behavior*. New York: Elsevier.

Barash, D. 1979. *The Whisperings Within*. London: Penguin.

Barash, D. 1982. "From Genes to Mind to Culture: Biting the Bullet at Last." *The Behavioral and Brain Sciences* 5, 7–8.

Barlow, G., and Silverberg, J., eds. 1980. *Sociobiology: Beyond Nature/Nurture?* Washington, D.C.: American Association for the Advancement of Science.

Bateson, P. 1978. "Sexual Imprinting and Optimal Outbreeding." *Nature* 273, 659–660.

Bateson, P. 1980. "Optimal Outbreeding and the Development of Sexual Preferences in Japanese Quail." *Zeitschrift für Tierpsychologie* 53, 231–244.

Bateson, P., ed. 1982a. *Current Problems in Sociobiology*. Cambridge: Cambridge University Press.

Bateson, P. 1982b. "Preferences for Cousins in Japanese Quail." *Nature* 295, 236–237.

Beatty, J. 1985. "The Hardening of the Synthesis." In P. Asquith and P. Kitcher, eds., *PSA 1984*. East Lansing: Philosophy of Science Association.

Bengtsson, B. 1978. "Avoiding Inbreeding: At What Cost?" *Journal of Theoretical Biology* 73, 439–444.

Bernds, W., and Barash, D. 1979. "Early Termination of Parental Investment in Mammals, Including Humans." In Chagnon and Irons 1979, 487–506.

Bertram, B. 1976. "Kin Selection in Lions and in Evolution." In Clutton-Brock and Harvey 1979a, 160–182.

Bertram, B. 1982. "Problems with Altruism." In Bateson 1982a, 251–267.

Bishop, D., and Cannings, C. 1978. "A Generalised War of Attrition." *Journal of Theoretical Biology* 70, 85–124.

Block, N., and Dworkin, G. 1976a. *The IQ Controversy*. New York: Pantheon.

Block, N., and Dworkin, G. 1976b. "IQ, Heritability, and Inequality." In Block and Dworkin 1976a, 410–540.

Bock, K. 1980. *Human Nature and History*. New York: Columbia University Press.

Bodmer, W., and Cavalli-Sforza, L. 1976. *Genetics, Evolution, and Man*. San Francisco: Freeman.

Bonner, J. T. 1980. *The Evolution of Culture in Animals*. Princeton: Princeton University Press.

Boorman, S., and Levitt, P. 1973. "Group Selection at the Boundary of a Stable Population." *Theoretical Population Biology* 4, 85–128.

Brockmann, H. J. 1984. "The Evolution of Social Behaviour in Insects." In Krebs and Davies 1984, 340–361.

Bull, J. 1979. "Evolution of Male Haploidy." *Heredity* 43, 361–381.

Caplan, A., ed. 1978. *The Sociobiology Debate*. New York: Harper and Row.

Carlson, E. 1966. *The Gene: A Critical History*. Philadelphia: Saunders.

Cavalli-Sforza, L., and Feldman, M. 1981. *Cultural Transmission: A Quantitative Approach*. Princeton: Princeton University Press.

Chagnon, N. 1968. *Yanomamo: The Fierce People*. New York: Holt, Rinehart, and Winston.

Chagnon, N. 1974. *Studying the Yanomamo*. New York: Holt, Rinehart, and Winston.

Chagnon, N. 1976. "Fission in a Yanomamo Tribe." *The Sciences* 16, 14–18.

Chagnon, N. 1977. *Yanomamo: The Fierce People*. 2d ed. New York: Holt, Rinehart, and Winston.

Chagnon, N. 1982. "Sociodemographic Attributes of Nepotism in Tribal Populations: Man the Rule-Breaker." In Bateson 1982a, 291–318.

Chagnon, N., and Bugos, P. 1979. "Kin Selection and Conflict: An Analysis of a Yanomamo Ax Fight." In Chagnon and Irons 1979, 213–238.

Chagnon, N., and Irons, W., eds. 1979. *Evolutionary Biology and Human Social Behavior: An Anthropological Perspective*. North Scituate, Massachusetts: Duxbury.

Chagnon, N., Flinn, M., and Melancon, T. 1979. "Sex-Ratio Variation among the Yanomamo Indians." In Chagnon and Irons 1979, 290–320.

Charlesworth, B. 1980. *Evolution in Age-Structured Populations*. Cambridge: Cambridge University Press.

Charnov, E. 1983. *The Theory of Sex Allocation*. Princeton: Princeton University Press.

Cheney, D. 1983a. "Intergroup Encounters among Old-World Monkeys." In Hinde 1983, 233–241.

Cheney, D. 1983b. "Proximate and Ultimate Factors Related to the Distribution of Male Migration." In Hinde 1983, 241–249.

Clutton-Brock, T., and Harvey, P., eds. 1979a. *Readings in Sociobiology*. San Fransisco: Freeman.

Clutton-Brock, T., and Harvey, P. 1979b. "Evolutionary Rules and Primate Societies." In Clutton-Brock and Harvey 1979a, 293–310.

Clutton-Brock, T., and Harvey, P. 1979c. "Primate Ecology and Social Organization." In Clutton-Brock and Harvey 1979a, 342–383.

Clutton-Brock, T., Guinness, F., and Albon, S. 1982. *Red Deer: The Behavior and Ecology of Two Sexes*. Chicago: University of Chicago Press.

Crow, J., and Kimura, M. 1970. *Introduction to Population Genetics Theory*. New York: Harper and Row.

Darwall, S. 1983. *Impartial Reason*. Ithaca: Cornell University Press.

Darwin, C. 1859. *The Origin of Species.*London: John Murray. Facsimile of 1st edition, edited by Ernst Mayr. 1967. Cambridge, Massachusetts: Harvard University Press.

Darwin, C. 1862. *On the Various Contrivances by Which British and Foreign Orchids Are Fertilised by Insects.* London: John Murray.

Darwin, C. 1871. *The Descent of Man.* London: John Murray.

Darwin, F., ed. 1888. *Life and Letters of Charles Darwin.* 3 vols. London: John Murray.

Darwin, F., ed. 1903. *More Letters of Charles Darwin.* 2 vols. London: John Murray.

Davies, N. 1978. "Territorial Defence in the Speckled Wood Butterfly (*Pararge aegeria*). The Resident Always Wins." *Animal Behaviour* 26, 138–147.

Dawkins, R. 1976. *The Selfish Gene.* Oxford: Oxford University Press.

Dawkins, R. 1982. *The Extended Phenotype.* San Francisco: Freeman.

Dickemann, M. 1979. "Female Infanticide, Reproductive Strategies and Social Stratification: A Preliminary Model." In Chagnon and Irons 1979, 321–367.

Dobzhansky, T. 1970. *Genetics of the Evolutionary Process.* New York: Columbia.

Dunbar, R. 1982. "Adaptation, Fitness, and the Evolutionary Tautology." In Bateson 1982a, 9–28.

Dunbar, R. 1983. "Relationships and Social Structure in Gelada and Hamadryas Baboons." In Hinde 1983, 299–307.

Durham, W. 1976. "The Adaptive Significance of Cultural Behavior." *Human Ecology* 4, 89–121.

Durham, W. 1979. "Toward a Coevolutionary Theory of Human Biology and Culture." In Chagnon and Irons 1979, 39–59.

Dworkin, G. 1970. "Acting Freely." *Nous* 4, 367–383.

Ehrman, L., and Parsons, P. 1981. *Behavior Genetics and Evolution.* New York: McGraw-Hill.

Eibl-Eibesfeldt, I. 1971. *Love and Hate: The Natural History of Behavior.* New York: Holt, Rinehart, and Winston.

Eldredge, N., and Cracraft, J. 1980. *Phylogenetic Patterns and the Evolutionary Process.* New York: Columbia University Press.

Emlen, S. T. 1978. "The Evolution of Cooperative Breeding in Birds." In J. R. Krebs and N. B. Davies, eds., *Behavioral Ecology: An Evolutionary Approach.* Oxford: Blackwell, 245–281.

Emlen, S. T. 1984. "Cooperative Breeding in Birds and Mammals." In Krebs and Davies 1984, 305–339.

Estep, D., and Bruce, K. 1981. "The Concept of Rape in Non-Humans: A Critique." *Animal Behaviour* 29, 1272–1273.

Fagen, R. 1981. *Animal Play Behavior.* New York: Oxford University Press.

Fagen, R. 1982. "Skill and Flexibility in Animal Play Behavior." *The Behavioral and Brain Sciences* 5, 162.

Feinberg, J. 1981. "Psychological Egoism." In J. Feinberg, ed., *Reason and Responsibility.* Belmont, California: Wadsworth.

Feller, W. 1968. *An Introduction to Probability Theory and Its Applications.* New York: Wiley.

Fisher, R. A. 1930. *The Genetical Theory of Natural Selection.* 1st ed. Oxford: Oxford University Press.

Fisher, R. A. 1958. *The Genetical Theory of Natural Selection.* 2d ed. New York: Dover.

Flanagan, O. 1984. *The Science of the Mind.* Cambridge, Massachusetts: MIT Press.

Frankfurt, H. 1970. "Freedom of the Will and the Concept of a Person." *Journal of Philosophy* 68, 5–20.

Futuyma, D., and Risch, S. 1984. "Sexual Orientation, Sociobiology, and Evolution." *Journal of Homosexuality* 9, 157–168.

Geist, V. 1971. *Mountain Sheep.* Chicago: University of Chicago Press.

Ghiselin, M. 1969. *The Triumph of the Darwinian Method.* Berkeley: University of California Press.

Gilpin, M. 1975. *Group Selection in Predator-Prey Communities.* Princeton: Princeton University Press.

Gould, S. J. 1977. "Biological Potentiality vs Biological Determinism." In S. J. Gould, *Ever Since Darwin.* New York: Norton, 251–259.

Gould, S. J. 1980a. "Sociobiology and the Theory of Natural Selection." In Barlow and Silverberg 1980, 257–269.

Gould, S. J. 1980b. *The Panda's Thumb.* New York: Norton.

Gould, S. J. 1981. *The Mismeasure of Man.* New York: Norton.

Gould, S. J. 1983. *Hen's Teeth and Horses' Toes.* New York: Norton.

Gould, S. J., and Lewontin, R. C. 1979. "The Spandrels of San Marco and the Panglossian Paradigm: A Critique of the Adaptationist Programme." *Proceedings of the Royal Society of London* B 205, 581–598. Reprinted in E. Sober, ed., *Conceptual Issues in Evolutionary Biology.* Cambridge, Massachusetts: MIT Press, 1984.

Gowaty, P. 1982. "Sexual Terms in Sociobiology: Emotionally Evocative and, Paradoxically, Jargon." *Animal Behaviour* 30, 630–631.

Grafen, A. 1982. "How Not to Measure Inclusive Fitness." *Nature* 298, 425–426.

Grafen, A. 1984. "Natural Selection, Kin Selection, and Group Selection." In Krebs and Davies 1984, 62–84.

Grafen, A., and Sibly, R. 1978. "A Model of Mate Desertion." *Animal Behaviour* 26, 645–652.

Greene, P. 1978. "Promiscuity, Paternity, and Culture." *American Ethnologist* 5, 151–159.

Gross, M., and Shine, R. 1981. "Parental Care and Mode of Fertilization in Ectothermic Vertebrates." *Evolution* 35, 775–793.

Hamilton, W. D. 1964a. "The Genetical Evolution of Social Behavior I." In Williams 1971, 23–43.

Hamilton, W. D. 1964b. "The Genetical Evolution of Social Behavior II." In Williams 1971, 44–89.

Harris, M. 1979. *Cultural Materialism: The Struggle for a Science of Culture.* New York: Random House.

Heinrich, B. 1979. *Bumblebee Economics.* Cambridge, Massachusetts: Harvard University Press.

Hinde, R. 1982. *Ethology.* Oxford: Oxford University Press.

Hinde, R., ed. 1983. *Primate Social Relationships.* Oxford: Blackwell.

Hölldobler, B. 1966. "Futterverteilung durch Männchen in Ameisenstaat." *Zeitschrift für Vergleichende Physiologie* 52, 430–455.

Hrdy, S. 1981. *The Woman Who Never Evolved.* Cambridge, Massachusetts: Harvard University Press.

Hull, D., ed. 1974. *Darwin and His Critics.* Cambridge, Massachusetts: Harvard University Press.

Jarman, P. 1982. "Prospects for Interspecific Comparison in Sociobiology." In Bateson 1982a, 323–342.

Jenkin, F. 1867. Review of *The Origin of Species* from *The North British Review.* Reprinted in Hull 1974, 303–344.

Jensen, A. 1969. "How Much Can We Boost IQ and Scholastic Achievement?" *Harvard Educational Review* 39, 1–123.

Kaffman, M. 1977. "Sexual Standards and Behavior of the Kibbutz Adolescent." *American Journal of Orthopsychiatry* 47, 207–217.

Kamin, L. 1976. "Heredity, Intelligence, Politics, and Psychology: I." In Block and Dworkin 1976a, 242–264.
Kettlewell, H. 1973. *The Evolution of Melanism*. Oxford: Oxford University Press.
Kitcher, P. 1981. "Explanatory Unification." *Philosophy of Science* 48, 507–531.
Kitcher, P. 1982a. *Abusing Science*. Cambridge, Massachusetts: MIT Press.
Kitcher, P. 1982b. "Genes." *British Journal for the Philosophy of Science* 33, 337–359.
Kitcher, P. 1984. "1953 and All That: A Tale of Two Sciences." *Philosophical Review* 93, 335–373.
Kitcher, P. 1985. "Darwin's Achievement." In N. Rescher, ed., *Reason and Rationality in Science*. Washington, D.C.: University Press of America.
Kleiman, D. 1977. "Monogamy in Mammals." *Quarterly Review of Biology* 52, 39–69.
Krebs, J. R., and Davies, N. 1978. *Behavioural Ecology: An Evolutionary Approach*. Oxford: Blackwell.
Krebs, J. R., and Davies, N. 1981. *An Introduction to Behavioural Ecology*. Oxford: Blackwell.
Krebs, J. R., and Davies, N. 1984. *Behavioural Ecology: An Evolutionary Approach*. 2d ed. Sunderland, Massachusetts: Sinauer.
Kruuk, H. 1972. *The Spotted Hyena*. Chicago: University of Chicago Press.
Kuhn, T. S. 1970. *The Structure of Scientific Revolutions*. 2d ed. Chicago: University of Chicago Press.
Kummer, H. 1971. *Primate Societies*. Chicago: Aldine.
Kurland, J. 1976. "Sisterhood in Primates: What to Do with Human Males." Paper presented at the 1976 meetings of the American Anthropological Association, Washington, D.C.
Kurland, J. 1979. "Paternity, Mother's Brother, and Human Sociality." In Chagnon and Irons 1979, 145–180.
Lakatos, I. 1969. "Falsification and the Methodology of Scientific Research Programmes." In I. Lakatos and A. Musgrave, eds., *Criticism and the Growth of Knowledge*. Cambridge: Cambridge University Press, 91–195.
Laudan, L. 1977. *Progress and Its Problems*. Berkeley: University of California Press.
Leeds, A., and Dusek, V., eds. 1983. *Sociobiology: The Debate Evolves. The Philosophical Forum* 13.
Lewontin, R. C. 1974. *The Genetic Basis of Evolutionary Change*. New York: Columbia University Press.
Lewontin, R.C. 1976. "The Analysis of Variance and the Analysis of Causes." In Block and Dworkin 1976a, 179–193.
Lewontin, R. C. 1983a. "Organism as Subject and Object of Evolution." *Scientia* 118, 65–82.
Lewontin, R. C. 1983b. Review of Lumsden and Wilson, *Genes, Mind, and Culture. The Sciences* (Proceedings of the New York Academy of Science).
Lewontin, R. C., and White, M. J. D. 1960. "Interaction between Inversion Polymorphisms of Two Chromosome Pairs in the Grasshopper, *Moraba scurra*." *Evolution* 14, 116–129.
Lewontin, R. C., Rose, S., and Kamin, L. 1984. *Not in Our Genes*. New York: Pantheon.
Livingstone, F. 1980. "Cultural Causes of Genetic Change." In Barlow and Silverberg 1980, 307–329.
Lorenz, K. 1966. *On Aggression*. New York: Harcourt Brace Jovanovich.
Lott, A. 1984. "Intraspecific Variation in the Social Systems of Wild Vertebrates." *Behaviour* 87, 266–325.
Low, B. 1979. "Sexual Selection and Human Ornamentation." In Chagnon and Irons 1979, 462–487.

Lumsden, C., and Wilson, E. O. 1981. *Genes, Mind, and Culture.* Cambridge, Massachusetts: Harvard University Press.
Lumsden, C., and Wilson, E. O. 1983a. *Promethean Fire.* Cambridge, Massachusetts: Harvard University Press.
Lumsden, C., and Wilson, E. O. 1983b. "Genes, Mind, and Ideology." *The Sciences* (Proceedings of the New York Academy of Science).
Macevicz, S., and Oster, G. 1976. "Modeling Social Insect Populations II: Optimal Reproductive Strategies in Annual Eusocial Insect Colonies." *Behavioral Ecology and Sociobiology* 1, 265–282.
McKinney, F., Derrickson, S., and Mineau, P. 1983. "Forced Copulation in Waterfowl." *Behaviour* 86, 250–293.
Mattern, R. 1978. "Altruism, Ethics, and Sociobiology." In Caplan 1978, 462–475.
May, R. M. 1979. "When to Be Incestuous." *Nature* 279, 192–194.
Maynard Smith, J. 1964. "Group Selection and Kin Selection." *Nature* 201, 1145–1147.
Maynard Smith, J. 1976a. "Group Selection." In Clutton-Brock and Harvey 1979a, 20–30.
Maynard Smith, J. 1976b. "Parental Investment: A Prospective Analysis." In Clutton-Brock and Harvey 1979a, 98–114.
Maynard Smith, J. 1978. *The Evolution of Sex.* Cambridge: Cambridge University Press.
Maynard Smith, J. 1982a. *Evolution and the Theory of Games.* Cambridge: Cambridge University Press.
Maynard Smith, J. 1982b. "Introduction." In Bateson 1982a, 1–3.
Maynard Smith, J., and Warren, N. 1982. Review of Lumsden and Wilson, *Genes, Mind, and Culture. Evolution* 36.
Mayr, E. 1983. "How to Carry Out the Adaptationist Program." *American Naturalist* 121, 324–334.
Mech, L. 1970. *The Wolf: The Ecology and Behavior of an Endangered Species.* Garden City: Natural History Press.
Meehl, P. 1984. "Consistency Tests in Estimating the Completeness of the Fossil Record: A Neo-Popperian Approach to Statistical Paleontology." In J. Earman, ed., *Testing Scientific Theories.* Minneapolis: University of Minnesota Press, 413–473.
Michod, R. 1982. "The Theory of Kin Selection." *Annual Review of Ecology and Systematics* 13, 23–55.
Midgley, M. 1978. *Beast and Man.* Ithaca: Cornell University Press.
Milkman, R. 1982. *Perspectives on Evolution.* Sunderland, Massachusetts: Sinauer.
Mills, S., and Beatty, J. 1979. "The Propensity Interpretation of Fitness." *Philosophy of Science* 46, 263–286.
Money, J., and Ehrhardt, A. 1972. *Man and Woman, Boy and Girl.* Baltimore: Johns Hopkins University Press.
Montagu, A. 1980. *Sociobiology Examined.* New York: Oxford University Press.
Morris, D. 1967. *The Naked Ape.* New York: McGraw-Hill.
Nelson, G., and Platnick, N. 1981. *Systematics and Biogeography: Cladistics and Vicariance.* New York: Columbia University Press.
Newton, I. [1687] 1962. *Mathematical Principles of Natural Philosophy and His System of the World,* translated by A. Motte and F. Cajori. 2 vols. Berkeley: University of California Press.
Nozick, R. 1981. *Philosophical Explanations.* Cambridge, Massachusetts: Harvard University Press.
Orians, G. 1969. "On the Evolution of Mating Systems in Birds and Mammals." In Clutton-Brock and Harvey 1979a, 115–132.
Oster, G., and Wilson, E. O. 1978. *Caste and Ecology in the Social Insects.* Princeton: Princeton University Press.

Packer, C. 1977. "Reciprocal Altruism in Olive Baboons." In Clutton-Brock and Harvey 1979a, 227–232.
Packer, C. 1979. "Inter-troop Transfer and Inbreeding Avoidance in *Papio anubis.*" *Animal Behaviour* 27, 1–36.
Packer, C., and Pusey, A. 1982. "Cooperation and Competition within Coalitions of Male Lions: Kin Selection or Game Theory?" *Nature* 296, 740–742.
Parker, G. 1978. "Searching for Mates." In Krebs and Davies 1978, 214–244.
Parker, G., and MacNair, M. 1978. "Models of Parent-Offspring Conflict. I. Monogamy." *Animal Behaviour* 26, 97–110.
Parker, G. 1974. "Assessment Strategy and the Evolution of Fighting Behaviour." In Clutton-Brock and Harvey 1979a, 271–292.
Plotkin, H., and Odling-Smee, F. 1981. "A Multiple-Level Model of Evolution and Its Implications for Sociobiology." *The Behavioral and Brain Sciences* 4, 225–268.
Pulliam, R., and Caraco, T. 1984. "Living in Groups: Is There an Optimal Group Size?" In Krebs and Davies 1984, 122–147.
Pusey, A. 1979. "Intercommunity Transfer of Chimpanzees in Gombe National Park." In D. A. Hamburg and E. R. McCown, eds., *The Great Apes.* Menlo Park: Benjamin/Cummings, 465–479.
Pusey, A. 1980. "Inbreeding Avoidance in Chimpanzees." *Animal Behaviour* 28, 543–552.
Quinn, P. 1981. *Divine Commands and Moral Requirements.* New York: Oxford University Press.
Raup, D., and Stanley, S. 1978. *Principles of Paleontology.* San Francisco: Freeman.
Rawls, J. 1951. "Outline of a Decision Procedure for Ethics." *Philosophical Review* 60, 177–197.
Rawls, J. 1971. *A Theory of Justice.* Cambridge, Massachusetts: Harvard University Press.
Rawls, J. 1980. "Kantian Constructivism in Moral Theory." *Journal of Philosophy* 77, 515–572.
Richardson, J., and Kroeber, A. L. 1940. "Three Centuries of Women's Dress Fashions. A Quantitative Analysis." *University of California Anthropological Records* 5, 111–153.
Richerson, P. J., and Boyd, R. 1978. "A Dual Inheritance Model of the Human Evolutionary Process I: Basic Postulates and a Simple Model." *Journal of Social and Biological Structures* 1, 127–154.
Roughgarden, J. 1979. *Theory of Population Genetics and Evolutionary Ecology: An Introduction.* New York: Macmillan.
Rubenstein, D. 1982. "Complexity in Evolutionary Processes." In Bateson 1982a, 87–89.
Ruse, M. 1979. *Sociobiology: Sense or Nonsense?* Dordrecht: D. Reidel.
Ruse, M. 1982. "Is Human Sociobiology a New Paradigm?" In Leeds and Dusek 1983, 119–143.
Sahlins, M. 1976. *The Use and Abuse of Biology.* Ann Arbor: University of Michigan Press.
Schaller, G. 1972. *The Serengeti Lion.* Chicago: University of Chicago Press.
Schelling, T. 1978. *Micromotives and Macrobehavior.* New York: Norton.
Schull, W. J., and Neel, J. V. 1965. *The Effects of Inbreeding on Japanese Children.* New York: Harper and Row.
Schuster, P., and Sigmund, K. 1981. "Coyness, Philandering, and Stable Strategies." *Animal Behaviour* 29, 186–192.
Seger, J. 1983. "Partial Bivoltivism May Cause Alternating Sex-Ratio Biases that Favour Eusociality." *Nature* 301, 59–62.
Seyfarth, R. 1978. "Social Relationships among Adult Male and Female Baboons. I.

Behaviour during Sexual Consortship. II. Behaviour throughout the Female Reproductive Cycle." *Behaviour* 64, 204–226, 227–247.

Shepher, J. 1971. "Mate Selection among Second-Generation Kibbutz Adolescents and Adults: Incest Avoidance and Negative Imprinting." *Archives of Sexual Behavior* 1, 293–307.

Singer, P. 1981. *The Expanding Circle*. New York: Farrar, Straus, and Giroux.

Smith, P. 1982. "Does Play Matter? Functional and Evolutionary Aspects of Animal and Human Play." *The Behavioral and Brain Sciences* 5, 139–155.

Smole, W. J. 1976. *The Yanomamo Indians*. Austin: University of Texas Press.

Sober, E. 1984. *The Nature of Selection*. Cambridge, Massachusetts: MIT Press.

Sober, E. 1985. "Methodological Behaviorism, Evolution, and Game Theory." *Synthese* (forthcoming).

Spielman, R. S., Neel, J. V., and Li, F. H. 1977. "Inbreeding Estimation from Population Data: Models, Procedures, and Implications." *Genetics* 85, 355–371.

Stammbach, E. 1978. "On Social Differentiation in Groups of Captive Female Hamadryas Baboons." *Behaviour* 67, 322–338.

Stamps, J., and Metcalf, R. 1980. "Parent-Offspring Conflict." In Barlow and Silverberg 1980.

Stamps, J., Metcalf, R., and Krishnan, V. 1978. "Genetic Analysis of Parent-Offspring Conflict." *Behavioral Ecology and Sociobiology* 3, 369–392.

Stanley, S. 1980. *Macroevolution: Pattern and Process*. San Francisco: Freeman.

Templeton, A. 1982. "Adaptation and the Integration of Evolutionary Forces." In Milkman 1982, 15–31.

Thoday, J. M. 1953. "Components of Fitness." *Symposium of the Society for Experimental Biology* 7, 96–113.

Thornhill, R., and Alcock, J. 1983. *The Evolution of Insect Mating Systems*. Cambridge, Massachusetts: Harvard University Press.

Tiger, L., and Fox, R. 1971. *The Imperial Animal*. New York: Holt, Rinehart and Winston.

Tinbergen, N. 1968. "On War and Peace in Animals and Man." In Caplan 1978, 76–99.

Trivers, R. 1971. "The Evolution of Reciprocal Altruism." In Clutton-Brock and Harvey 1979a, 189–226.

Trivers, R. 1972. "Parental Investment and Sexual Selection." In Clutton-Brock and Harvey 1979a, 52–97.

Trivers, R. 1974. "Parent-Offspring Conflict." In Clutton-Brock and Harvey 1979a, 233–257.

Trivers, R., and Willard, D. 1973. "Natural Selection and Parental Ability to Vary the Sex Ratio of Offspring." *Science* 179, 90–92.

Turnbull, C. 1972. *The Mountain People*. New York: Simon and Schuster.

van den Berghe, P. 1979. *Human Family Systems*. New York: Elsevier North Holland.

van den Berghe, P. 1980. "Incest and Exogamy: A Sociobiological Reconsideration." *Ethology and Sociobiology* 1, 151–162.

van den Berghe, P. 1982. "Resistance to Biological Self-Understanding." *The Behavioral and Brain Sciences* 5, 27.

van den Berghe, P. 1983. "Human Inbreeding Avoidance: Culture in Nature." *The Behavioral and Brain Sciences* 6, 91–123.

van den Berghe, P., and Mesher, G. 1980. "Royal Incest and Inclusive Fitness." *American Ethnologist* 7, 300–317.

Vehrencamp, S., and Bradbury, J. 1984. "Mating Systems and Ecology." In Krebs and Davies 1984, 251–278.

Washburn, S. L. 1980. "Human Behavior and the Behavior of Other Animals." In Montagu 1980, 254–282.

Watson, G. 1975. "Free Agency." *Journal of Philosophy* 72, 205–220.

Williams, G. C. 1966. *Adaptation and Natural Selection.* Princeton: Princeton University Press.

Williams, G. C., ed. 1971. *Group Selection.* Chicago: Aldine.

Williams, G. C. 1975. *Sex and Evolution.* Princeton: Princeton University Press.

Williams, G. C., and Williams, D. C. 1957. "Natural Selection of Individually Harmful Social Adaptations among Sibs with Special Reference to Social Insects." *Evolution* 11, 32–39.

Wilson, D. S. 1980. *The Natural Selection of Populations and Communities.* Menlo Park: Benjamin/Cummings.

Wilson, D. S. 1983. "Individual and Group Selection: A Historical and Conceptual Review." *Annual Review of Ecology and Systematics* 14, 159–188.

Wilson, E. O. 1971. *The Insect Societies.* Cambridge, Massachusetts: Harvard University Press.

Wilson, E. O. 1975a. *Sociobiology: The New Synthesis.* Cambridge, Massachusetts: Harvard University Press.

Wilson, E. O. 1975b. "Human Decency is Animal." *New York Times Magazine,* October 12, 38–50.

Wilson, E. O. 1976. "Academic Vigilantism and the Political Significance of Sociobiology." In Caplan 1978, 291–303.

Wilson, E. O. 1978. *On Human Nature.* Cambridge, Massachusetts: Harvard University Press.

Wilson, E. O. 1980a. "The Relation of Science to Theology." *Zygon* 15, 425–434.

Wilson, E. O. 1980b. "Comparative Social Theory." *Tanner Lecture,* University of Michigan.

Wittenberger, J., and Tilson, R. 1980. "The Evolution of Monogamy: Hypothesis and Evidence." *Annual Review of Ecology and Systematics* 11, 197–232.

Wolf, A. P., and Huang, C. 1980. *Marriage and Adoption in China 1845–1945.* Stanford: Stanford University Press.

Wolf, L. 1975. " 'Prostitution' Behavior in a Tropical Hummingbird." *Condor* 77, 140–144.

Wolf, S. 1980. "Asymmetrical Freedom." *Journal of Philosophy* 77, 151–166.

Woolfenden, G. 1975. "Florida Scrub Jay Helpers at the Nest." *Auk* 92, 1–15.

Woolfenden, G., and Fitzpatrick, J. 1978. "The Inheritance of Territory in Group-Breeding Birds." *Bioscience* 28, 104–108.

Index

The index contains three types of entries. There are technical terms, whose page references are given in italics to show where the meanings of the terms are explained. There are names of people and kinds of organisms, whose page references include all pertinent discussions of the people or kinds of organisms in question. Finally, there are entries intended to help the reader locate quickly discussions of particular aspects of the sociobiology controversy.